INTRODUCTORY STATISTICS AND PROBABILITY

A Basis for Decision Making

INTRODUCTORY STATISTICS

A Basis for Decision Making

AND PROBABILITY

DAVID W. BLAKESLEE
SAN FRANCISCO STATE COLLEGE

WILLIAM G. CHINN
CITY COLLEGE OF SAN FRANCISCO

EDITORIAL ADVISERS

ANDREW M. GLEASON
HARVARD UNIVERSITY

ALBERT E. MEDER, JR.
RUTGERS UNIVERSITY

HOUGHTON MIFFLIN COMPANY · BOSTON

NEW YORK · ATLANTA · GENEVA, ILLINOIS · DALLAS · PALO ALTO

Copyright © 1971 by Houghton Mifflin Company

All rights reserved. No part of this work may be reproduced or transmitted in any form or by any means, electronic or mechanical, including photocopying and recording, or by any information storage or retrieval system, without permission in writing from the publisher. Printed in the U.S.A.

Library of Congress Catalog Card Number: 77-128707

ISBN: 0-395-11063-7

Preface

In writing this book, our aim has been to provide a basic course in statistics and probability suitable for several different types of readers. First, the extensive use of statistical methods in the modern world makes it imperative that the educated citizen understand the fundamental ideas that underlie decisions which are reached by these methods. Fortunately a sound and useful introduction to statistics and probability can be developed by using only the mathematical tools of high school algebra.

This text is also aimed at the student who plans to enter such fields as economics, business, education, psychology, sociology, biology, and medicine, which now require statistics for their effective pursuit. We believe that successful completion of a course based on this textbook will be adequate preparation for the usual college courses offered in these fields.

Any student who is preparing for study in mathematics, physical science, or the engineering sciences will eventually find need for one or more courses in mathematical statistics. These courses require several semesters or quarters of calculus, but the necessary fundamental background in probability and statistics is presented in this text.

The following outline of the chapters will indicate the structure of our development:

In Chapters 1, 2, and 3 the fundamental ideas of probability are discussed. Some elementary topics concerning the handling of numerical data are presented in Chapter 4. The treatment of random variables in Chapters 5 and 6 is unusually thorough for a beginning course. We believe that the material is presented in a way which makes it accessible to the reader who has had some familiarity with functions from a course in algebra.

Chapters 7 and 8 deal with the important special cases of the binomial and normal random variables. After a brief introduction to sampling in Chapter 9, the remaining chapters are devoted to the use of statistical methods in determining confidence intervals and significance levels. Including a discussion of statistical hypotheses, tests, and Type I and Type II errors in Chapter 11 is also unusual for an introductory course. Again, we feel that the material is not too difficult and that it is sufficiently important to decision making to warrant its inclusion.

There are several special features of this text which should be noted before its study is begun. The exercises include problems of three types: (1) those marked with a ● we consider to be essential to an understanding of the material. Many of these are referred to in the text and in later exercises. (2) Those marked with a ★ are to be considered optional. They are designed to be more challenging and to provide certain insights which are not exploited in the text. (3) The unmarked problems are designed to provide extra practice.

We have provided solutions for all problems marked with a ●. It has been our experience that simply giving the answer to a problem in probability is sometimes worse than useless. It does the reader little good to discover that the probability of a particular event is $\frac{3}{8}$ unless he is able to see how the answer is obtained. This feature of the text helps to make it self-contained and hence appropriate for individual as well as classroom usage. In addition, answers are provided for the odd-numbered problems not marked with a ●.

The sections labeled "What Do You Think?" are designed to allow the reader to develop certain ideas for himself. They often lead to material which is presented in more detail in subsequent sections.

Sections marked with a ★ are to be considered as optional or supplementary. Their omission will not interfere with the flow of the text.

David W. Blakeslee
William G. Chinn

Contents

Chapter 1		**Introduction**	**1**
	1.1	Decisions in the Face of Uncertainty	2
	1.2	Statistics and Probability	5
	1.3	Experimental and Historical Data	8
	1.4	Organizing Numerical Data	10
	1.5	A Word about Words	12
	1.6	What Do You Think?	13
Chapter 2		**Probability—Preliminary Ideas**	**15**
	2.1	Theoretical (*A Priori*) and Experimental (Empirical) Probabilities	15
	2.2	Sets of Possible Outcomes of Experiments	17
	2.3	Assigning Probabilities	22
	2.4	Events and Their Probabilities	26
	2.5	Summary of Probabilities of Events	31
	2.6	The Probability of the Union of Two Events	34
	2.7	Proving Some Theorems	40
	2.8	What Do You Think?	42
		Chapter Summary	44
		Chapter Review	45
Chapter 3		**Conditional Probability**	**47**
	3.1	A Formula for Conditional Probability	47
	3.2	Experiments Having Several Steps	53
	3.3	Experiments with Assigned Probabilities	61
	3.4	Independent Events	64
	3.5	Repeated Independent Trials	69
	★ 3.6	Bayes' Formula	73
		Chapter Summary	78
		Chapter Review	78
Chapter 4		**Numerical Data**	**81**
	4.1	Frequency Distributions	81
	4.2	Grouped Data	88
	4.3	The Mean	92
	4.4	The Variance and the Standard Deviation	96
		★ Computer Investigations	103

	4.5	What Do You Think?	104
		Chapter Summary	106
		Chapter Review	107

Chapter 5   Random Variables and Probability Functions   109

5.1	Random Variables	109
5.2	Probability Functions and Their Graphs	113
5.3	Continuous Probability Distributions	120
5.4	The Standard Normal Distribution	125
5.5	The Expected Value of a Random Variable	128
5.6	What Do You Think?	133
	Chapter Summary	136
	Chapter Review	136

Chapter 6   Functions of Random Variables   139

6.1	$E(aX + b)$, $E(X^2)$	140
6.2	Variance and Standard Deviation	147
6.3	Sums of Random Variables	152
6.4	A Standardized Random Variable	158
★ 6.5	What Do You Think?	162
	★ Computer Investigations	173
	Chapter Summary	173
	Chapter Review	175

Chapter 7   The Binomial Distributions   177

7.1	Definition of a Binomial Distribution	177
7.2	Permutations and Combinations	185
	★ Computer Investigations	189
7.3	Binomial Coefficients	190
7.4	More about Binomial Distributions	193
7.5	What Do You Think?	199
	Chapter Summary	201
	Chapter Review	202

Chapter 8   Using Continuous Distributions   203

8.1	Standardizing a Normal Distribution	203
8.2	Normal Approximation to a Binomial Distribution	206
	★ Computer Investigations	212
★ 8.3	Poisson Approximation to a Binomial Distribution	212
8.4	What Do You Think?	214
	Chapter Summary	217
	Chapter Review	217

Chapter 9		Sampling	219
	9.1	Population Random Variable	220
	9.2	Random Sampling	222
	9.3	The Sample Mean, $\overline{X}$	226
	9.4	The Law of Large Numbers	230
	9.5	More about the Distribution of $\overline{X}$	232
	9.6	What Do You Think?	235
		Chapter Summary	238
		Chapter Review	239

Chapter 10		Estimation	241
	10.1	Sample of Size One	241
	10.2	Point Estimation	243
	10.3	Confidence Intervals	244
	10.4	Using Confidence Intervals	248
		★ Computer Investigations	253
	10.5	Significance	253
	10.6	What Do You Think?	258
		Chapter Summary	260
		Chapter Review	261

Chapter 11		Decision Making	263
	11.1	Statistical Hypotheses	264
	11.2	Two-Sided Statistical Tests	266
	11.3	One-Sided Statistical Tests	270
	11.4	Risk; Type I Errors	273
	11.5	Alternate Hypotheses; Type II Errors	275
	11.6	Decisions, Decisions	278
		Chapter Summary	279
		Chapter Review	279

Appendix	Note on Summation Notation	281
	List of Formulas	284
	Selected Solutions and Answers	286

Tables	Table B Binomial Distributions $b(n, p; x)$	339
	Table N The Cumulative Standard Normal Distribution	344
	Table P Poisson Distributions	345
	Table R Squares and Square Roots	347

Index		353

Chapter 1

Introduction

Before we begin our study of statistics and probability, we shall take a brief look ahead at the ground we shall cover. This kind of preview may be likened to an investigation before starting on a trip. Then we may ask such questions as these:

Where are we going?

Why?

What preparations are necessary?

What difficulties may we meet?

What profit or enjoyment may we expect to gain from this trip?

Similarly, we may now ask:

What is the purpose of studying statistics and probability?

In particular, we may ask:

To what uses can this knowledge be put?

What kinds of problems can be solved?

Is this subject of any immediate relevance to our lives?

What mathematical background do we need in order to study it?

In this introductory chapter, we shall suggest some answers to these questions.

1.1 Decisions in the Face of Uncertainty

Our earlier experience with mathematics may have led us to expect that it is a subject in which a proposed answer to a problem must be either right or wrong. From this it might seem to follow that a decision based on careful mathematical analysis must be correct. This is, of course, the case when all the necessary information is available. For example, the decision as to the proper orbital height of a particular satellite can be made precisely on the basis of the appropriate mathematical and physical ideas.

Many times, however, decisions must be made in the face of uncertainty—when information is incomplete. Examine the following situations:

1. A college student needs to decide whether or not to buy collision insurance on his sports car. He wishes to protect his investment, but the premiums would place a heavy burden on his budget.

2. A company has developed a new product and wishes to decide on the most effective and economical way to advertise it. Television is expensive but reaches a large audience.

3. A doctor needs to decide whether or not to use a certain treatment on one of his patients. The treatment may be helpful, but sometimes the treatment produces undesirable effects.

Clearly, if the student *knew* that he would be in an accident which would damage his car, he would buy the insurance. If the company *knew* that television was the best medium, it would choose that method of advertising. If the doctor *knew* that the side effects would not develop in his patient, he would start the treatment.

Can mathematics help us in cases such as those listed above?

Of course, no mathematical procedures can guarantee that a "correct" decision can be made in such cases. Our task in this text will be to try to develop a mathematical theory and technique which will aid in making decisions that will have a *better chance* of being correct.

Naturally, the types of decisions which will concern us here are those which can be analyzed mathematically. Many situations involve personal, moral, social, or psychological factors that are outside the realm of mathematics.

Notice that we said above that we would seek decisions that would have a *better chance* of being correct. The study of the mathematics of chance began with questions raised in connection with the so-called "games of chance." How a pair of dice will fall on the next throw, whether a tossed coin will fall heads or tails, where a spinner will stop, and which card will be dealt first from a shuffled deck are obvious examples of situations in which the outcome is uncertain.

Perhaps the first attempt at an organized treatment of the subject was given by Girolamo Cardano (1501–1576), an Italian algebraist, when he wrote a kind of gambler's manual. A particular gaming question* asked by a French nobleman led Blaise Pascal (1623–1662) and Pierre de Fermat (1601–1665) to consider carefully the mathematics of chance. Other famous men who contributed greatly to the early history of this topic were the Dutch scientist Christian Huygens (1629–1695), the great Swiss mathematician Jacques Bernoulli (1654–1705), and the famous French scientist Pierre S. Laplace (1749–1827).

Because simple games of chance are relatively easy to understand and to analyze, we shall use games involving dice, coins, spinners, and cards as illustrations in much of this text. A *die* (singular of "dice") is a cube with

the six faces marked "1" through "6,"

and we shall assume that one of these faces will be uppermost when the die is tossed. The number appearing on the uppermost face is said to be "the number thrown." We shall assume that if a coin is tossed, it will land either "heads" or "tails," not on its edge. A regular deck of playing cards consists of

four suits (spades, hearts, diamonds, clubs)

of 13 cards each:

ace, king, queen, jack, 10, 9, . . . , 2

We shall have occasion to refer to coins or dice as *fair*, or *unbiased*. By a *"fair coin,"* we shall mean one that is *equally likely* to turn up heads or tails when it is tossed. Similarly, with a fair die, all six faces are *equally likely* to appear uppermost in a toss. We shall also assume that a deck of cards is well shuffled so that all 52 cards are *equally likely* to be drawn.

EXERCISES 1.1†

In Problems 1–14, you are supposed to guess intelligently. Later we shall develop a mathematical treatment for such problems.

By a "game" here, we shall mean a situation with two players, exactly **one** *of whom "wins." A game is "fair" if each player has an equal chance of winning.*

Decide whether or not each of the following is a game as defined above, and if it is a game, whether or not it is fair. (Assume that the coins and dice are fair and that the cards are well shuffled.)

- 1. A coin is thrown. You win if "heads" shows; you lose and your opponent wins if "tails" shows.

    2. Two coins are thrown. You win if they "match"; you lose otherwise.

*See Problem 9, Exercises 3.5.

†All problems marked with ● are considered to be essential to the development of the course. Solutions of all problems so marked, as well as answers to all odd-numbered problems not so marked, are given at the back of the book.

3. A die is thrown. You win if the number thrown is divisible by 3; you lose otherwise.

4. A die is thrown. You win if a 1 or 2 shows; you lose if a 3, 4, or 5 shows; if a 6 shows, the die is thrown again.

5. A die is thrown. You win if the number showing is divisible by 3; you lose if the number showing is divisible by 2; if neither is the case, the die is thrown again.

6. A die is thrown. You win if the number shown is a prime; your opponent wins if the number is even.

7. A die is thrown. You win if the number shown is prime; you lose otherwise.

8. A die is thrown. You win if the number shown is even; your opponent wins otherwise.

9. Two dice are thrown. You win if both dice show prime numbers; you lose otherwise.

10. Two dice are thrown. You win if exactly one die shows a prime number; you lose otherwise.

11. A card is drawn. You win if a spade is drawn; you lose otherwise.

12. A card is drawn. You win if the card is an "honor" card (10, jack, queen, king, ace); you lose otherwise.

13. A dial for a spinner is marked as shown at the left below. You win if the spinner stops on red; you lose if it stops on blue. (Assume that the pointer does not stop on a line.)

(13) 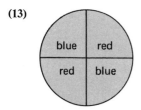 (14)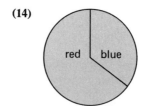

14. Repeat Problem 13 for the dial shown at the right above.

15. A roulette wheel has 38 positions in which the ball may stop. Eighteen of these are colored red, 18 black, and 2 white. You win if the ball stops in a red position; you lose otherwise. (You may think of this as a spinner with 38 equally likely positions.)

16. Cardano, Pascal, Fermat, Huygens, Jacques Bernoulli, and Laplace have been mentioned in connection with the history of probability. Try to find out what other mathematical or scientific contributions were made by one or more of these men.

1.2 Statistics and Probability

Our goal in this text, as we stated in Section 1.1, is to develop mathematical techniques to aid us in making decisions in situations involving uncertainty. The methods which we shall develop are those of *statistical inference* (Chapters 9–11). This phrase refers to the drawing of conclusions, or to the making of predictions, based on the analysis of available numerical data (statistics). In general:

The study of numerical data is called *statistics*.

Sometimes the data which are available in a given situation lead to rather obvious conclusions and hence to a reasonably simple choice among alternate courses of action. More often, the evidence at hand does not offer an obvious choice. What is needed then is an underlying mathematical theory to which we may appeal.

In Section 1.1 we noted that in problems involving uncertainty, we would try to make decisions that would have a *better chance* of being correct. In general:

The mathematical study of chance occurrences is called *probability*.

We all have some feeling or intuition about *probability*. We have been exposed to *probabilistic* statements about uncertain situations in connection with such familiar things as weather forecasting, games of chance, betting odds on sporting events, and election predictions. We often feel its influence in our daily living. For instance, our preparations for wet weather are different if the forecast (prediction) is "90% chance of rain" from those we make if it is "10% chance of rain."

Weather forecasting offers a good example of the use of both statistics and probability in estimating a result. The predictions are based on current data together with long-term, or long-run, information. By the prediction "90% chance of rain," the forecaster indicates that, in the past, it has rained 90% of the time that the present atmospheric conditions have occurred.

In very much the same way, if we say, "the probability that a coin will fall heads is $\frac{1}{2}$," we are expressing our belief that if the coin were tossed many times, the ratio of heads to the total number of tosses would tend to become close to $\frac{1}{2}$.

These examples illustrate the idea that a measure can be assigned to *the probability of a particular event*—that we can use a *number* (from 0 to 1, as it turns out) to express a *degree of belief* that a certain thing will happen. From this basic idea, a mathematical *theory of probability* can be built up. Chapters 2, 3, 5, and 6 are devoted to an introduction to this theory, while the final chapters deal with its application to statistical inference and decision making.

What mathematical background do we need in order to begin our study? Advanced courses in probability require a great deal of "higher" mathematics, including the calculus. Fortunately, for an introductory course we need mainly some skill in arithmetic, some techniques of elementary algebra, and a familiarity with the language (and notation) of sets. Of course, we also need to have had some experience with mathematical (deductive) methods of reasoning.

There will be places in the text (especially in Chapters 5, 8, and 9) where we shall wish to use conclusions that require more advanced mathematical ideas for their proof. When we arrive at these places, we shall try to make the results appear reasonable through examples and informal arguments. Then we shall simply point out that the necessary proofs must be postponed until we have had further mathematical training.

Remark 1

Whether a numerical expression of a degree of belief can be applied properly *only* to long range situations has been a matter for considerable debate among mathematicians and philosophers. Some would argue that probability theory cannot be applied in connection with single, unique, nonrepeatable situations. Suppose that a particular, perhaps bent or dented, coin were to be tossed only once and then discarded. Is there any meaning to a statement of the probability that the coin will fall heads on that one toss?

Remark 2

There is also the point of view that probability (degree of belief) is a personal, subjective matter. Those holding this view agree that one's belief is influenced by facts and observations. They argue, however, that different individuals, exposed to the same evidence, can properly arrive at different probability judgments. As a simple example, consider the differences of opinions held by qualified sports authorities about the outcome of some future athletic event. On the other hand, there are many who would assert that probability is objective. They feel that the useful probability judgments are those on which reasonable men would agree. As an illustration, suppose that a bag contains 50 red and 50 blue marbles. The marbles are thoroughly mixed and then one marble is drawn without looking. It can be argued that in this case "everyone" would agree that there is the same probability of drawing a red marble as a blue marble.

The preceding remarks may be borne in mind as we study probability theory. At this point, it would be wise not to try to decide which of the viewpoints to adopt.

EXERCISES 1.2

Here are some situations in which data are given and conclusions drawn. Judge each conclusion as to your degree of belief as to its truth according to the following scale:

(1) *feel strongly that it is true*
(2) *believe it is true*
(3) *feel it is just as likely to be true as false*
(4) *believe it is false*
(5) *feel strongly that it is false*

There are no "right" answers. Our goal is to gain an awareness of how it seems easier to draw conclusions in some cases than in others.

In Problems 1–4, we simply present a list of numbers without explaining where the numbers come from. We can pretend that they came from some experimental situations.

- **1.** Data: 1, 2, 4, 8, 16, 32, __?__
  Conclusions: The next number to be obtained will be:
  a. positive   **b.** larger than 32
  c. exactly 64   **d.** exactly 33.89

  2. Data: 47, −3, −8, 9, −23, 14, __?__
  Conclusions: The next number to be obtained will be:
  a. positive   **b.** negative
  c. less than 1000   **d.** less than −1000

- **3.** Data: 1, 1, 1, 1, 1, 1, __?__
  Conclusions: The next number to be obtained will be:
  a. 1   **b.** 17
  c. different from 1   **d.** a whole number

  4. Data: 1, 0, 0, 1, 0, 1, __?__
  Conclusions: The next number to be obtained will be:
  a. either 0 or 1   **b.** 0
  c. 1   **d.** −3.14159

- **5.** The approximate population of the United States (in millions) for each of the last six decades was, respectively, 92, 106, 123, 132, 151, 179.
  Conclusions: In the next decade, the population of the United States (in millions) will be:
  a. less than 179   **b.** between 179 and 200
  c. more than 200   **d.** exactly 216

- **6.** In the years 1910, 1930, and 1950, the population of San Francisco (in thousands) was, respectively, 417, 634, and 775.
  Conclusions: After the next twenty years, the San Francisco population (in thousands) will be:
  a. greater than 775   **b.** less than 775
  c. exactly 800   **d.** between 900 and 1,000

1.3 Experimental and Historical Data

We have explained that we are concerned with finding ways of making decisions in uncertain situations, basing our decisions on numerical data. How do we obtain such data?

Sometimes data are obtained from *experiments*.

In the physical, biological, behavioral, or social sciences, the necessary data are frequently collected by means of planned, controlled experiments. A botanist, for example, might plant many seedlings, provide them with the same type of soil, the same amount of moisture and sunlight, but vary the type of fertilizer used. He would then record the growth rate of the plants. After subjecting his data to statistical analysis, he might be able to decide the relative value of the different types of fertilizer.

Another kind of example is the familiar one of deciding whether or not a particular coin is "fair," that is, as likely to fall heads as it is to fall tails. Expressed a bit differently: Is the *probability of heads* equal to the *probability of tails?* No amount of study of the physical characteristics of the coin could determine the answer, although such a study would perhaps enable us to make a reasonable guess. One method of obtaining information relevant to the question is to experiment—to toss the coin many times.

As an experiment, suppose that we have two coins, and suppose that we toss them both 10,000 times with the following results:

	Heads	Tails
Coin 1	5,084	4,916
Coin 2	8,347	1,653

Let us consider Coin 2. Our evidence leads us, without any complicated mathematical reasoning, to be very sure that Coin 2 is not fair.

For Coin 1, the situation is a bit different. The number of heads and tails are about the same. We might well conclude that either the coin is fair or that it is slightly more likely to fall heads than tails. If we had to choose "fair" or "not fair," we would probably vote for "fair." With how much confidence would we vote? We shall return to questions like this in later chapters.

This example illustrates an interesting principle: If a proposition, such as "the coin is fair," is greatly in error, then experimentation will rather easily reveal its falsity. On the other hand, if the proposition is indeed true, it is difficult, if not impossible, to determine its truth by experimentation. It has been said that "Nature shouts NO, but whispers YES."

We have been discussing the possibility of using experiments to obtain numerical data which can be used to draw conclusions or to make decisions. The problems which arise in designing an experiment for a particular purpose may be very complicated. Later in this text, we shall touch upon some of these problems.

Sometimes data are drawn from *historical records.*

In practice, in situations in the "real world," it is not always possible to design the required controlled experiments. In many cases, data must be obtained from historical evidence—records of what has happened in the past in the same or similar situations. This is the method used by the weather forecaster, the insurance statistician, the economic theorist, and some social scientists. Historical data are inherited, rather than obtained from planned experiments.

Either kind of information, experimental and historical, can be subjected to mathematical analysis and used as a basis for statistical inference.

Remark

Throughout the text we shall often propose "mind" or "thought" experiments. In Remark 2 on page 6, we said: "Suppose that a bag contains 50 red and 50 blue marbles. The marbles are thoroughly mixed and one marble is to be drawn without looking." We can imagine such an experiment being performed without actually doing it ourselves.

EXERCISES 1.3

- 1. Refer to the two coins mentioned in this section (see page 8). Let us agree that our data lead us to conclude that Coin 1 is fair and that Coin 2 is biased in favor of heads. In a game using that Coin 2, the player choosing heads apparently has an advantage over the player choosing tails.

    Can you suggest rules for a game using both coins that might produce results making it closer to being a "fair game"?

  2. a. What data would you collect if you want to decide whether it is true that Mr. Jones is an easier grader than Miss Smith?
     b. Suppose that you wonder whether or not the students in Mr. Jones's classes are smarter than those in Miss Smith's classes, what changes would you want to make in the collection of data?

In some situations, missing data might be "retrieved" by continuing or extending the experiment. In other situations, missing data cannot be retrieved because conditions in an experiment cannot be duplicated without repeating the entire experiment.

Some data are missing in each situation below. State whether or not you believe that the data can be retrieved (how?). Would retrieval be important, or could we just ignore the missing information?

- **3.** The experiment consists of tossing a coin 10,000 times to determine the ratio of heads to tails. Data for five of these tosses are missing.
- **4.** The experiment consists of tossing a coin 10,000 times to determine the number of "runs" of four heads in a row. Data for five of these tosses are missing.
  5. The experiment consists of comparing the arithmetic skills of students in two schools during the third week in the semester. During that week, one school had 25% of the students absent due to a flu epidemic; the other school had only 2% of the students absent.
  6. The experiment consists of comparing the growth of liverworts in lunar dust with growth in regular earth soil. The labels dropped off some of the containers and it was uncertain which containers had the lunar dust.
- **7.** A study was to be made of paths of hurricanes since the turn of the century. Details of the tracks were missing for five of these years.

1.4 Organizing Numerical Data

Chapter 4 is devoted to an introduction to the treatment of numerical data. At this point, we make only preliminary remarks.

Whether the information we are considering comes from designed experiments or from historical records, it is essential that the material be *organized* in an orderly way. If one has just a "bucket of numbers," it is difficult to make sense out of the information.

Sometimes it is most useful to arrange the numbers in order of size. Chronological or alphabetical listings are often useful. For example, if we were dealing with population statistics of the states of the United States, we might wish to list the states either in order of size or alphabetically.

At other times we group certain pieces of information together. If we gave a 100-point test to a large number of college freshmen, we might wish to group together the scores lying in intervals 1–10, 11–20, ..., 91–100.

In any event, careful organization of data is a practical first step in its analysis.

A less obvious remark concerns the nature of the data that are at hand. It is sometimes useful to distinguish between *discrete* and *continuous* data. It is not possible to give a precise definition of these terms at this moment. The following example, a nonmathematical one, should help to illustrate the distinction between discrete and continuous information.

Suppose that we are interested in recording the progress of some action, say a footrace. Using an appropriate camera, we could take a series of k still photographs. If we label the photographs 1, 2, 3, ..., k in the order in which we have snapped them, then we have a set of pictures—each separate and distinct. The actual race is a continuous action; the ordered set of photographs gives us discrete information about the race.

This example can be carried a bit farther. If we have a movie camera (which really is a very fast still camera), we can, by viewing the resulting film, obtain a very close approximation to the actual continuous action.

One additional point should be recognized. In some investigations or experiments, it is possible to obtain *all* the relevant information. If we wish to investigate some physical characteristic, say height, of a small group of men who have just reached their eighteenth birthday, we can simply measure them all. Our findings will apply directly to that group. We might find, for instance, that the average height was 5' 9". On the other hand, we might be interested in the average height of all 18-year-olds in the United States; we would be unable to measure them all. To conduct such a study, we would have to be content to measure the heights of *some* (perhaps very many) 18-year-olds. We would realize that our information was not complete, only partial. Hopefully, the subset of the individuals that we measured would in some sense represent the whole set of 18-year-olds. The entire set under consideration is called the *population*. The selected subset is a *sample* of the population. In Chapter 9 we shall discuss the idea of a sample more carefully.

EXERCISES 1.4

For each of the following data, decide whether it "makes sense" to reorder the numbers. If so, propose a reasonable reordering.

- **1.** The numbers: 1, 16, 128, 2, 32, 4, 8, 64, ...

- **2.** The numbers: 95, 95, 97, 100, 98, 96, 99, 102, 104, ...

- **3.** The heights of several students in a class:

Ann	5' 3"	Doris	5' 11"	Grace	5' 5"
Bob	6' 1"	Everett	4' 11"	Henry	6' 0"
Charles	5' 7"	Freda	5' 3"	Irwin	5' 8"

- **4.** The hourly record of pressure in pounds per square inch.

5:00 A.M.	95	8:00 A.M.	100	11:00 A.M.	99
6:00 A.M.	95	9:00 A.M.	98	12:00 NOON	102
7:00 A.M.	97	10:00 A.M.	96	1:00 P.M.	104

- **5.** The annual average temperature (in degrees Fahrenheit) of certain cities in the United States:

Atlanta	61.4	Memphis	61.5	Phoenix	69.0
Chicago	50.8	Miami	75.1	New York	54.5
Denver	49.5	Minneapolis	43.7	San Francisco	56.8

- **6.** Give a common example in which a sampling is offered to represent the measurement for a large population.

1.5 A Word about Words

Mathematicians often adopt ordinary English words and give them special meanings. In elementary mathematics courses, words such as *commutative*, *associative*, and *distributive* apply in a very special way to properties of numbers. We are used to the mathematical meaning of *function*, *union* (of sets), *factor*, *product*, and a host of other words.

The possibility of confusing the ordinary with the mathematical meaning of words is particularly great in a study of statistics and probability. Already we have introduced some words which will have special meaning for us. In the preceding section, we discussed (but did not formally define) *discrete*, *continuous*, *population*, and *sample*. As we proceed with our study, we shall encounter a great many more such words. It is necessary to remember that one must judge from the context whether such words are being used in the general or the mathematical sense.

Later, we shall also give special meanings to the words *outcome*, *event*, *independent*, and *significance*.

One special term to be used in this text requires some explanation. In considering actual or mind experiments, we often wish to refer to drawing a ball, marble, or numbered slip of paper from a container. The container might be a bag or a box. In common language we sometimes speak of "drawing from a hat." In any event, the idea should be clear: The container must be constructed so that the person making the drawing cannot peek. It is customary in probability courses to call this kind of container an *urn*. Problems connected with drawings from urns are called *urn problems*.

There are two distinct types of urn problems. Sometimes, in making several draws, we return the marble (or other object) to the urn before making the next draw. Other times we do not return the marble; we simply draw the next time from the remaining marbles. These types of urn problems are distinguished by using the phrases "drawing *with* replacement" and "drawing *without* replacement." In either of these cases, we are thinking of drawing *at random;* that is, no marble is especially favored or unfavored, and each is *equally likely* to be drawn.

EXERCISES 1.5

1. Find ordinary dictionary definitions for the following words. You might want to *contrast* these meanings with their mathematical definitions as we introduce these concepts.

   continuous population sample
   outcome event independent
   significance

- 2. Suppose that there are 100 marbles in a box and that 5 marbles are to be drawn in one drawing. Explain how you might consider this to be an urn problem in which repeated drawings are made. Would this be with or without replacement?

1.6 What Do You Think?

We shall include, at the end of most chapters, a section with this title. We hope to offer some special problems which, though varied in nature, will offer a challenge to the intuition and imagination. Many of these problems you will not be able to "answer" mathematically at the time they are presented. They are designed to stimulate discussion. You are supposed to make your decision on the basis of intuition and intelligent guessing. We include such sections in the belief that by actively engaging in an attempt to think through these problems, you will gain considerably more insight into the topics under discussion. As we progress, and as our familiarity with probability and statistical inference increases, we shall reexamine some of these problems. Eventually, we shall develop the necessary mathematical tools to attack them systematically.

EXERCISES 1.6

- 1. Suppose that we have a coin which we *know* is fair. In each experiment described below, which result do you think is more likely?

	Experiment	Result I	Result II
a.	Toss once	1 head	0 heads
b.	Toss five times	3 heads	5 heads
c.	Toss five times	3 heads	2 heads
d.	Toss ten times	3 heads	5 heads

- **2.** Again, assume that we have a fair coin. For each situation, which result do you think is more likely?
    - **a.** (I) Toss 10 times, get 5 heads, or
        (II) toss 100 times, get 50 heads.
    - **b.** (I) Toss 10 times, get from 4 to 6 heads, or
        (II) toss 100 times, get from 40 to 60 heads.
    - **c.** (I) Toss 100 times, get from 45 to 55 heads, or
        (II) toss 1000 times, get from 450 to 550 heads.
    - **d.** (I) Toss 10 times, get 8 or more heads, or
        (II) toss 100 times, get 80 or more heads.

- **3.** An urn contains 10 marbles. We are told that there are either (I) 2 red, 8 blue, or (II) 8 red, 2 blue marbles. We stir, draw a marble, record its color, replace it, stir again, draw again. In each case decide whether (I) or (II) is more likely:
    - **a.** One draw; result: red
    - **b.** Three draws; results: red, red, red
    - **c.** Three draws; results: red, blue, blue
    - **d.** Four draws; results: red, blue, blue, red.

Chapter 2

Probability— Preliminary Ideas

In this chapter, we shall lay the groundwork for our development of the theory of probability. The approach here is to use informal arguments based on some simple ideas of sets rather than to attempt a more rigorous development, which would require a rather thorough knowledge of set theory.

Particularly important in this chapter is the introduction of some of the technical language mentioned in Section 1.5.

2.1 Theoretical (*A Priori*) and Experimental (Empirical) Probabilities

Often to estimate (in the face of uncertainty) the probability, or measure of the degree of belief, that some given situation will occur, we can perform experiments. However, there are some situations in which we can arrive at a measure of a degree of belief even before any experimental results have been obtained.

For example, suppose that an urn contains 6 red and 2 blue marbles indistinguishable by texture, size, or weight. It is reasonable to argue that if a ball is drawn *at random* (recall Section 1.5), it is three times as likely to be red as blue. In this case, the measure of our degree of belief in the color of the marble to be drawn is based on the following deductive reasoning. If a ball is drawn "at random," we assume that the marbles are thoroughly mixed and that the draw is made in such a way that no marble is especially favored or unfavored. Thus, we conclude that all marbles are *equally likely* to be drawn. Since there are three times as many red as blue marbles, we arrive logically at the result stated above. This type of reasoning is called *a priori* reasoning, and the probability assigned is called an *a priori* probability.

In the real world, of course, such clear-cut situations are rather rare. For the urn example, we knew in advance the ratio of red to blue marbles, and we made estimates on the outcomes of the drawings. In practical problems, the situation is often quite the reverse; that is, we wish to estimate the ratio of the "mix" from observed results of "drawings."

For instance, a manufacturer of electronic equipment knows that not every item produced by his plant will meet specifications. There are many places in the manufacturing process where slight imperfections may occur. How can he determine what fraction of his manufactured parts may be faulty? One method, of course, would be to test every item. Unfortunately, testing is always expensive and time-consuming. Furthermore, in some cases, it may be damaging or destructive to the part that is being tested. (If an electric fuse is designed to blow under a 20-ampere load, how does one test whether or not it will blow without destroying the fuse?)

The solution is to test *some* of the manufactured parts and, from these results, make estimates about the entire lot of manufactured parts. Suppose that testing a lot of 100 pieces shows 2 pieces to be defective. If there is reason to believe that the tested items are representative of the total production, then it can be estimated that the ratio of good to faulty parts is 98 to 2. That is, if a customer buys one of the parts, the probability that the part is good is 49 times the probability that the part is faulty. (In most manufacturing processes, the percent of faulty parts is quite a bit less than 2%.) Such probabilities are called **empirical,** in contrast to the *a priori* probabilities mentioned at the beginning of this section. Empirical probabilities may also be based in "historical data" (Section 1.3).

What, if any, is the relation between these two kinds of probability? We mention here one important idea. Suppose that we wish to determine whether or not a given coin is fair. We might make the *a priori* assumption that heads and tails are *equally likely*. On the basis of this assumption we predict, before testing, that the coin will fall heads in about half of all tosses. We can then test the coin by making numerous tosses. The actual results—the experimental evidence—can then be compared with the predicted results. Finally, we can make a decision as to whether or not our *a priori* assumption was correct. (Recall that in Section 1.3 we mentioned the difficulty of deciding whether an assumption is, in fact, true.) Such a procedure is an example of *hypothesis testing*, and we shall devote Chapter 11 to a discussion of such testing.

In short, we can think of an *a priori* probability estimate as a kind of "model" with which experimental evidence can be compared.

EXERCISES 2.1

Some of the questions in this set of exercises do not have a "right" answer. Try to make an "intelligent guess," giving some justification for your guess.

- **1.** In a production of 100,000 items, a random selection of 100 of these were tested and 2 were found to be defective.
  - **a.** What is the estimated number of defective items in the entire production?
  - **b.** Suppose that every one of the 100 items tested was found to be defective. If you were the manufacturer, what kinds of action might this discovery cause you to take?
  - **c.** Suppose that every one of the 100 items tested was found to be defective, and when another ten were tested none of these were found to be defective. If you were the manufacturer, what questions might you ask about the testing procedures? What explanation might account for such results?
  - **d.** To ensure "randomness" in selecting 100 out of 100,000 items, a foreman suggests selecting every 1000th item for testing. Comment on the possible weakness in such a program for selection.

- **2.** In a testing program to find out the general ability of students in a particular field, various school districts throughout a large state were selected at random and every student in each of these chosen districts was tested.
  - **a.** Comment on the possible weakness in such a testing program.
  - **b.** Suggest ways to improve on the proposed scheme.

2.2 Sets of Possible Outcomes of Experiments

Let us return to the example (Section 2.1) of the urn containing 6 red and 2 blue marbles. Consider drawing twice, with replacement (recall Section 1.5). What are the possible results of this experiment? The answer depends on what things interest us. Examine the following list of some questions that we might ask.

1. Do the marbles in the two draws match in color?
2. How many red marbles were drawn?
3. What were the colors on each draw?

We shall concentrate on this list, but it should be noticed that this is only a partial list of questions that could be asked. The point is that the same experiment (in this case, a repeated drawing with replacement) can be the framework for a number of different questions. For example, we might be interested in the color of the second marble drawn, or whether the number of red marbles drawn is greater than the number of blue marbles, and so on.

Associated with each question is a **set of possible outcomes:**

Question	Set of Possible Outcomes
1. Do the marbles in the two draws match in color?	{match, do not match}
2. How many red marbles were drawn?	{2, 1, 0}
3. What were the colors on each draw?	{rr, rb, br, bb}

Here we have used abbreviations to indicate the results. By the notation "rb," for example, we mean the ordered pair

"a red marble on the first draw, a blue marble on the second draw."

Although the sets of possible outcomes listed above are different, they share two common features:

First, in each case, the elements of the set *exhaust* (form a complete list of) *the possibilities* of what will happen if the experiment were performed. The colors either match or do not match; we obtain exactly 2, 1, or 0 red marbles; rr, rb, br, bb describe all the possibilities of the color of the marbles drawn on the two drawings.

Second, the elements in each set are *distinct*; they do not overlap. Thus, the colors cannot both match and not match; thus too, if we get exactly two red marbles, we cannot get exactly one red; and so on.

The two features stated above are incorporated in the following description.

By a *set of outcomes of an experiment,* we mean a set of distinct possibilities which exhaust the possible results of the experiment.

We have seen that a particular experiment may have more than one set of outcomes associated with it. Which set we decide to work with depends on our interest in making the experiment.

If we examine the set

{rr, rb, br, bb},

we see that these elements have been chosen by considering the 2-draw experiment in two steps. First, we list what can happen on the first draw. Either a red marble or a blue marble can be obtained. This situation can be visualized as in Figure 2.1.

Figure 2.1

First draw

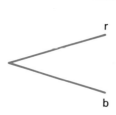

No matter which color appears on the first draw, there are again two choices for the second draw. This is illustrated by extending Figure 2.1 to Figure 2.2.

Figure 2.2

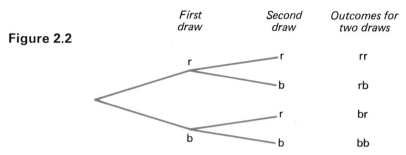

The diagram of Figure 2.2 is called a **tree diagram.** Such a diagram may be very useful in helping to determine a set of outcomes.

As another example, suppose that a coin is to be tossed and then a die is to be thrown. One possible result is to have the coin fall heads and a 3 to be thrown on the die. Let us agree to represent this by the notation h3. Using a tree diagram, we see that there are 12 possible outcomes of this type. Again, we think of the experiment in two steps as shown in Figure 2.3.

Figure 2.3

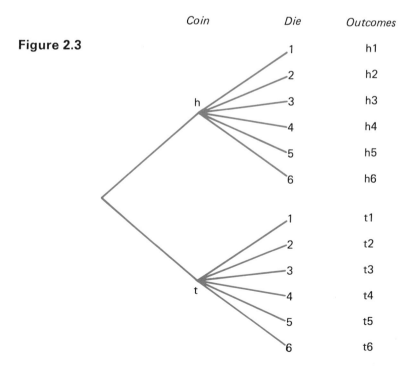

Our examples have illustrated the following:

MULTIPLICATION PRINCIPLE If one step of an experiment can happen in exactly *m* ways, and if, for each of these, a second step can happen in exactly *n* ways, then the experiment which consists of these two steps can happen in exactly $m \cdot n$ ways.

It is clear that we may extend the multiplication principle to more than two steps. If a die is thrown three times, one set of possible outcomes contains $6 \cdot 6 \cdot 6$, or 216, members. In this case, we would not like to draw the complete tree diagram, but we might imagine or visualize it.

In our examples, all of the sets of outcomes have been finite. Consider now a simple "experiment" consisting merely of choosing a real number in the interval from 0 to 10. Here, the set of possible outcomes is no longer finite, and different theories apply to the probabilities. However, it is often possible to approximate such an infinite set by a finite one (see Problem 15, in the following set of Exercises). This idea is somewhat the same as the idea of approximating continuous data by discrete data which was mentioned in Section 1.4.

EXERCISES 2.2

- 1. Eight playing cards, ace and 8 of hearts and ace, 6, 5, 4, 3, and 2 of spades, are shuffled and placed face down on a table. One card is to be selected at random.
    a. Describe a set of outcomes having two elements if our interest is in the suit of the card selected.
    b. Describe a set of outcomes having two elements if our interest is in whether or not an ace is drawn.
    c. Describe a possible set of outcomes with eight elements.
    d. For which of these sets is it reasonable to argue that the individual outcomes are equally likely?
- 2. A penny and a dime are to be tossed. Describe (list) three sets of possible outcomes.
- 3. Three coins are to be tossed.
    a. Describe a set of possible outcomes based on the number of heads that may appear.
    b. Describe a set of possible outcomes having eight members. (You may wish to use a tree diagram.)
- 4. Four coins are to be tossed. A tree diagram leads us to 16 outcomes. How many outcomes would we have if:
    a. 5 coins are to be tossed?
    b. *n* coins are to be tossed?

- 5. Suppose that a die is thrown and then a coin is tossed. Draw a tree diagram for this two-step experiment. Show that there is a one-to-one correspondence between the set of outcomes you obtain and the set indicated in Figure 2.3.
- 6. Suppose that a coin and a die are to be thrown at the same instant (simultaneously). Find a set of possible outcomes having 12 members.
- 7. For the urn problem of Section 2.1 we recorded on page 18 three sets of possible outcomes:

$$S_1 = \{\text{match, do not match}\}$$
$$S_2 = \{2, 1, 0\}$$
$$S_3 = \{\text{rr, rb, br, bb}\}$$

  - a. Show that each element of S_1 may be identified with a subset of S_2.
  - b. Similarly for S_1 and S_3.
  - c. Similarly for S_2 and S_3.
  - d. If we know that rb occurred, then we can answer the questions, "Do they match in color?" and "How many red marbles were drawn?" Give an argument that it might be more useful to use S_3 rather than S_1 or S_2 as a set of outcomes for this experiment.
- 8. Again we start with an urn with 6 red and 2 blue marbles. This time we make three draws, one after another, with replacement.
  - a. Draw a tree diagram for the outcomes of this experiment considered as being composed of three steps.
  - b. How many outcomes are there?
  - c. Does the multiplication principle apply?
- 9. Still using an urn with 6 red and 2 blue marbles, we again make three draws, but this time *without replacement*.
  - a. Draw a tree diagram.
  - b. How many outcomes are there?
  - c. Does the multiplication principle apply here?
- 10. There are two urns. Urn I contains 1 red, 1 blue, and 1 green marble. Urn II contains 2 red marbles and 1 blue marble. One of the urns is selected and then a marble is drawn from that urn. Draw a tree diagram, and give a set of possible outcomes for this experiment based on the diagram.
- 11. An urn contains 3 identical envelopes, of which one contains a penny, one contains a nickel, and one contains a dime. Three drawings with replacement are to be made. Draw or visualize a tree diagram.
  - a. From the tree diagram, explain whether you think that the possible outcomes are equally likely; if not, which seems to be the most likely?
  - b. How many branches are there in this tree diagram?
  - c. What is the set of possible outcomes in terms of the dollar value of the coins drawn? List the possible outcomes for this. Are these *equally likely?*

12. An urn contains envelopes containing a penny, a nickel, and a dime as in Problem 11. Drawings with replacement are to be made. How many possible outcomes are there in
    a. 1 draw?   b. 2 draws?   c. 3 draws?
13. Each step of an experiment can happen in m ways. How many ways are there after:
    a. 2 steps?   b. 3 steps?   c. n steps?
14. Three sports car drivers, Ames, Brown, and Carlson, compete in a special competition. How many possible orders of finish are there if we assume no ties? (Use a tree diagram.)
- 15. "Choose a real number in the interval 0 to 10" leads to an infinite set of outcomes. Let us modify this experiment a bit. After choosing a real number, we might round it off to the nearest whole number, the nearest tenth, and so on. If we do this rounding off, the resulting set of outcomes is finite. For each modification, determine the number of elements in the corresponding set of outcomes.
    a. Round off to the nearest whole number.
    b. Round off to the nearest tenth.
    c. Round off to the nearest hundredth.
    d. Round off to the nearest 10^{-n}.
16. A single die is to be tossed. Describe four sets of possible outcomes.
17. Two marbles are to be drawn from an urn containing 3 red marbles, 3 blue, and 3 green.
    a. Suggest a set of possible outcomes which will specify the number of green marbles drawn.
    b. Describe a nine-element set of possible outcomes.
18. A coin is tossed. If heads turns up, the coin is tossed again. If tails turns up, a die is thrown. Describe a set of possible outcomes.

2.3 Assigning Probabilities

In this section we shall concentrate on the *a priori* type of probability.

In Sections 2.1 and 2.2 we considered experiments with an urn containing 6 red and 2 blue marbles. In the case of drawing a single marble from this urn, we agreed in Section 2.1 that whatever probability we associate with "red," it should be three times the probability associated with "blue." Now our goal is to assign to each outcome a number as a measure of our degree of belief. This number will be called the *probability of the outcome*.

Before we settle on a scheme for assigning this number, we shall consider some of the characteristics we would like it to have. Unless we have reason to believe otherwise, it makes sense to look for a measure that expresses closely our intuitive feelings, such as the agreement mentioned in the preceding paragraph. That is, we shall wish to assign a larger number to an outcome in which we have a greater degree of belief.

Referring to our example of the 6 red and 2 blue marbles, we indicated earlier that if we assign any weight, w, to red, then we must assign only a third as much, or $\frac{1}{3}w$, to blue. One possibility would be to assign 6 to red and 2 to blue. There are disadvantages in this, however.

For instance, suppose that we think about an urn with 12 red and 4 blue marbles and assign the number 12 to red and 4 to blue. As was the case for the 6-to-2 mixture, intuition leads us to think that the measure of our degree of belief for red must be 3 times that for blue. So the degree of belief for red is the same in the two situations. However, different numbers (6 in one case, 12 in the other) are assigned to the same degree of belief. Likewise for our degree of belief for blue. What shall we do?

In both situations, the red marbles are $\frac{3}{4}$ of the total number of marbles. If we agree to use $\frac{3}{4}$ as a measure of our degree of belief in red, it follows that $\frac{1}{4}$ is the appropriate measure for our degree of belief in blue. Using this procedure, we would have the same probability, $\frac{3}{4}$, for drawing a single red marble in any urn problem in which the mixture of red and blue is in the ratio 3:1.

Example 2.3.1 Suppose that we consider an urn containing 5 red, 3 blue, and 2 yellow marbles. If we follow the procedure suggested above, we assign 0.5, 0.3, and 0.2 as the probabilities of the outcomes red, blue, and yellow, respectively.

Example 2.3.2 Four cards—the ace, king, queen, and jack of spades—are placed face down and one card is selected at random. Since each card has the same chance of being selected as any other (that is, the outcomes are *equally likely*), we would assign the same *a priori* probability (namely, $\frac{1}{4}$) to each.

Example 2.3.3 If an urn contains only red and blue marbles, we agree that {red, blue} is an appropriate set of outcomes. Suppose that someone suggests {red, blue, yellow} as a possible set of outcomes. We could, of course, reject the suggestion. If we do agree to include yellow, what probability should we assign to this outcome? It is reasonable to assign 0 as the measure of our degree of belief.

Example 2.3.4 If an urn contains only red marbles, we agree that there is only one possible outcome, {red}, and that the measure of our degree of belief, in this case, is 1.

In the examples that we have considered, it is relatively simple to agree on a set of outcomes and on the associated probabilities. In more complicated situations, matters might not be as straightforward. On the basis of these few examples, however, we can say the following:

Given a finite set of outcomes, *the probability of each outcome is a number between 0 and 1 (inclusive) assigned in such a way that the sum of the probabilities of all outcomes in the set is 1;* equally likely outcomes shall be assigned the same probability.

In mathematical terminology, these rules may be compactly stated thus:

For a finite set of n outcomes, $S = \{s_1, s_2, \ldots, s_n\}$,

(1) each $s_i \in S$ is assigned a unique number (probability) $f(s_i)$;*

(2) for each i,
$$0 \leq f(s_i) \leq 1;$$

(3) $f(s_1) + f(s_2) + \cdots + f(s_n) = \sum_{i=1}^{n} f(s_i) = 1;$†

(4) If $s_1, s_2, \ldots, s_n$ are equally likely, then for each i,
$$f(s_i) = \frac{1}{n}.$$

Note that the assignment of a unique number, $f(s_i)$, to each outcome, s_i, is a *function*. We shall discuss this further in our later development (Section 2.5).

EXERCISES 2.3‡

- 1. If a fair coin is tossed, what probability should be assigned to "heads"? (Recall that by fair, we mean heads and tails are equally likely (Section 1.1)).
- 2. If a balanced (fair) die is thrown, what probability would be assigned to "1"? to "4"?
- 3. If a fair coin is tossed and a balanced die is thrown, we may consider a set of outcomes having 12 members. What probability should be assigned to each?
- 4. If an experiment leads to k outcomes, each of which is equally likely, what probability should be assigned to each?
  5. A card is to be dealt from a shuffled deck. What probability should be associated with the outcome "ace of spades"?

*"each $s_i \in S$" means "each s_i belonging to S."
†See the note on summation notation in the Appendix.
‡Items marked with ★ are to be considered optional.

6. The junior class has 700 men and 500 women. John is a member of the class. One junior student is selected at random.
   a. What is the probability that John is selected?
   b. Suppose that the selection is made only from the men students. In that case, what is the probability that John is selected?

7. For the eight-member set of outcomes of Problem 1c, Exercises 2.2, what probability should be assigned to each?

8. An urn contains a penny, a nickel, and a dime in identical envelopes. Two drawings with replacement are to be made. In dollar value, a set of outcomes is {0.02, 0.06, 0.10, 0.11, 0.15, 0.20}. What is the probability for each member of this set of outcomes? (Hint: Draw a tree diagram.)

9. You and a friend each choose a letter of the alphabet at random. What is the probability that both of you choose the same letter?

10. Suppose that a spinner, with dial marked as shown, is spun once. (We have omitted the pointer in the illustration.)

    a. $f(x) = $ __?__
    b. $f(y) = $ __?__
    c. $f(z) = $ __?__

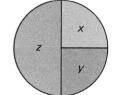

11. a. What is wrong with the following argument?
    "In the United States, a person is chosen at random. The probability that he is a resident of a state whose name begins with the letter M is equal to the probability that he is a resident of a state whose name begins with the letter N. (There are 8 M-states and 8 N-states.)"
    b. How would you determine the probability for each (an M-stater and an N-stater)?

12. The automobile license numbers for a certain state consist of some letters followed by three digits (from 000 to 999).
    a. What is the probability that a license number has all 3 digits the same?
    b. What is the probability that a license number has 3 successive digits in sequence, that is 012, ..., 789?
    ★ c. What is the probability that a license number has two similar digits and one different digit?
    ★ d. What is the probability that a license number is symmetric (reads the same forward or backward)?

13. A lot of 100 manufactured items have been placed in a bin after production. It is known that five of these items are defective. If one item is removed at random from the bin, what is the probability that this item is defective?

★ 14. The state referred to in Problem 12 decides that the digit "0" cannot appear in automobile license numbers. Revise the probabilities in parts a through d.

2.4 Events and Their Probabilities

"When, in the course of human events, . . ."

In the opening words of the Declaration of Independence, quoted above, the noun "event" has its ordinary, dictionary meaning. In this sense, an event is an occurrence, an incident, a happening. For our purposes, however, we wish to give the word a special mathematical meaning (recall Section 1.5), and we shall do that in this section. In what follows we are restricting ourselves to sets of outcomes that are finite.

The seven words of the phrase quoted above contain a total of 28 letters. Suppose that we select one word from this phrase in such a way that the probability that any word is selected is proportional to the number of letters in the word. Thus, because "course" contains 6 letters and "in" contains 2, the word "course" is three times as likely to be selected as is the word "in." This situation would occur, for example, if the words were spelled out evenly along the edge of the dial of a spinner as shown in Figure 2.4. If we think of this selection as an experiment, we have, as a set of outcomes,

Figure 2.4

$$S = \{\text{when, in, the, course, of, human, events}\}.$$

Our method of selecting a member of this set leads us to assign the following probabilities:

Word selected (outcome)	when	in	the	course	of	human	events
Probability of selecting that word	$\frac{4}{28}$	$\frac{2}{28}$	$\frac{3}{28}$	$\frac{6}{28}$	$\frac{2}{28}$	$\frac{5}{28}$	$\frac{6}{28}$

Let us examine some particular subset of S, say,

$$\{\text{course, human}\}.$$

This particular subset may be described verbally as "the subset of S consisting of all words in the quotation containing the letter u." We may now ask:

What is the probability that the selected word is a member of this subset?

Our subset consists of two outcomes, "course" and "human." The corresponding probabilities are $\frac{6}{28}$ and $\frac{5}{28}$. The subset {course, human} may be indicated as in Figure 2.5.

Figure 2.5

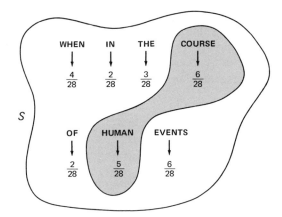

The total of the probabilities of the elements of S is 1. The word "course" contributes $\frac{6}{28}$ to that total, and "human" contributes $\frac{5}{28}$. Together, these elements contribute $\frac{6}{28} + \frac{5}{28} = \frac{11}{28}$. We have, then,

probability of {course, human} = $\frac{11}{28}$,

or using a more compact notation,

$$P(\{\text{course, human}\}) = \tfrac{11}{28}.$$

As we have noted before, {course, human} is a subset of S, the set of possible outcomes. Frequently we shall have occasion to refer to this concept "*a subset of a set of possible outcomes.*" To replace this awkward expression with a simpler one, we make the following definition:

An *event* is a subset of a set of possible outcomes of an experiment.

Further, since a subset of S consists of distinct elements of S, each with an assigned probability, an event has a probability:

The *probability of an event* equals the sum of the probabilities of its individual outcomes.

Formally, let S be a set of outcomes of an experiment:

$$S = \{s_1, s_2, \ldots, s_n\}$$

Let $A = \{s_1, s_2, \ldots, s_m\}$, $m \leq n$, be a subset of S. Then, recalling (Section 2.3) that by the notation $f(s_i)$ we mean the probability that s_i occurs, the above statement may be expressed as:

$$P(A) = \sum_{i=1}^{m} f(s_i)$$

So, at this point, two kinds of things have probabilities: outcomes and events. We can simplify things a bit if we identify each outcome with the corresponding event having just that one outcome as a member. For example, the outcome "course" is thought of as the event {course}. We shall call such an event an **elementary event**. Hence,

$$P(\{\text{course}\}) = \tfrac{6}{28}.$$

With this agreement, we can state that *events, and only events, have probabilities*.

In order to avoid excessive notation, it is often convenient to omit the braces and write, for example,

$$P(\text{course}) = \tfrac{6}{28}, \qquad P(\text{course, human}) = \tfrac{11}{28}.$$

Further, it is common practice to use a capital letter to designate a set. If we let $C = \{\text{course}\}$, then $P(C) = \tfrac{6}{28}$.

EXERCISES 2.4

- 1. This problem refers to the example of Section 2.4: "When, in the course of human events, ..." We give verbal descriptions of the events. List the elements of each event and find its probability.
  a. E is the event "the word selected contains the letter e."
  b. W is the event "the word selected begins with the letter w."
  c. V is the event "the word selected ends in a vowel."
  d. H is the event "the word selected contains the letter h."
  e. T is the event "the word selected has exactly two letters."

- 2. A balanced die is to be thrown. (Assume that there are six equally likely outcomes.) Find the probability that the number shown on the die is:
  a. 5   b. an even number   c. a prime number
  d. divisible by 3   e. less than 5   f. 5 or more

- 3. A card is to be chosen at random from a shuffled deck. Find the probability that the card drawn is:
  a. the ace of spades   b. an ace   c. a spade
  d. an honor card (The honor cards are ace, king, queen, jack, 10.)
  e. black (Spades and clubs are black.)

- 4. A red and a green die are to be thrown. We may describe a set S of 36 equally likely outcomes. (Think of the tree diagram.) Let (3, 2), for example, stand for "3 on the red die, 2 on the green die."
  a. List the 36 elements of S.
  b. For each element of S, examine the *sum* of the numbers. (If (3, 2) occurs, the sum is 5.) Complete the following table.

Sum	2	3	4	5	6	7	8	9	10	11	12
Number of outcomes	1	2				5					

c. Let X represent the sum for any particular throw. Complete the following table.

Possible value of X	2	3	4	5	6	7	8	9	10	11	12
Probability that X has that value	$\frac{1}{36}$	$\frac{2}{36}$									

d. What is the probability of obtaining a sum of 7 *or* 11?

e. Look at the 36 outcomes of part a. What is the probability that at least one die shows a 6?

- 5. The dial of a spinner for a game is divided into 9 equal sections and marked as shown. Assuming that the pointer does not stop on a dividing line, find the probability of spinning:

    a. 1; a number greater than 1

    b. an odd number; an even number

    c. less than 3; 3 or more

    d. 4; either 1, 2, 3, 5, or 6

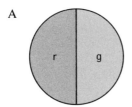

- 6. A player "wins" if he guesses correctly the color of the region in which the pointer of a spinner stops. Two choices of spinners are available for playing, with dials colored as labeled in Figures A and B (r for red, g for green):

    A   B

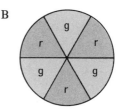

    Dave maintains that he would rather play on Spinner A because the wider bands of the region make it easier for the pointer to stop at the color of his choice. Explain whether or not he is justified in his belief.

- 7. Which of the following statements is (are) correct interpretation(s) for the forecast "the probability for rain tomorrow is 20%."

    a. It will rain one minute out of every five.

    b. During the 24-hour interval, it will rain approximately 20% (4 hours 48 minutes) of the time.

    c. Historically, for such weather conditions, it has rained 1 time out of every 5.

    d. Weather conditions seldom duplicate exactly those previously recorded. Some conditions appear at some times; others appear at other times. From these past performances, a sort of compromise "percentage" of rain is reached.

    e. History has nothing to do with the forecast; the weatherman considers that he will be correct in forecasting rain 20% of the time.

8. Two requirements were made in Section 2.2 on a set of possible outcomes. Suppose that two coins were tossed.
   a. Explain whether or not {both heads, not both heads} meets the requirements for a set of possible outcomes.
   b. Explain whether or not {both heads, both tails} meets the requirements for a set of possible outcomes.
   c. Are the outcomes in the event {both heads, not both heads} equally likely? That is, if you were to guess "both heads" or "not both heads," would you have equal chance of being correct taking either choice? (Hint: Use a tree diagram.)
   d. $P(\text{both heads}) = \underline{\ ?\ }$ $P(\text{not both heads}) = \underline{\ ?\ }$

9. Suppose that the events "son" and "daughter" are equally likely. For families having two children (exactly), what is the probability for both children to be sons?

10. An automobile manufacturer took a survey and found that 50% of the potential customers surveyed preferred car tops to be painted green and 50% preferred the lower half of the car to be painted red.
   a. Explain whether or not the manufacturer is justified in producing 50% two-tone cars, green on the top and red on the bottom.
   b. If the 50% figures given above are changed to 60%, explain why the manufacturer might decide to produce 20% "green top, red bottom" cars.

11. Consider the following sentence and select one word in such a way that the probability that any word is selected is proportional to the number of letters in the word: "A card is to be chosen at random from a shuffled deck." Describe the elements in each event and determine its probability.
   a. C is the event "the word selected contains the letter c."
   b. F is the event "the word selected contains exactly four letters."
   c. T is the event "the word selected contains at least three letters."

12. What is the probability of selecting the following at random from the set of spades and clubs of a shuffled deck of cards:
   a. a jack
   b. the queen of spades
   c. an honor card (see Problem 3d)

13. In Problem 4, what is the probability that the sum of the two numbers thrown is:
   a. odd   b. even   c. a prime number

14. Considering the toss of two dice:
   a. explain whether or not {both odd, both even} meets the requirements for a set of possible outcomes.
   b. explain whether or not {both odd, not both odd} meets the requirements for a set of possible outcomes.
   c. explain whether or not {at least one odd, neither odd} meets the requirements for a set of possible outcomes.
   d. $P(\text{at least one odd}) = \underline{\ ?\ }$; $P(\text{neither odd}) = \underline{\ ?\ }$.

2.5 Summary of Probabilities of Events

We shall now examine carefully some special subsets of a set of outcomes,
$$S = \{s_1, s_2, \ldots, s_n\}.$$
(We might call these "special events"!)

First, we shall make formal statements about some properties suggested by the discussion in Section 2.3.

Set S, of course, is a subset of itself. Hence S is also an event. What is the probability of this event? Our definition of the probability of an event (Section 2.4) gives us
$$P(S) = f(s_1) + f(s_2) + \cdots + f(s_n) = \sum_i f(s_i).*$$

By rule (3) near the end of Section 2.3, $\sum_i f(s_i) = 1$. Therefore,
$$P(S) = 1.$$

You may recall from your earlier work with sets that the empty set, $\emptyset$, is a subset of every set. Now $\emptyset$ contains no members; so if we attempt to add the probabilities of the members of $\emptyset$, we see that
$$P(\emptyset) = 0.$$
This is surely an appropriate measure of our degree of belief in this event.

From these results, together with those of Section 2.4, we see that if A is any subset of A,
$$0 \leq P(A) \leq 1.$$

Finally, let
$$S = \{a_1, a_2, \ldots, a_r, b_1, b_2, \ldots, b_t\}, \quad r + t = n.$$
If $A = \{a_1, a_2, \ldots, a_r\}$, then the *complement* of A in S is the event $\{b_1, b_2, \ldots, b_t\}$. Let us denote the complement of A by $\sim A$. If we know the probability of A, it is easy to find the probability of $\sim A$:

$1 = P(S)$
$\quad = \underbrace{f(a_1) + f(a_2) + \cdots + f(a_r)}_{P(A)} + \underbrace{f(b_1) + f(b_2) + \cdots + f(b_t)}_{P(\sim A)}$

Hence
$$P(\sim A) = 1 - P(A).$$

*For convenience, we are using the symbol $\sum_i f(s_i)$ in place of $\sum_{i=1}^{n} f(s_i)$. We do this on the assumption that there is no confusion here about the summation. In fact, where we feel that there is not likely to be confusion, we shall further, for convenience, drop the summation index i, writing, for example, simply $\sum f(s_i)$ or even $\sum f(s)$.

We can summarize our development of probability up to this point as follows:

Given an experiment and a set of possible outcomes, S:

1. An event is a subset of S. Events, and only events, have probabilities.
2. $P(S) = 1$.
3. $P(\emptyset) = 0$.
4. If A is an event, then $0 \leq P(A) \leq 1$.
5. If A is an event, then $\sim A$ is also an event, and
$$P(\sim A) = 1 - P(A).$$

Since a unique real number in the interval $[0, 1]$* is associated with the event, such an association was noted to be a function (Section 2.3). This *probability function* may be depicted thus:

an event $A \to$ a real number in $[0, 1]$

In the language of functions, we say that each event A of S corresponds to a real number in the interval $[0, 1]$. The *domain* of this function is the set of possible events in S. The *range* is a subset of $[0, 1]$.

Notice that the probability of an event or an outcome is defined *relative to the set of possible outcomes*. For example, the event "a king is drawn from a (shuffled) deck of playing cards" has probability $\frac{1}{13}$. A pinochle deck, on the other hand, has 48 cards, double the ordinary playing cards from nines to aces. Thus, there are 8 kings, and so the event "a king is drawn from a pinochle deck" has probability $\frac{1}{6}$. We might use the notation $P(\text{king})$ for both cases, but the number assigned to $P(\text{king})$ depends on the particular experiment and the associated set of outcomes.

In the development so far, we have relied mainly on informal arguments, and no attempt has been made to give a rigorous treatment. In Section 2.7, we shall find that statements 3 and 5 above can be proved as theorems.

EXERCISES 2.5

- **1.** This problem refers to the quotation given at the beginning of Section 2.4, "When, in the course of human events, ...," together with the procedure described there for establishing probabilities. Find the probability for the event specified:

    a. if E is the event "the word selected contains the letter e," the probability of $\sim E$

    b. if H is the event "the word selected contains the letter h," $P(\sim H)$

*"$a \in [0, 1]$" means "$0 \leq a \leq 1$."

c. if the event G is "the word ends in g," $P(G)$
d. $P(D)$, where D is the event "the word selected contains fewer than 10 letters."

● 2. Consider the same quotation and same procedure for probabilities as in Problem 1.
   a. Let F be the event "the word selected contains at least one of the first five letters of the alphabet." Is it easier to find F or $\sim F$? What is $P(F)$?
   b. Let L be the event "the word selected contains at least one of the last three letters of the alphabet." $P(L) = \underline{\quad ? \quad}$

● 3. A red and a green die are to be thrown (Problem 4, Exercises 2.4). Let G be the event "the number on the green die is greater than the number on the red die."
   a. Describe $\sim G$ in words. (There are several ways of doing this.)
   b. $P(G) = \underline{\quad ? \quad}$, $P(\sim G) = \underline{\quad ? \quad}$.
   c. What is the probability that the sum of two numbers thrown is more than 12? Less than 12? 12 or less?
   d. What is the probability that the sum of two numbers thrown is 1?

● 4. Two dice (one red, one green) are to be thrown (Problem 4, Exercise 2.4).
   a. Find the probability of the event, A, that both numbers thrown are even.
   b. Find the probability of the event B that both numbers are odd.
   c. Why isn't $P(A) + P(B) = 1$?

5. One day on the New York Stock Exchange the number of stocks showing an advance (rise from the previous day) was 651; the number of declines, 726; the number unchanged, 255. If a stock were picked on the NYSE at random that day, what is the probability that it showed no gain?

6. The fifteen most active stocks that day had closing prices (to the nearest dollar) as follows:

   50, 45, 29, 21, 134, 18, 29, 30, 24, 163, 24, 42, 26, 24, 62.

   a. What is the probability of the event "a stock selected from this list sells for less than $100"?
   b. What is the probability of the event "a stock from this list is less than $200"?
   c. What is the probability of the event "a stock from this list is in the $20 to $30 price range"? (More than $20, less than or equal to $30.)
   d. What is the probability of the event "a stock from this list is in the $80 to $100 price range"?

7. A card is selected at random from a shuffled deck. Let D be the event "a diamond is selected." What is $P(\sim D)$?

8. A card is selected at random from a shuffled deck. Let H be the event "an honor card is selected." Find $P(\sim H)$.

Odds. *Suppose that the probability of a given event A is p. Then $P(\sim A) = 1 - p$. The ratio $p:(1-p)$ is often called the "odds on A." For example, if $P(A) = \frac{3}{5}$, then $P(\sim A) = \frac{2}{5}$. The ratio $\frac{3}{5}:\frac{2}{5}$, or $3:2$, expresses the fact that the odds on A are $3:2$ or "the odds for A are $3:2$." We could also say, "the odds against A are $2:3$."*

9. What are the odds on throwing heads on a single toss of a fair coin?
10. What are the odds on drawing a green marble from an urn containing 3 green, 1 red marble?
11. If the odds on an event are $1:100$, what is the probability of the event?
12. If $P(A) = \frac{1}{100}$, what are the odds on A?
13. If the odds of an event A are $a:b$, what is $P(A)$?
14. If the probability of an event is 1, does it make any sense to speak about odds?
15. If the odds on an event are $n:n$, for whatever positive integer, n, what can you say about the probability of this event? of its complement?
16. What are the odds against drawing the ace of spades from a shuffled deck of playing cards?

2.6 The Probability of the Union of Two Events

Suppose that we have two events, A and B, which are subsets of a given set of outcomes S. The union, $A \cup B$, of A and B is a subset of S and is therefore also an event. If we know $P(A)$ and $P(B)$, it is reasonable to ask whether we can determine $P(A \cup B)$ from our knowledge of A and B. Is there a formula which gives $P(A \cup B)$ in terms of $P(A)$ and $P(B)$? Two examples help to throw some light on the answer.

Example 2.6.1 In the problem of Section 2.4, let $T = \{\text{in, of}\}$ and $H = \{\text{when, the, human}\}$. Then $P(T) = \frac{4}{28}$ and $P(H) = \frac{12}{28}$. Now, $T \cup H = \{\text{in, of, when, the, human}\}$ and $P(T \cup H) = \frac{16}{28}$.

$$P(T) + P(H) = \frac{4}{28} + \frac{12}{28} = \frac{16}{28},$$

and so we have

$$P(T \cup H) = P(T) + P(H).$$

Example 2.6.2 Referring to the same problem, if $E = \{\text{the, course}\}$, then $P(E) = \frac{9}{28}$. $H \cup E = \{\text{when, the, human, course}\}$ and $P(H \cup E) = \frac{18}{28}$. But

$$P(H) + P(E) = \frac{12}{28} + \frac{9}{28} = \frac{21}{28},$$

and so for this example,

$$P(H \cup E) \neq P(H) + P(E).$$

The difference between Example 2.6.1 and Example 2.6.2 becomes quite clear on examination. In Example 2.6.1, the sets T and H are **disjoint**; that is, their intersection is the empty set:

$$T \cap H = \emptyset$$

$$P(T \cap H) = 0$$

On the other hand, in Example 2.6.2,

$$H \cap E = \{\text{the}\} \neq \emptyset.$$

To describe the first case, a special phrase is used: If two events have no outcomes in common, the events are said to be **mutually exclusive**. A diagram representing two mutually exclusive sets, A and B, is shown in Figure 2.6.

Figure 2.6

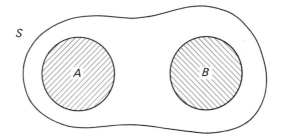

In general:

Let A and B be two mutually exclusive events of a finite set of outcomes, S, with

$$A = \{a_1, a_2, \ldots, a_r\}$$

and

$$B = \{b_1, b_2, \ldots, b_t\},$$

where a_i and $b_j \in S$ and $A \cap B = \emptyset$. Hence

$$A \cup B = \{a_1, a_2, \ldots, a_r, b_1, b_2, \ldots, b_t\}$$

and

$$P(A \cup B) = \underbrace{f(a_1) + f(a_2) + \cdots + f(a_r)}_{} + \underbrace{f(b_1) + f(b_2) + \cdots + f(b_t)}_{}$$

$$= \sum_{i=1}^{r} f(a_i) \quad + \quad \sum_{j=1}^{t} f(b_j)$$

$$= P(A) \quad + \quad P(B).$$

Of course, this is a straightforward result of our agreement that the probability of an event equals the sum of the probabilities of its individual outcomes.

Example 2.6.3 Problem 4, Exercises 2.4, deals with 36 equally likely outcomes of the experiment of throwing two dice. We shall now consider part d of that problem in detail. Let N be the event the sum of the dice is 7, and E be the event the sum of the dice is 11. Then

$$N = \{(1, 6), (6, 1), (2, 5), (5, 2), (3, 4), (4, 3)\}$$

and

$$E = \{(5, 6), (6, 5)\}.$$

Here, $N \cap E = \emptyset$, and

$$\begin{aligned}P(N \cup E) &= P(\{(1, 6), (6, 1), (2, 5), (5, 2),\\ &\qquad (3, 4), (4, 3), (5, 6), (6, 5)\})\\ &= \tfrac{8}{36} = \tfrac{6}{36} + \tfrac{2}{36} = P(N) + P(E).\end{aligned}$$

The probability of throwing a total of 7 *or* 11 is $\tfrac{8}{36}$, or $\tfrac{2}{9}$.

On the other hand, if two events are not mutually exclusive, then they will have some outcomes in common. This situation is illustrated in Figure 2.7.

Figure 2.7

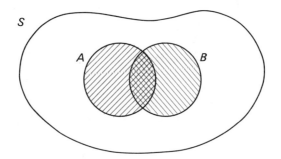

In general:

Let

$$A = \{a_1, a_2, \ldots, a_q, c_1, c_2, \ldots, c_k\}$$

and

$$B = \{b_1, b_2, \ldots, b_r, c_1, c_2, \ldots, c_k\},$$

where $c_1, c_2, \ldots, c_k$ are the outcomes that the subsets A and B have in common. Then

$$A \cup B = \{a_1, a_2, \ldots, a_q, c_1, c_2, \ldots, c_k, b_1, b_2, \ldots, b_r\}.$$

Hence,

$$\begin{aligned}P(A \cup B) &= \underbrace{f(a_1) + \cdots + f(a_q)}_{\sum f(a)} + \underbrace{f(c_1) + \cdots + f(c_k)}_{\sum f(c)} + \underbrace{f(b_1) + \cdots + f(b_r)}_{\sum f(b)}\\ &= \sum f(a) \quad + \quad \sum f(c) \quad + \quad \sum f(b).\end{aligned}$$

But we have:
$$P(A) = f(a_1) + \cdots + f(a_q) + f(c_1) + \cdots + f(c_k)$$
$$= \sum f(a) + \sum f(c)$$
$$P(B) = f(b_1) + \cdots + f(b_r) + f(c_1) + \cdots + f(c_k)$$
$$= \sum f(b) + \sum f(c)$$

Therefore,
$$P(A) + P(B) = \sum f(a) + \sum f(c) + \sum f(b) + \sum f(c)$$
$$= P(A \cup B) + \sum f(c).$$

Now $\sum f(c)$ is exactly the probability of the event $\{c_1, c_2, \ldots, c_k\}$, namely, the subset consisting of *all* the outcomes common to A and B. That is,
$$\sum f(c) = P(A \cap B).$$

Hence,
$$P(A) + P(B) = P(A \cup B) + P(A \cap B),$$

or, equivalently,
$$P(A \cup B) = P(A) + P(B) - P(A \cap B).$$

This result was to be expected, since in the sum $P(A) + P(B)$, the contribution of the outcomes in $A \cap B$ was counted twice.

Example 2.6.4 Suppose that a die is to be thrown. Let A be the event "a prime number is shown" and B be the event "an odd number is shown." Then $A = \{2, 3, 5\}$ and $B = \{1, 3, 5\}$. We see that
$$A \cap B = \{3, 5\} \neq \emptyset$$
and
$$A \cup B = \{1, 2, 3, 5\}.$$

The probability of throwing a prime number *or* an odd number is $P(A \cup B)$:
$$P(A \cup B) = P(A) + P(B) - P(A \cap B)$$
$$= \tfrac{3}{6} + \tfrac{3}{6} - \tfrac{2}{6} = \tfrac{4}{6} = \tfrac{2}{3}.$$

From the argument above, we conclude that in order to compute $P(A \cup B)$, we need more information than the values of $P(A)$ and $P(B)$. We also need the value of $P(A \cap B)$. It would seem natural to turn our attention then to the development of methods of computing $P(A \cap B)$. However, we shall postpone that until Section 3.1.

EXERCISES 2.6

1. Enrollment figures at State College show the following:

	Male	Female
Freshmen	200	200
Sophomores	150	125
Juniors	250	200
Seniors	200	175
TOTAL	800	700

   What is the probability that a student, chosen at random, will be:
   a. a freshman?
   b. a male?
   c. a male freshman?
   d. a male *or* a freshman?
   e. a female upper classman (junior or senior)?

2. In Problem 1, suppose that we limit our selection to members of the junior class. What is the probability that a female is selected?

3. An urn contains 6 red, 4 blue, 7 green, and 3 yellow marbles. If a marble is drawn at random, find the probability that the marble drawn is:
   a. blue
   b. some color other than blue
   c. green or yellow
   d. green, yellow, or red

4. A red and a green die are to be thrown. Find the following probabilities, assuming that the 36 outcomes are equally likely.
   a. The red die shows 4.
   b. The sum of the numbers thrown is 10.
   c. The red die shows 4 and the sum is 10.
   d. The red die shows 4 or the sum is 10.

5. Continuing Problem 4, find the probability that:
   a. the sum of the numbers thrown is 7.
   b. the sum is 7 and the red die shows 6.
   c. the sum is 5 and the red die shows 6.
   d. the sum is 5 or the red die shows 6.

6. Let A, B, be events and complete the following.
   a. If $P(A) = 0.2$, $P(B) = 0.7$, $P(A \cap B) = 0.3$, then $P(A \cup B) = $ __?__.
   b. If $P(A) = 0.4$, $P(B) = 0.5$, $P(A \cup B) = 0.8$, then $P(A \cap B) = $ __?__.
   c. If $P(A) = 0.15$, $P(B) = 0.45$, $P(A \cup B) = 0.6$, then $P(A \cap B) = $ __?__.
   d. If $P(A) = \frac{1}{3}$, $P(A \cap B) = \frac{1}{6}$, $P(A \cup B) = \frac{2}{3}$, then $P(B) = $ __?__.

7. Consider the experiment "select an integer from 1 to 100 at random." Then let the set of possible outcomes, S, consist of all the integers from 1 to 100 inclusive, let A be the event "the perfect squares in S" ($A = \{1, 4, 9, \ldots, 100\}$), and let B be the event "the perfect cubes in S." Suppose that it is known that
$$P(A) = \tfrac{1}{10},$$
$$P(B) = \tfrac{1}{25},$$
and
$$P(A \cap B) = \tfrac{1}{50}.$$
   a. Find $P(A \cup B)$ without listing the outcomes in any of the sets.
   b. Verify by listing the outcomes in A and B that the result obtained in part a is correct.

8. Suppose that 53% of a sophomore class are boys ($P(B) = 0.53$). If 40% of this class are honor students and 73% are either honor students or boys, what is the probability that a student chosen at random is a male honor student?

9. In a survey conducted at a men's college, it was found that 94% of the students like blonde girls, that 91% like redhead girls, and that 87% like both kinds of girls.
   a. What is the probability that a student chosen at random at this college likes either kind of girls?
   b. What is the probability that a student chosen at random likes neither kind of girls?

10. In a class of 500 students, 150 take mathematics, 200 take a science, and 50 take both mathematics and a science. If a student is to be chosen at random from this class, what is the probability that the person chosen takes mathematics or a science?

11. A card is to be selected at random from a shuffled deck of playing cards. Find the probability that:
    a. an ace is drawn
    b. a spade is drawn
    c. the ace of spades is drawn
    d. an ace or a spade is drawn

12. A single die is thrown. What is the probability that a prime number or an even number shows?

13. A survey of 100 people showed that 26 preferred foreign cars to American models, 48 preferred sports cars to standard models, and 50 liked foreign cars or sports cars. What is the probability that a person chosen at random from this group likes foreign sports cars?

14. Statistics in a certain city reveal that $\tfrac{2}{3}$ of its population is made up of persons under 25 years of age. It is also known that $\tfrac{3}{5}$ of the population is female, and that $\tfrac{5}{8}$ of the population is male or 25 years of age or older. What is the probability that a person selected at random from this city is male and at least 25 years old?

2.7 Proving Some Theorems

Let us now take a second look at the ideas that have been introduced in this chapter.

We analyze below the results that we have obtained up to now, distinguishing definitions, assumptions, and theorems.

DEFINITIONS

D_1: Associated with a given experiment is a **set, S, of outcomes**. The outcomes are distinct and exhaust all possibilities. (Section 2.2)

D_2: An **event** is a subset of S. (Section 2.4)

D_3: If $a \in S$, then $\{a\}$ is an **elementary event**. (Section 2.4)

D_4: A function, P, which maps the collection of all events of S into the real numbers is a **probability function** if the function satisfies the conditions A_1 to A_3 below. (Section 2.5)

D_5: Given an event A and a probability function, then $P(A)$ is the **probability** of the event A. Only events have probabilities. (Sections 2.4, 2.5)

D_6: If $A \cup B = S$ and $A \cap B = \emptyset$, then A and B are **complementary events**. We write $B = \sim A$. (Section 2.5)

Note that this is a restatement of the following idea brought out in Section 2.5: Given a subset A of S, if B is the set of all outcomes of S that are not in A, then B is the complement of A in S.

D_7: Two events, A and B, are **mutually exclusive** if $A \cap B = \emptyset$. (Section 2.6)

ASSUMPTIONS

A_1: For every event A, $0 \leq P(A) \leq 1$. (Section 2.5)

A_2: $P(S) = 1$. (Section 2.5)

A_3: If A and B are mutually exclusive events, then $P(A \cup B) = P(A) + P(B)$. (Section 2.6)

Remark

While our arguments have developed these results for a finite set of outcomes, we state these as assumptions which apply to infinite sets of outcomes also.

THEOREMS

T_1: $P(\emptyset) = 0$. (Section 2.5)

T_2: If $P(A) = p$, then $P(\sim A) = 1 - p$. (Section 2.5)

T_3: If $A_1, A_2, \ldots, A_k$ are events and if $A_i \cap A_j = \emptyset$, $(i \neq j)$, then

$$P(A_1 \cup A_2 \cup \cdots \cup A_k) = P(A_1) + P(A_2) + \cdots + P(A_k).$$

T_4: If A and B are events, then

$$P(A \cup B) = P(A) + P(B) - P(A \cap B). \quad \text{(Section 2.6)}$$

In our discussions, we have made the results stated here as T_1 to T_4 seem plausible by way of examples. With the assumptions and definitions that we have listed above, these can indeed be proved. We demonstrate the proofs for two of these, T_1 and T_4, and leave the proofs for T_2 and T_3 as exercises (Problems 1 and 2, Exercises 2.7) for the reader.

THEOREM T_1: $P(\emptyset) = 0$.

Proof: $A \cap \emptyset = \emptyset$ and so A and $\emptyset$ are mutually exclusive (D_7). Since $A \cup \emptyset = A$, we have $P(A) = P(A \cup \emptyset)$. From A_3,
$$P(A \cup \emptyset) = P(A) + P(\emptyset).$$
Therefore,
$$P(A) = P(A) + P(\emptyset).$$
Subtracting $P(A)$, we have
$$0 = P(\emptyset).$$

THEOREM T_4: If A and B are events, then
$$P(A \cup B) = P(A) + P(B) - P(A \cap B).$$

Proof: The figure below illustrates that we may write A as the union of two mutually exclusive (disjoint) sets:
$$A = (A \cap B) \cup (A \cap {\sim}B)$$

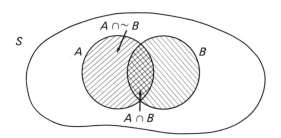

From A_3, we have
$$P(A) = P(A \cap B) + P(A \cap {\sim}B)$$
or
$$P(A \cap {\sim}B) = P(A) - P(A \cap B). \qquad (1)$$
We may also write $A \cup B$ as the union of two mutually exclusive sets:
$$A \cup B = (A \cap {\sim}B) \cup B$$
Again using A_3, we have
$$P(A \cup B) = P(A \cap {\sim}B) + P(B). \qquad (2)$$
Substituting from (1) into (2), we have:
$$P(A \cup B) = P(A) - P(A \cap B) + P(B)$$
$$= P(A) + P(B) - P(A \cap B)$$

Theorem T_3 applies to a finite number of mutually exclusive events. If we wished to extend this result to the infinite case, an additional assumption would be needed, stating in effect that for such cases the probability of mutually disjoint (no two events intersecting) events equals the sum of the probabilities. As it turns out, we shall not use this assumption in our work.

EXERCISES 2.7

- **1.** Prove Theorem T_2.
- ★ **2.** Prove Theorem T_3. (Hint: Use mathematical induction.)

2.8 What Do You Think?

In ordinary conversation, we occasionally hear reference to the "law of averages." For example, a basketball coach was overheard making the following statement to his star performer at half-time. "Joe, you only hit 3 of 12 shots in the first half but you have averaged about 50% during the season. Shoot more in the second half; the law of averages is in your favor." This may be good psychology, but our interest in this remark is mathematical.

By what assumptions has the "law of averages" been invoked? Is the conclusion justified from a mathematical point of view? In later chapters we shall develop the mathematical theory of independent, repeated trials and discuss the mathematical basis of the "law of averages" in that light. For the present, we shall merely look at a common misconception regarding it.

Coin tossing provides a simple illustration. Let us assume that we have a fair coin. On a single toss,

$$P(\text{heads}) = P(\text{tails}) = \tfrac{1}{2}.$$

The long-run interpretation of this probability statement is that if the coin were tossed *many* times, we would obtain heads on about $\tfrac{1}{2}$ the total number of tosses. If we begin tossing the coin and get 9 heads in a row, something rather unusual or rare has happened. (Unexpected things do happen. The New York Mets, for example, won the National League Championship and the World Series in 1969, although before the season started, the probability of their winning was judged to be extremely small.) The occurrence of 9 heads at the beginning of a series of tosses of a fair coin is not exactly unexpected. In fact, we shall see in Chapter 7 that if we conduct 10,000 series of 9 tosses of a fair coin, we would "expect" about 20 of these series to consist of 9 heads.

If we *do* start with 9 heads, the interesting question has to do with the 10th toss. What is the probability that the 10th toss will fall heads? Many people argue that the probability of tails is now more than $\tfrac{1}{2}$, since the "law of averages" says that the number of tails must catch up to the number of heads. Their idea is: the probability of 10 heads in a row is quite small;

since we already have 9 heads, the probability of a 10th head must be very small.

It is quite correct that 10 heads in a row has small probability, but nevertheless the conclusion stated above is quite false. Ten heads in a row is unusual, but *not* if we already have 9 heads. In fact, the probability of 10 heads, given 9 heads, is exactly $\frac{1}{2}$. We might think of it this way. On the 10th toss, the coin does *not* say to itself, "I have fallen heads 9 times; I had better strain my molecules so as to fall tails. I mustn't violate the law of averages."

The usual wording for the argument of the last paragraph is "probability has no memory." More accurately, the coin has no memory.

We have pointed out a common misconception about the law of averages. What do you think is a *proper* interpretation of the law of averages?

EXERCISES 2.8

In the following, you are not expected to answer every question precisely. We are trying to develop a "feel" for the law of averages.

- **1.** Twenty tosses of a coin are made which yield 16 heads, 4 tails. How would you judge the following conclusions?
   a. The coin is not fair; heads are more probable than tails.
   b. The coin is fair; something rather unusual has occurred.
- **2.** A player at a roulette table watches "red" occur three times in a row. He now bets "black," arguing to himself that black is more likely than red by the law of averages. Comment on this. (A roulette wheel has 38 "pockets," of which 18 are red, 18 are black, and 2 are white.)
- **3.** Consider 10 tosses of a *fair* coin. Problem 4 of Exercises 2.2 tells us that there are $2^{10} = 1024$ possible outcomes.
   a. Are these equally likely?
   b. We agree that the sequence *hhhhhhhhhh* is unusual because its probability is small, namely $\frac{1}{1024}$. Is the sequence, *hththththt*, unusual?
   c. Is the sequence *htthththtt* unusual?
   d. Make a generalization.
- **4.** Examine Problem 2a, Exercises 1.6. We shall learn in Chapter 7 that $P(5 \text{ heads, } 10 \text{ tosses}) \doteq 0.25$ and in Chapter 8 that $P(50 \text{ heads, } 100 \text{ tosses}) \doteq 0.08$. Do these results contradict the long-run interpretation of the law of averages?
- **5.** Examine Problems 2b and 2c, Exercises 1.6, and comment.
- **6.** Comment on Problem 2d, Exercises 1.6.
- **7.** One apparently unusual event in the real world is the real world itself. In particular, the precise circumstances that make life as we know it possible—the amount of solar radiation, the size of the earth, the composition of the atmosphere, the presence of large bodies of water—are very unusual in the universe. The part that probability plays in these considerations can be highlighted by the differences of opinions on the matter of

life elsewhere in the universe. One school of thought holds that the circumstances are *so* unusual as to be impossible by accident. Thus there must be a cause of some kind. Others hold that while rare, there are so many stars and so many possible "earths" that our circumstances had to occur elsewhere in space as well. Does mathematics (and the law of averages) help either argument?

Chapter Summary

1. a. An *a priori* probability is arrived at by logical reasoning.
   b. An *empirical* probability is assigned as a result of experiments or historical evidence.
2. a. A *set of outcomes* of an experiment is a set of distinct possibilities which exhaust the possible results. A *tree diagram* may be used to picture a set of outcomes resulting from several steps.
   b. *Multiplication Principle:* If one step of an experiment can happen in exactly m ways, and if, for each of these, a second step can happen in exactly n ways, then the experiment which consists of these two steps can happen in exactly $m \cdot n$ ways.
3. Let $S = \{s_1, s_2, \ldots, s_n\}$ be a finite set of all possible outcomes. Then:
   (1) each s_i is assigned a unique number, its *probability*, $f(s_i)$;
   (2) for each i, $0 \leq f(s_i) \leq 1$;
   (3) $\sum_{i=1}^{n} f(s_i) = 1$;
   (4) if $s_1, s_2, \ldots, s_n$ are equally likely, then for each i, $f(s_i) = \dfrac{1}{n}$.

   The assignment of a unique number, $f(s_i)$, to each outcome s_i is a *function*.
4. a. An *event* is a subset of a set of possible outcomes of an experiment.
   b. The *probability of an event* equals the sum of the probabilities of its individual outcomes. By interpreting an outcome as an event with only one member, we can say that *only events have probabilities*.
5. a. If S is a set of outcomes, then $P(S) = 1$; $P(\emptyset) = 0$.
   b. If A is an event, then $0 \leq P(A) \leq 1$.
   c. The *complement* of A in S is the set of all outcomes in S that are not in A; it is denoted by $\sim A$. $P(\sim A) = 1 - P(A)$.
   d. The *probability function* may be denoted by the symbolism

   $$\text{an event } A \longrightarrow \text{a real number in } [0, 1].$$

   The *domain* is the set of possible events in S. The *range* is a subset of $[0, 1]$.

6. a. If A and B are events, then
$$P(A \cup B) = P(A) + P(B) - P(A \cap B).$$
b. If events C and D have no outcomes in common, that is, $C \cap D = \emptyset$, then they are said to be *mutually exclusive*. In this case,
$$P(C \cup D) = P(C) + P(D).$$

Chapter Review

1. For each of the equipments and sets of possible outcomes listed below, describe an experiment yielding the outcomes. (The experiment may involve more than one stage.)
   a. Two coins: {0, 1, 2}
   b. Two coins: {0, 1, 2, 3, 4}
   c. One die: {2, 3, 4, 5, 6, 7, 8, 9, 10, 11, 12}
   d. One die: a set consisting of two outcomes

2. A die is thrown. Give two reasons why the following is *not* a set of possible outcomes:
   {a number divisible by 2, a number divisible by 3}

3. An urn contains 6 red and 2 blue marbles. A marble is drawn. Without replacing the marble that is drawn, a marble of the opposite color is added to the urn. What is the probability of red on the second draw if the first marble drawn is:
   a. red?   b. blue?

4. Using a tree diagram, decide whether or not the following game is fair. If not fair, whom does the rule favor?
   a. Two coins are tossed. You win if both coins match. Your opponent wins if they do not match.
   b. Three coins are tossed. You win if all three coins match. Your opponent wins if none of the coins match.

5. There are two urns marked H and T. Urn H contains 2 red marbles and 1 blue marble. Urn T contains 1 red and 2 blue marbles. A coin is to be tossed. If it lands heads, a marble is drawn from Urn H. If it lands tails, a marble is drawn from Urn T.
   a. Draw a tree diagram associated with the experiment which leads to six equally likely outcomes. Describe this set by listing.
   b. Draw a tree diagram associated with the experiment which leads to four outcomes that are *not* all equally likely. Describe this set by listing.

6. Two dice are thrown. Find the probability that one of the dice shows a prime number and the other shows an even number.

7. Two dice, one red and one green, are thrown. Find the probability that the red die shows a prime number and the green die shows an even number.
8. A card is drawn. Find the probability that it is either a 2, 3, 5, 7, jack, king, or a black ace.
9. A marble is drawn from an urn containing blue marbles, red marbles, green marbles. If $P(\text{blue}) = \frac{1}{2}$ and $P(\text{red}) = \frac{1}{3}$, find the probability of drawing a marble that is either red or blue.
10. The total enrollment in a school is 2000. Of these, 1200 students take French, 900 take German, and 350 take both French and German. What is the probability that a student selected at random from this school takes either French or German?
11. If A and B are events in a set of possible outcomes with the probabilities,

$$P(A) = \tfrac{3}{4}, \quad P(B) = \tfrac{1}{3}, \quad P(A \cup B) = \tfrac{2}{3},$$

find $P(A \cap B)$.

12. Let S be a set of possible outcomes. If A and B are to be complementary events in this set, one of two conditions that must be met is $A \cup B = S$. What is the other condition?

Chapter 3

Conditional Probability

We have seen two methods of constructing probability functions. A series of experiments may lead us to the conclusion that we are dealing with a situation involving uncertainty. We would then assign probabilities to the various observed outcomes in accordance with the experimental evidence. On the other hand, we may be able to assign *a priori* probabilities having in mind some "model." Our urn problems have illustrated this second approach.

In assigning *a priori* probabilities, we begin with a certain set, S, of possible outcomes. Since any event, A, is a subset of S, we cannot determine $P(A)$ unless S is specified. Consider this problem: An urn contains three colored marbles. One marble is to be drawn at random. Let R be the event "a red marble is drawn." Is it possible to assign a meaningful *a priori* probability to R? The answer is "No, not without knowing how many of the marbles are red."

The probability of an event depends on the set of outcomes and on the probability function that we establish. In a sense, then, the probability of any event is *conditional;* it depends, first of all, on the set of possible outcomes. Thus, if S is a set of outcomes of an experiment and if A is an event in S, then we could speak of "the probability of A, given the set S." In practice, we usually know S and so the qualification "given S" is omitted.

The idea that probability is conditional, however, has important implications. In particular, we saw in Section 2.6 that in order to compute the probability of $A \cup B$ where A and B are events, we needed to know $P(A \cap B)$. It turns out that we can attack the problem of finding $P(A \cap B)$ by examining the notion of conditional probability more closely.

3.1 A Formula for Conditional Probability

Let us consider an experiment and a corresponding set of outcomes. Further, let us assume that we have constructed a probability function, and so for any event A, $P(A)$ is known. After the experiment has been performed, we may receive information about the results. If we know precisely what has happened, then we know that either A has occurred or it hasn't. An interesting situation is one for which we receive some partial information.

Let us start with some examples.

Example 3.1.1 A joint committee of the two houses of a state legislature consists of 2 Republicans and 4 Democrats from the upper house, and 3 Republicans and 3 Democrats from the lower house. This information may be depicted more vividly in the form of a table such as shown below:

	Upper house (A)	Lower house (B)	Total
Republicans (R)	2	3	5
Democrats (D)	4	3	7
Total	6	6	12

A chairman is to be selected. In actual practice, legislative committees are chaired by a member of the majority party. For the sake of our example, however, let us assume that the committee decides to choose a chairman by lot (at random). Then:

Let A be the event "the chairman is a member of the upper house."
Let B be the event "the chairman is a member of the lower house."
Let R be the event "the chairman is a Republican."
Let D be the event "the chairman is a Democrat."

We may consider our set of outcomes either as the set consisting of 12 equally likely outcomes (individuals) or as the set

$$S = \{R \cap A, R \cap B, D \cap A, D \cap B\},$$

where the outcomes are not equally likely. In this latter case, the associated probabilities are

$$\tfrac{2}{12}, \tfrac{3}{12}, \tfrac{4}{12}, \text{ and } \tfrac{3}{12},$$

respectively, as can be seen from the table, since the total is 12.

a. To determine the probability that the chairman is a Republican, we have

$$P(R) = P(R \cap A) + P(R \cap B)$$
$$= \tfrac{2}{12} + \tfrac{3}{12} = \tfrac{5}{12}.$$

b. Suppose now that we have some reliable information that a member of the *upper house* has been chosen as chairman. We are now concerned only with the *reduced* set of outcomes,

$$S' = \{R \cap A, D \cap A\},$$

limiting our attention to the first column of the table, or subset A. With reference to *this* set, we would agree that the probability that R has occurred, given that A has occurred, is

$$\tfrac{2}{6} = \tfrac{1}{3}.$$

Thus, our partial information has reduced the probability of a Republican chairman's being chosen from $\tfrac{5}{12}$ to $\tfrac{1}{3}$.

Example 3.1.2 Suppose instead that the information we get in the situation described in Example 3.1.1 is that a member of the *lower house* has been chosen as chairman. An argument similar to that in part b of Example 3.1.1 leads us to the conclusion that the probability of "R given B" is

$$\tfrac{3}{6} = \tfrac{1}{2}.$$

With this information, then, the probability of a Republican's being chosen has increased from $\tfrac{5}{12}$ to $\tfrac{1}{2}$.

We were concerned in the preceding examples with computing probabilities with reference to a reduced set of outcomes. Let us introduce a notation to describe this situation. By the notation

$$P(R \mid A)$$

we shall mean *the probability of the event R, given that the event A has occurred.*

Using this notation, we write the following for comparison:

In Example 3.1.1b, we found: In Example 3.1.2, we found:

$$P(R \mid A) = \tfrac{1}{3} \qquad\qquad P(R \mid B) = \tfrac{1}{2}$$

Moreover: $P(R \cap A) = \tfrac{1}{6}$ Moreover: $P(R \cap B) = \tfrac{1}{4}$

$$P(A) = \tfrac{1}{2} \qquad\qquad\qquad P(B) = \tfrac{1}{2}$$

Do you see a pattern? It appears that

$$P(R \mid A) = \frac{P(R \cap A)}{P(A)} = \frac{\tfrac{1}{6}}{\tfrac{1}{2}} = \frac{1}{3}, \qquad P(R \mid B) = \frac{P(R \cap B)}{P(B)} = \frac{\tfrac{1}{4}}{\tfrac{1}{2}} = \frac{1}{2}.$$

We now return to Example 3.1.1 and consider the set of 12 equally likely outcomes. The probability of each of these is $\tfrac{1}{12}$. These are represented by Figure 3.1a. In Figure 3.1b attention is focused on subset A, and the portion of R that overlaps A (namely, $R \cap A$).

Figure 3.1

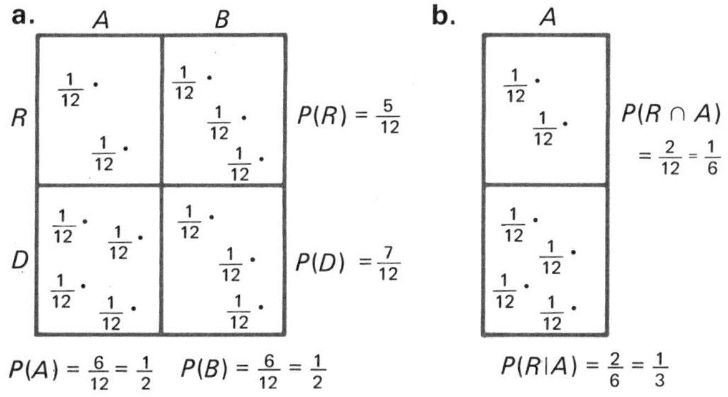

a. $P(R) = \tfrac{5}{12}$ $P(D) = \tfrac{7}{12}$ $P(A) = \tfrac{6}{12} = \tfrac{1}{2}$ $P(B) = \tfrac{6}{12} = \tfrac{1}{2}$

b. $P(R \cap A) = \tfrac{2}{12} = \tfrac{1}{6}$ $P(R \mid A) = \tfrac{2}{6} = \tfrac{1}{3}$

Following the pattern developed above, we see from Figure 3.1b that

$$P(D \mid A) = \frac{P(D \cap A)}{P(A)} = \frac{\tfrac{1}{3}}{\tfrac{1}{2}} = \frac{2}{3}.$$

Notice that

$$P(R \mid A) + P(D \mid A) = \tfrac{1}{3} + \tfrac{2}{3} = 1,$$

since the classifications "Republicans" and "Democrats" are mutually exclusive and exhaust the subset A.

We are led to the following *definition:*

If E and G are events from a set of outcomes S and if $P(G) \neq 0$, then the *conditional probability* of E, given that G has occurred, is

$$P(E \mid G) = \frac{P(E \cap G)}{P(G)}.$$

Example 3.1.3 Six cards—ace, king, queen of hearts; ace, king of diamonds; and king of clubs—are placed face down on a table. One card is to be selected at random. If A is the event "an ace is drawn" and H is the event "a heart is drawn", we have

$$P(A) = \tfrac{2}{6} = \tfrac{1}{3},$$

and, since $P(A \cap H) = \tfrac{1}{6}$ and $P(H) = \tfrac{3}{6} = \tfrac{1}{2}$,

$$P(A \mid H) = \frac{\tfrac{1}{6}}{\tfrac{1}{2}} = \frac{1}{3}.$$

Note that in this case $P(A) = P(A \mid H)$; that is, the information that a heart has been selected does not change the probability that an ace is selected.

In Section 2.6, we found that to compute the probability of the union of two events, we needed to know the probability of their intersection. It happens that frequently we can easily determine $P(E \mid G)$ directly, and from this, obtain $P(E \cap G)$:

$$P(E \cap G) = P(G) \cdot P(E \mid G).$$

Remark 1

If $P(E) \neq 0$, then the roles of E and G are interchangeable because $E \cap G = G \cap E$. It follows that

$$P(E \cap G) = P(E) \cdot P(G \mid E) \quad \text{or} \quad P(G \mid E) = \frac{P(E \cap G)}{P(E)}.$$

Remark 2

As a very special case, it is possible that E and G are mutually exclusive events. (See Problem 2b, Exercises 3.1.) Then, since $P(E \cap G) = 0$,

$$P(E \mid G) = P(G \mid E) = 0 \quad \text{if} \quad P(E) \neq 0, \quad P(G) \neq 0.$$

Remark 3

The notation, $P(E \mid G)$, while commonly used, is rather unfortunate. Events, and only events, have probabilities. Now, "$E \mid G$" is not a set (not an event). $P(E \mid G)$ indicates that we are considering the probability of E *relative to* the event G.

EXERCISES 3.1

1. A penny, a nickel, and a dime are to be tossed. Assume that the coins are fair and find the probability of:
   a. obtaining 3 heads
   b. obtaining 3 heads, given that the penny shows heads
   c. obtaining 3 heads, given that at least one of these coins shows heads
   d. obtaining 3 heads, given that the penny and nickel both show heads
   e. having the dime show heads (without any condition). Compare this result with the one in part d.

2. Using the data of Example 3.1.1, suppose that it is known that the chairman is a Republican. What are the probabilities that he is:
   a. a member of the upper house?
   b. a member of the lower house?
   (Note: The sum of the probabilities in parts a and b is 1.)

3. Refer to Example 3.1.3. Find the probability:
   a. of an ace, given that a diamond was turned
   b. of an ace, given that a club was turned
   c. of a heart, given that an ace was turned
   d. of a club, given that a king was turned
   e. Explain why it is true that $P(A \mid C) = P(C \mid A)$.

4. A pair of fair dice, one red, one green, are to be tossed. Let A be the event "the numbers thrown total 7," B the event "the numbers thrown total 10," and C the event "the red die shows a 4." Find:
   a. $P(A), P(B), P(C), P(A \cap C), P(B \cap C)$
   b. $P(A \mid C)$ c. $P(B \mid C)$ d. $P(C \mid A)$ e. $P(C \mid B)$

5. a. Given $P(E) = 0.3$, $P(F) = 0.4$, and $P(E \cap F) = 0.2$, find $P(E \mid F)$, $P(F \mid E)$, $P(E \cup F)$.
   b. Given $P(A \mid B) = 0.6$ and $P(B) = 0.5$, find $P(A \cap B)$.
   c. Given $P(R \mid T) = 0.4$ and $P(R \cap T) = 0.3$, find $P(T)$.

6. If A and B are events from S with $P(A) \neq 0, P(B) \neq 0$, and $P(A \cap B) \neq 0$, prove that $P(A \cap B) \leq P(A \mid B)$. Under what conditions does the equality sign hold?

7. Let S be the set of outcomes
   $$\{\text{when, in, the, course, of, human, events}\}.$$
   If H, E, V, and T are the events described in Problem 1 of Exercises 2.4:
   a. Show that $P(E) = \dfrac{P(H \cap E)}{P(H \mid E)}$.
   b. Explain why $P(T) \neq \dfrac{P(V \cap T)}{P(V \mid T)}$.

8. The table below shows the number and type of cars competing in a sports car rally.

	Stick (G)	Automatic (H)
American (A)	5	2
Foreign (F)	3	2

Let A be the event "an American car is chosen."
Let F be the event "a foreign car is chosen."
Let G be the event "the chosen car has a stick shift."
Let H be the event "the chosen car has an automatic shift."
  a. Is $P(G \mid A) = P(A \mid G)$?
  b. Find $P(A \cup F), P(A), P(F)$.
  c. Without referring to the table, tell what the results in part b show about $P(F \mid A)$.

9. We are told of the following experiment. A coin is tossed. If it lands heads, a marble is drawn from Urn H, which contains 1 red and 1 blue marble. If it lands tails, a marble is drawn from Urn T, which contains 2 red marbles and 1 blue marble.
  a. We are told that the coin fell heads. What is the probability that a blue marble was drawn? That is, find $P(B \mid H)$.
  b. Find $P(B \mid T)$.
  c. Suppose that we are informed that a blue marble was drawn. Find $P(H \mid B)$.
  d. Is it true that $P(H \mid B) = P(B \mid H)$?

10. If $P(A \cap C) = P(B \cap C)$ and $P(C) \neq 0$, show that $P(A \mid C) = P(B \mid C)$.

11. Let S be the set of possible outcomes with events E, F, and G such that E is a subset of F. Show that:
  a. $P(E \mid E) = 1$
  b. $P(\emptyset \mid G) = 0$
  c. $P(E \mid G) \leq P(F \mid G)$
  d. $P(F \mid E) = 1$
  e. How is $P(\sim E \mid G)$ related to $P(E \mid G)$?
  f. Express $P(E \mid S)$ in terms of $P(E)$.

12. Four players are in a game of cards. Each player has been dealt four cards and is to be dealt another from the deck.
  a. If the four cards one of the players got were all spades, what is the probability of his getting a fifth spade?
  b. Suppose that the fifth card turned out to be a heart and the player is allowed to trade that for another card from the deck. What is his probability of getting a fifth spade?
  c. Suppose that out of the five cards, each player had to expose (turn face up) three of his cards. The player who got 4 spades and a heart exposed 3 of his spades. He noticed that there are 3 other spades exposed in addition to his. If he were allowed to trade his heart for another card from the deck, what is his probability of getting a fifth spade?

13. A spinner, with dial marked as shown, is spun once. Each portion of the dial is identified by a color (R for red, Y for yellow, B for blue, G for green) as well as by a number (1, 2, 3, 4). Find the probability that the spinner points to:
    a. 2, given that the region is red.
    b. yellow, given that the region is labeled 3.
    c. an odd number, given that the region is green.

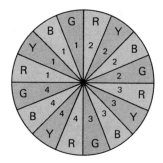

14. A card is to be chosen at random from a shuffled deck. Find the probability that the card drawn is
    a. the queen of clubs, given that it is an honor card.
    b. a red four, given that the face number is even.
    c. a seven, given that the face number is a prime.

15. An urn contains 6 red, 4 blue, 7 green, 3 yellow marbles. If a marble is drawn at random, find the probability that the marble drawn is
    a. red, given that it is not blue.
    b. green, yellow, or red, given that it is not blue.

16. Suppose that the population of a certain town is made up of 45% men and 55% women. Of the men, 40% wear glasses, and, of the women, 20% wear glasses. Given that a person chosen at random from this town wears glasses, what is the probability that the person is a woman?

3.2 Experiments Having Several Steps

In Chapter 2 we used tree diagrams to assist us in counting the outcomes of experiments having more than one step. Again we find it convenient to use an urn problem as an illustration:

Suppose that we have two urns. Urn H contains 3 red and 2 blue marbles. Urn T contains 2 red, 4 blue, and 2 green marbles. Suppose then that a coin is tossed. If the coin falls heads, a marble is drawn from Urn H. If it falls tails, a marble is drawn from Urn T (all drawings at random).

Figure 3.2

Urn H: 3 red, 2 blue

Urn T: 2 red, 4 blue, 2 green

Let R be the event "a red marble is drawn." We wish to find

$$P(R).$$

Since there are 13 marbles in all and 5 of them are red, we might be tempted to guess that $P(R) = \frac{5}{13}$. (Of course, we might make other guesses, such as $\frac{1}{2}$ or $\frac{2}{5}$ or even $\frac{17}{40}$.) Let us see.

Clearly, we may think of this experiment in two steps: first, select an urn, and then draw a marble. A tree diagram shows that a set of five outcomes is appropriate (see Figure 3.3).

Figure 3.3

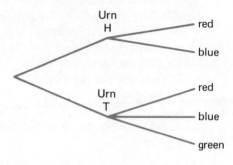

Let us use H to represent the event "the coin falls heads," or, what amounts to the same thing, "the marble is drawn from Urn H." If R represents the event "a red marble is drawn," then the intersection

$$H \cap R$$

is the event "a red marble is drawn from Urn H." We may interpret

$$H \cap B, \quad T \cap R, \quad T \cap B, \quad \text{and} \quad T \cap G,$$

in a similar manner. We may consider then (see Figure 3.4) the set of possible outcomes

$$S = \{H \cap R, H \cap B, T \cap R, T \cap B, T \cap G\}.$$

Figure 3.4

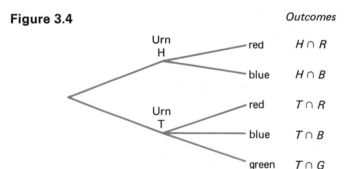

If the coin is assumed to be fair, we have

$$P(H) = \tfrac{1}{2} \quad \text{and} \quad P(T) = \tfrac{1}{2}.$$

$P(R \mid H)$ is easily determined, since each of the 5 marbles in Urn H is equally likely to be drawn; so we have (Figure 3.2)

$$P(R \mid H) = \tfrac{3}{5}.$$

Also,

$$P(B \mid H) = \tfrac{2}{5}.$$

Similarly,

$$P(R \mid T) = \tfrac{2}{8} = \tfrac{1}{4}, \quad P(B \mid T) = \tfrac{4}{8} = \tfrac{1}{2}, \quad P(G \mid T) = \tfrac{2}{8} = \tfrac{1}{4}.$$

We are interested in the event R, and we notice that exactly two branches of our tree lead us to an outcome in R. That is,

$$R = (H \cap R) \cup (T \cap R).$$

Now $H \cap R$ and $T \cap R$ are mutually exclusive events, and so, using the results of Section 2.6, we have

$$P(R) = P(H \cap R) + P(T \cap R).$$

To compute $P(H \cap R)$, we may apply the formula of Section 3.1, using the probabilities computed above:

$$P(H \cap R) = P(R \cap H) = P(H) \cdot P(R \mid H)$$
$$= \tfrac{1}{2} \cdot \tfrac{3}{5} = \tfrac{3}{10}.$$

In a similar manner, we find

$$P(T \cap R) = P(R \cap T) = P(T) \cdot P(R \mid T)$$
$$= \tfrac{1}{2} \cdot \tfrac{2}{8} = \tfrac{1}{8}.$$

Therefore,

$$P(R) = P(H \cap R) + P(T \cap R)$$
$$= \tfrac{3}{10} + \tfrac{1}{8} = \tfrac{17}{40}.$$

Hence, we see that our first guess of $\tfrac{5}{13}$ for $P(R)$ was wrong, and our "unlikely" guess of $\tfrac{17}{40}$ was right!

Example 3.2.1 For the urn problem we have been discussing, we can find other probabilities. For example:

$$P(B) = P(H \cap B) + P(T \cap B)$$
$$= P(H) \cdot P(B \mid H) + P(T) \cdot P(B \mid T)$$
$$= \tfrac{1}{2} \cdot \tfrac{2}{5} + \tfrac{1}{2} \cdot \tfrac{4}{8} = \tfrac{1}{5} + \tfrac{1}{4} = \tfrac{9}{20}$$
$$P(G) = P(H \cap G) + P(T \cap G)$$
$$= 0 + P(T) \cdot P(G \mid T)$$
$$= \tfrac{1}{2} \cdot \tfrac{2}{8} = \tfrac{1}{8}$$

It is satisfying to observe that

$$P(R) + P(B) + P(G) = \tfrac{17}{40} + \tfrac{9}{20} + \tfrac{1}{8}$$
$$= \tfrac{17}{40} + \tfrac{18}{40} + \tfrac{5}{40} = 1,$$

since we know that $\{R, B, G\}$ is also a set of possible outcomes for this experiment.

If we now return to our tree diagram (Figure 3.4), we may attach probabilities to the various branches. The first step is to label the branches leading to H and to T, as shown in Figure 3.5.

Figure 3.5

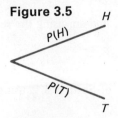

The branches leading from "Urn H" indicate the possible results of drawing from that urn. Thus, the branch from "Urn H" to "red" indicates "a red marble is drawn from Urn H." The probability to be attached to this branch is simply $P(R \mid H)$. Completing the labelling, we have Figure 3.6.

Figure 3.6

Outcomes	*Probabilities*
$H \cap R$	$P(H) \cdot P(R \mid H)$
$H \cap B$	$P(H) \cdot P(B \mid H)$
$T \cap R$	$P(T) \cdot P(R \mid T)$
$T \cap B$	$P(T) \cdot P(B \mid T)$
$T \cap G$	$P(T) \cdot P(G \mid T)$

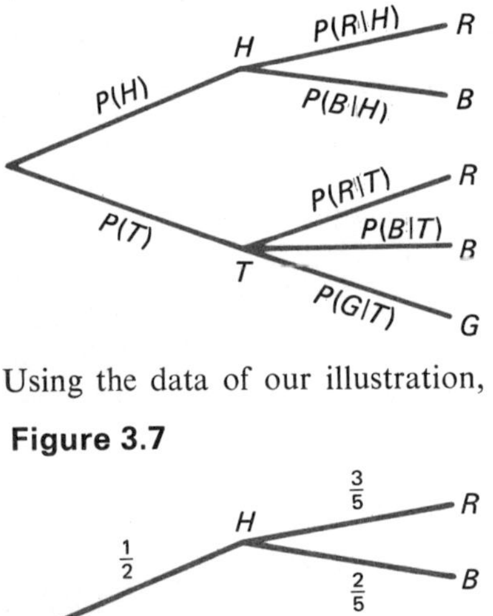

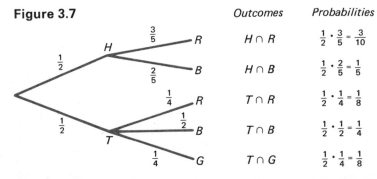

Using the data of our illustration, we display our results in Figure 3.7.

Figure 3.7

Outcomes	*Probabilities*
$H \cap R$	$\frac{1}{2} \cdot \frac{3}{5} = \frac{3}{10}$
$H \cap B$	$\frac{1}{2} \cdot \frac{2}{5} = \frac{1}{5}$
$T \cap R$	$\frac{1}{2} \cdot \frac{1}{4} = \frac{1}{8}$
$T \cap B$	$\frac{1}{2} \cdot \frac{1}{2} = \frac{1}{4}$
$T \cap G$	$\frac{1}{2} \cdot \frac{1}{4} = \frac{1}{8}$

The probabilities associated with the various branches leading from any branch point are called **conditional probabilities.** Notice that *the sum of the conditional probabilities leading from each branch point is 1:*

From Figures 3.6 and 3.7 we see that

$$P(H) + P(T) = \tfrac{1}{2} + \tfrac{1}{2} = 1$$

and that

$$P(R \mid H) + P(B \mid H) = \tfrac{3}{5} + \tfrac{2}{5} = 1,$$

and

$$P(R \mid T) + P(B \mid T) + P(G \mid T) = \tfrac{1}{4} + \tfrac{1}{2} + \tfrac{1}{4} = 1.$$

Conditional Probability / 57

Moreover, we have

$$P(H \cap R) + P(H \cap B) + P(T \cap R) + P(T \cap B) + P(T \cap G)$$
$$= \tfrac{3}{10} + \tfrac{1}{5} + \tfrac{1}{8} + \tfrac{1}{4} + \tfrac{1}{8} = 1.$$

We can draw diagrams as shown in Figure 3.8 to picture the probabilities as we did earlier in Figure 3.1.

Figure 3.8 a.

	H	T	
R	$\tfrac{1}{10}\cdot\;\tfrac{1}{10}\cdot\;\tfrac{1}{10}\cdot$	$\tfrac{1}{16}\cdot\;\tfrac{1}{16}\cdot$	$P(R) = \tfrac{3}{10} + \tfrac{2}{16} = \tfrac{17}{40}$
B	$\tfrac{1}{10}\cdot\;\tfrac{1}{10}\cdot$	$\tfrac{1}{16}\cdot\;\tfrac{1}{16}\cdot\;\tfrac{1}{16}\cdot\;\tfrac{1}{16}\cdot$	$P(B) = \tfrac{2}{10} + \tfrac{4}{16} = \tfrac{9}{20}$
G		$\tfrac{1}{16}\cdot\;\tfrac{1}{16}\cdot$	$P(G) = \tfrac{2}{16} = \tfrac{1}{8}$

$P(H) = \tfrac{5}{10} = \tfrac{1}{2}$ $P(T) = \tfrac{8}{16} = \tfrac{1}{2}$

b.

	H	
R	$\tfrac{1}{10}\cdot\;\tfrac{1}{10}\cdot\;\tfrac{1}{10}\cdot$	$P(R \cap H) = \tfrac{3}{10}$
B	$\tfrac{1}{10}\cdot\;\tfrac{1}{10}\cdot$	$P(B \cap H) = \tfrac{2}{10} = \tfrac{1}{5}$
G		

$P(R|H) = \tfrac{3}{5}$, $P(B|H) = \tfrac{2}{5}$

c.

	T	
R	$\tfrac{1}{16}\cdot\;\tfrac{1}{16}\cdot$	$P(R \cap T) = \tfrac{2}{16} = \tfrac{1}{8}$
B	$\tfrac{1}{16}\cdot\;\tfrac{1}{16}\cdot\;\tfrac{1}{16}\cdot\;\tfrac{1}{16}\cdot$	$P(B \cap T) = \tfrac{4}{16} = \tfrac{1}{4}$
G	$\tfrac{1}{16}\cdot\;\tfrac{1}{16}\cdot$	$P(G \cap T) = \tfrac{2}{16} = \tfrac{1}{8}$

$P(R|T) = \tfrac{2}{8} = \tfrac{1}{4}$, $P(B|T) = \tfrac{4}{8} = \tfrac{1}{2}$

$P(G|T) = \tfrac{2}{8} = \tfrac{1}{4}$

Earlier in this section we used our knowledge of the conditional probabilities $P(R \mid H)$ and $P(R \mid T)$ to compute $P(R)$. Having obtained this answer, we can now use $P(R)$ to compute $P(H \mid R)$ and $P(T \mid R)$ as illustrated in the following example.

Example 3.2.2 Suppose that the experiment described above has been performed and we are told that the marble drawn is red. What is the probability that Urn H was chosen? Urn T? We first determine $P(H \mid R)$. We find

$$P(H \mid R) = \frac{P(H \cap R)}{P(R)} = \frac{\tfrac{3}{10}}{\tfrac{17}{40}} = \frac{12}{17}.$$

Similarly,

$$P(T \mid R) = \frac{P(T \cap R)}{P(R)} = \frac{\tfrac{1}{8}}{\tfrac{17}{40}} = \frac{5}{17}.$$

As a check, we observe that $P(H \mid R) + P(T \mid R) = 1$. (Since a red marble was drawn, it must have come from either Urn H or Urn T.)

We can also find the following conditional probabilities:

$$P(H \mid B) = \frac{\frac{1}{5}}{\frac{9}{20}} = \frac{4}{9} \quad \text{and} \quad P(T \mid B) = \frac{\frac{1}{4}}{\frac{9}{20}} = \frac{5}{9};$$

$$P(H \mid G) = \frac{0}{\frac{1}{8}} = 0 \quad \text{and} \quad P(T \mid G) = \frac{\frac{1}{8}}{\frac{1}{8}} = 1.$$

Of course, $P(H \mid B) + P(T \mid B) = 1$ and $P(H \mid G) + P(T \mid G) = 1$.

The probabilities found in Example 3.2.2 are represented in the diagrams of Figure 3.9.

Figure 3.9 a.

$$R \quad \begin{array}{|cccc|} \hline \frac{1}{10} \cdot & \frac{1}{10} \cdot & \frac{1}{10} \cdot & \frac{1}{16} \cdot \quad \frac{1}{16} \cdot \\ \hline \end{array} \quad P(R) = \frac{17}{40}$$

$$P(H \mid R) = \frac{\frac{3}{10}}{\frac{17}{40}} = \frac{12}{17}, \quad P(T \mid R) = \frac{\frac{2}{16}}{\frac{17}{40}} = \frac{5}{17}$$

b.

$$B \quad \begin{array}{|cccc|} \hline \frac{1}{10} \cdot & \frac{1}{10} \cdot & \frac{1}{16} \cdot \frac{1}{16} \cdot \\ & & \frac{1}{16} \cdot \frac{1}{16} \cdot \\ \hline \end{array} \quad P(B) = \frac{9}{20}$$

$$P(H \mid B) = \frac{\frac{2}{10}}{\frac{9}{20}} = \frac{4}{9}, \quad P(T \mid B) = \frac{\frac{4}{16}}{\frac{9}{20}} = \frac{5}{9}$$

The technique that we have used is generally applicable to any experiment that can be considered as consisting of two or more steps.

Example 3.2.3 A game is played with a fair coin and two spinners. A player tosses the coin. If it lands heads, he spins Spinner H; if it lands tails, he spins Spinner T. The player then receives the number of points indicated on the spinner dial. (We assume that the pointer does not stop on a line.)

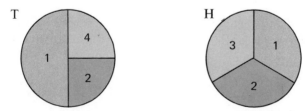

If we let X represent the number of points a player receives on his first play, then X has the possible values 1, 2, 3, and 4. The associated probabilities are:

$$P(X = 1) = P(H) \cdot P((X = 1) \mid H) + P(T) \cdot P((X = 1) \mid T)$$
$$= \tfrac{1}{2} \cdot \tfrac{1}{2} + \tfrac{1}{2} \cdot \tfrac{1}{3} = \tfrac{1}{4} + \tfrac{1}{6} = \tfrac{5}{12}$$

$$P(X = 2) = P(H) \cdot P((X = 2) \mid H) + P(T) \cdot P((X = 2) \mid T)$$
$$= \tfrac{1}{2} \cdot \tfrac{1}{4} + \tfrac{1}{2} \cdot \tfrac{1}{3} = \tfrac{1}{8} + \tfrac{1}{6} = \tfrac{7}{24}$$
$$P(X = 3) = P(T) \cdot P((X = 3) \mid T) = \tfrac{1}{2} \cdot \tfrac{1}{3} = \tfrac{1}{6}$$
$$P(X = 4) = P(H) \cdot P((X = 4) \mid H) = \tfrac{1}{2} \cdot \tfrac{1}{4} = \tfrac{1}{8}$$

Check: $\dfrac{5}{12} + \dfrac{7}{24} + \dfrac{1}{6} + \dfrac{1}{8} = \dfrac{10 + 7 + 4 + 3}{24} = 1.$

EXERCISES 3.2

1. Assume that male and female births are equally likely and assume further that the sex of the first born does not change the probability of the sex of the second born. In a family with two children, find the probability of:
   a. two boys
   b. exactly 1 boy, 1 girl
   c. two girls

2. Making the assumptions of Problem 1, find the probability that both children are boys, given that:
   a. at least one is a boy
   b. the first born is a boy

3. Urn I contains 2 red marbles and 1 blue marble. Urn II contains 1 red marble and 1 blue marble. Urn III contains 1 red and 2 blue marbles. An experiment consists of choosing an urn at random and then drawing a marble from that urn at random. Find:
   a. $P(R)$, where R is the event "a red marble is drawn." Note that there are *three* branches of the tree leading to the event R.
   b. $P(B)$
   Let I be the event "Urn I is chosen," and so on. Then find:
   c. $P(I \mid R)$ d. $P(II \mid R)$ e. $P(III \mid R)$

4. Two dice, one red and one white, are rolled. We are interested in T, the event "the numbers thrown total 10." Which of the following pieces of information is most favorable to the event T?
   a. We are told that "at least one die shows a six" (call this event A).
   b. We are told that "exactly one die shows a six" (call this event B).
   c. We are told that "the red die shows a six" (call this event R).

5. A fair coin is tossed until a tail appears or until three heads appear.
   a. Draw a tree diagram for the outcomes of this experiment. Stop each branch when the desired result is obtained.
   b. Find $P(\text{tails})$.
   c. Let X represent the number of heads that appear. Find $P(X = 0)$, $P(X = 1)$, $P(X = 2)$, $P(X = 3)$.
   d. Is $P(\text{tails})$ equal to $P(X = 0)$?

6. Two different integers are selected at random from the set $\{1, 2, 3, 4, 5\}$. (The order of selection is not considered, and so there are ten possible outcomes.) Find the probability that:
   a. the sum of the integers selected is even
   b. both integers selected are odd
   c. both integers selected are odd, given that the sum is even
   d. the sum is even, given that both integers selected are odd
   e. the sum is odd, given that both integers selected are even

7. The four aces from a deck of playing cards are lying face down on a table. We remove one card at a time until a red ace is found or until two black aces are found.
   a. Draw a tree diagram of the experiment, stopping each branch when the desired result is obtained.
   b. What is the probability of getting a red ace given that a black one was obtained on the first draw?
   c. Find the probability of obtaining a red ace.

8. Suppose that three different integers are selected at random from the set $\{1, 2, 3, 4, 5\}$ (compare Problem 6). (Order is not important; so there are 10 possible outcomes.) Find the probability that:
   a. the sum of the integers is even
   b. the three integers selected are odd
   c. the sum is odd, given that two of the integers are 1 and 3
   d. two of the integers are 3 and 5, given that the sum is 10

9. Consider two urns. There are 2 red marbles and 1 blue marble in Urn I, and 1 red marble and 1 blue marble in Urn II. The experiment consists of choosing an urn at random, drawing a marble from it, and placing it in the other urn. Then a marble is drawn from the second urn.
   a. Draw a tree diagram for the experiment.
   b. Find the probability of ending up with a red marble.
   c. Find the probability of ending up with a red marble given that Urn I was chosen.

10. Teams A and B are participating in a series of games. Team A has 30% chance of winning against B, team B has 50% chance of winning against A, and the chance of their tying is 20%.
    a. Draw a tree diagram of a two-game series.
    b. Find the probability of one tie game, given that A wins at least one of the games.

3.3 Experiments with Assigned Probabilities

In all our previous experiments with coins, we have assumed that the coin is fair. Hence, the probability of heads on each toss is $\frac{1}{2}$ and the probability of tails is $\frac{1}{2}$. Suppose now that the probability of heads on each toss is, instead, some number p, and the probability of tails is q. The technique shown in Section 3.2, making use of tree diagrams, will help us just as effectively in this more generalized picture.

We always assume that a coin falls either heads or tails, not both. That is, the events "heads" and "tails" are exhaustive and mutually exclusive; so

$$p + q = 1,$$

or

$$q = 1 - p.$$

Example 3.3.1 A coin is tossed twice in succession. Suppose that the probability of heads on the first toss is p. If we assume that the first toss does not change the characteristics of the coin, then the probability of heads on the second toss is also p. (The conditional probability of "heads on the second toss, given heads on the first toss" is still p.) We have the following diagram:

	Outcomes	Probabilities
	hh	$p \cdot p = p^2$
	ht	$p \cdot q = pq$
	th	$q \cdot p = qp$
	tt	$q \cdot q = q^2$

Our set of outcomes is

$$\{hh, ht, th, tt\}.$$

Here, for example, the notation "ht" means "heads on the first toss followed by tails on the second toss."

If we let X represent the number of heads that appear in the two tosses, we find:

$$P(X = 2) = P(\{hh\}) \quad = p \cdot p \quad\quad = p^2$$
$$P(X = 1) = P(\{ht, th\}) = p \cdot q + q \cdot p = 2pq$$
$$P(X = 0) = P(\{tt\}) \quad = q \cdot q \quad\quad = q^2$$

Check: $p^2 + 2pq + q^2 = (p + q)^2 = (1)^2 = 1$

Example 3.3.2 Two marbles are drawn, one after the other, *without* replacement from an urn containing 4 red and 2 blue marbles. If a red marble is drawn on the first draw, then there are only 3 red and 2 blue marbles left. Thus,

$$P(\text{red on second} \mid \text{red on first}) = \tfrac{3}{5}.$$

Completing the tree diagram in a similar manner, we have:

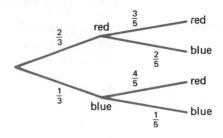

If we record the number of red marbles drawn, our set of outcomes is $\{2, 1, 0\}$. Our diagram shows the following information:

$$P(2 \text{ reds}) = \tfrac{2}{3} \cdot \tfrac{3}{5} = \tfrac{6}{15}$$
$$P(1 \text{ red}) = \tfrac{2}{3} \cdot \tfrac{2}{5} + \tfrac{1}{3} \cdot \tfrac{4}{5} = \tfrac{8}{15}$$
$$P(0 \text{ red}) = \tfrac{1}{3} \cdot \tfrac{1}{5} = \tfrac{1}{15}$$

Check: $\tfrac{6}{15} + \tfrac{8}{15} + \tfrac{1}{15} = 1$

EXERCISES 3.3

1. An urn contains red and blue marbles in the ratio 5:3. Two marbles are to be drawn, one after the other with replacement.
   a. Draw a tree diagram, attaching probabilities to the branches.
   b. Would the assignment of probabilities differ if the urn contains 5 red and 3 blue marbles or 10 red and 6 blue?

2. Suppose that two marbles are drawn from an urn containing red and blue marbles in the ratio 5:3, this time *without* replacement.
   a. What other information would be needed to determine the probabilities of the colors for the two marbles drawn?
   b. If X represents the number of red marbles obtained in the two draws, show that the probabilities $P(X = 2)$, $P(X = 1)$, and $P(X = 0)$ for 5 red and 3 blue marbles are different from those for 10 red and 6 blue marbles.
   c. Suppose that an urn contains 5 red and 3 blue marbles. If two marbles are drawn simultaneously, which of the methods—with or without replacement—is applicable for the probabilities here?

3. An urn contains r red and b blue marbles. If two marbles are to be drawn without replacement, show that:

a. $P(B \mid R) \neq P(R \mid B)$ if $r \neq b$, B is the event "a blue marble is drawn," and R is the event "a red marble is drawn."

   b. The probability of drawing a blue marble first and a red marble second is equal to the probability of red first and blue second.

- **4.** A college student is taking courses in mathematics and physics. An examination in mathematics is scheduled for Monday and one in physics for Friday. He estimates that his probability of passing mathematics is 0.80. If he does pass, his confidence carries over and his probability of also passing physics is 0.70. However, if he fails his mathematics examination, he judges that he has only probability 0.40 of passing physics. Draw a tree diagram with appropriate probabilities on the branches. What is the probability of his passing physics? (We assume that his judgments are correct.)

- **5.** In Problem 4, what is the probability that the student passed mathematics, given that he failed physics?

- **6.** Two men each shoot once at a target. Mr. A has probability of 0.6 of scoring a bull's-eye; Mr. B has probability of 0.8. If the success or failure of one man does not affect the other's chances for success, find the probability that:

   a. both hit the bull's-eye

   b. exactly one hits the bull's-eye

   c. at least one hits the bull's-eye

   (Hint: Draw a tree diagram for success and failure of first one man and then the other.)

   7. A shipment of 4 widgets is known to contain exactly one defective widget. If the four are tested one at a time and all four are to be tested in any case, find the probability that the defective one is the item:

   a. tested first **b.** tested second
   c. tested third **d.** tested fourth

   8. A container has 3 good items and 1 defective item. The items are drawn one at a time and tested until the defective item is found.

   a. Let X represent the number of the draw on which the defective item is found. Find the probability function of X. (Hint: Use a tree diagram.)

   b. Some of us may be surprised at the result of part a. Try to give an "intuitive" explanation.

   9. Assume that 80% of the drivers are rated "good" and 20% are rated "poor" drivers. Suppose that it is known that 3% of all "good" drivers receive a traffic ticket and 15% of all "poor" drivers receive a traffic ticket. What is the probability that a driver is "good," given that he received a traffic ticket?

- **10.** Suppose that a certain disease can be detected by laboratory blood test with probability 0.94. (That is, if a person has the disease, the probability is 0.94 that the test will reveal it.) Unfortunately, this test also gives a "false positive" in 1% of the healthy people tested. (That is, if a healthy

person takes the test, it will imply (falsely) that he has the disease with probability 0.01). Suppose that 1000 persons have been tested, of whom only 5 have the disease. If one of the 1000 were selected at random, find the probability that:
   a. he is healthy
   b. the test shows "positive"
   c. he is healthy, given that the test is "positive."
11. A professional football team wins with probability 0.7 when it is not snowing. However, on a snowy field, the probability of their winning is reduced to 0.4. The probability of snow in January is 0.5.
   a. What is the probability of their winning if they play a game on New Year's Day?
   b. Given that they won the game, what is the probability that they played on a snowy field?
12. A machine produces good items with probability p, and defective items with probability $1 - p$. Suppose that it produces 3 items, successively, which are then inspected.
   a. What is the probability that all three items are good?
   b. What is the probability that the first item is defective, given that the third item is defective?
   c. What is the probability that the first item is defective, given that exactly two are defective?

3.4 Independent Events

Starting with an experiment, a set of possible outcomes, and an *a priori* probability function, if E and G are events, we can find $P(E)$, $P(G)$, and $P(E \cap G)$. We may calculate the conditional probability $P(E \mid G)$ provided $P(G) \neq 0$:

$$P(E \mid G) = \frac{P(E \cap G)}{P(G)}$$

We have seen situations for which $P(E \mid G) = P(E)$; that is, the *a priori* probability of E equals the conditional probability, knowing the event G has occurred. Recall Example 3.1.3 as one illustration.

Other illustrations involve coin tossing or dice throwing. If a coin is thrown twice, the result of the second toss is not influenced by how the coin fell on the first toss. This conclusion is reasonable and is intuitively obvious; that is,

$P(\text{heads on 2d toss}) = P(\text{heads on 2d toss} \mid \text{heads on 1st toss})$.

If a red and a green die are tossed, a similar argument leads us to conclude, for example, that

$P(\text{red die shows 6}) = P(\text{red die shows 6} \mid \text{green die shows 3})$.

Another, only slightly less obvious, situation arises in connection with urn problems where successive draws are made with replacement. See the following example.

Example 3.4.1 An urn contains 5 red and 3 blue marbles. Two marbles are drawn, one after the other, with replacement after each drawing. The *a priori* probability of red on the first draw is $\frac{5}{8}$. Suppose that we do draw red on the first trial. Replacing the marble restores the contents of the urn to its original state. Thus, the probability of red on the second draw is again $\frac{5}{8}$. An appropriate tree diagram is as follows (recall Problem 1, Exercises 3.3):

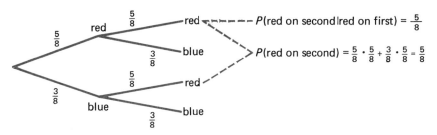

Notice that in this case

$$P(\text{red on second}) = P(\text{red on second} \mid \text{red on first}).$$

Example 3.4.2 Using the same urn as in Example 3.4.1, we now consider making two draws, this time *without* replacement (recall Problem 2, Exercises 3.3). If a red marble is drawn first and not replaced, then the urn contains 4 red and 3 blue marbles. If a blue marble is drawn first, a corresponding change is made in the contents of the urn. Our tree diagram shows the details together with the resulting probabilities:

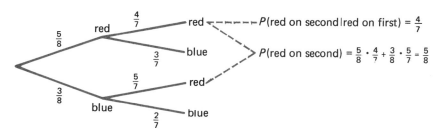

Here,

$$P(\text{red on second}) \ne P(\text{red on second} \mid \text{red on first}).$$

Most of us are surprised to learn that

$$P(\text{red on second}) = \tfrac{5}{8} = P(\text{red on first}),$$

whether the marble is replaced or not. (See Problem 6, Exercises 3.6 for more on this seeming paradox.)

The coin tossing illustration and Example 3.4.1 were both experiments in which it appeared "obvious" that $P(E \mid G) = P(E)$, where E and G represent the appropriate events. For the experiment without replacement in Example 3.4.2 it seemed equally "obvious" that $P(E \mid G) \neq P(E)$. Other situations are not so obvious. Consider the following examples.

Example 3.4.3 A red and a green die are tossed. Let E be the event "the red die shows an even number" and G the event "the sum of the faces is five." Using a set of 36 equally likely outcomes, we find that

$$P(E) = \tfrac{1}{2}, \qquad P(G) = \tfrac{1}{9}, \qquad P(E \cap G) = \tfrac{1}{18},$$

and so

$$P(E \mid G) = \frac{\tfrac{1}{18}}{\tfrac{1}{9}} = \frac{1}{2} = P(E).$$

Example 3.4.4 Let E be as in 3.4.3 and G the event "the sum of the faces is 6." Here

$$P(G) = \tfrac{5}{36}, \qquad P(E \cap G) = \tfrac{1}{18},$$

and so

$$P(E \mid G) = \frac{\tfrac{1}{18}}{\tfrac{5}{36}} = \frac{2}{5} \neq P(E).$$

If $P(E \mid G) = P(E)$, there is a "special relationship" between the event E and the event G. We remember that $P(E \mid G)$ has meaning only if $P(G) \neq 0$. If it is also true that $P(E) \neq 0$, then we have, from our multiplication formula,

$$P(E \cap G) = P(G) \cdot P(E \mid G) = P(E) \cdot P(G \mid E).$$

Since $P(E \mid G) = P(E)$, the last equality gives us

$$P(G) \cdot P(E) = P(E) \cdot P(G \mid E)$$

or

$$P(G \mid E) = P(G), \quad \text{since } P(E) \neq 0.$$

We conclude that whenever the "special relationship" exists between E and G, it also exists between G and E, provided $P(E) \neq 0$ and $P(G) \neq 0$.

To find a formula that will avoid the necessity for making the qualifications $P(E) \neq 0$ and $P(G) \neq 0$, we proceed as follows. If

$$P(E \mid G) = P(E),$$

then

$$P(E \mid G) = \frac{P(E \cap G)}{P(G)} = P(E)$$

and

$$P(E \cap G) = P(E) \cdot P(G).$$

We would reach the same result if we started with $P(G \mid E) = P(G)$.

Now we are ready to give a name to events having this "special relationship":

Two events *E* and *G* are called *independent events* if and only if
$$P(E \cap G) = P(E) \cdot P(G).$$

This definition avoids the specific use of conditional probability and is meaningful even without restrictions on $P(E)$ or $P(G)$.

The events of Examples 3.4.1 and 3.4.3 are independent events, while the events of Examples 3.4.2 and 3.4.4 are *not* independent events.

In the case of Examples 3.4.3 and 3.4.4, it may seem strange that E, "the red die shows an even number," and G, "the sum of the numbers is 5," are independent events, while E, "the red die shows an even number," and G, "the sum of the numbers is 6," are not independent. But a little reflection will show that the sum of the numbers can be 5 in just as many ways if the red die shows an odd number as if it shows an even number. Actually, they are (listing the throw of the red die first)

$$(1, 4), \quad (2, 3),$$
$$(3, 2), \quad (4, 1).$$

So the sum of the numbers, if 5, is independent of the oddness or evenness of the number thrown on either die.

But if the sum is 6, the possible throws are

$$(1, 5), \quad (2, 4),$$
$$(3, 3), \quad (4, 2),$$
$$(5, 1).$$

Hence, the sum is 6 more often if the number on the red die is odd than if it is even. So the evenness or oddness of the throw and a sum of 6 are not independent. (This reasoning may be extended to any odd sum (which is independent of the evenness of the numbers thrown) and any even sum (which is not).)

The mathematical definition of independent events serves as a check or verification of our intuitive feeling that certain events should be independent. We would suspect two events to be independent, for example, if the outcomes in one event was not physically connected with the outcomes of a second event. Sometimes the reverse argument is useful. If our mathematical "model" leads us to conclude that two events are independent, we might examine carefully the conditions of the experiment and see whether the events ought to be independent in the intuitive sense.

For simple examples (coin tossing, drawing with replacement), our definition coincides with our intuitive feeling that the occurrence of one event does not affect the probability of the other. Our formal, mathematical, definition now requires us to judge whether or not two events are independent *only* by deciding whether or not their probabilities satisfy our definition. Thus, we must check our intuition about E and G by computing $P(E)$, $P(G)$, and $P(E \cap G)$.

Remark

Some students, when first studying probability, confuse "mutually exclusive" events with "independent" events. To emphasize the difference between these concepts, see Problem 4 in the following set of exercises.

EXERCISES 3.4

1. One card is drawn from a shuffled deck of 52 cards. Let A be the event "an ace is drawn," D the event "a diamond is drawn," and H the event "an honor card (ace, king, queen, jack, or 10) is drawn." Which of the following pairs of events are independent events?
   a. A, D b. A, H c. D, H

2. Refer to Example 3.1.3 and Problem 3 of Exercises 3.1.
   a. Show that the events "ace" and "hearts" are independent events.
   b. Show that the events "ace" and "clubs" are not independent events.

3. In Problem 4, Exercises 3.1, which of the following pairs of events are independent events?
   a. A and B b. A and C c. B and C

4. a. Given that E and G are mutually exclusive events and $P(E) \neq 0$, $P(G) \neq 0$. Prove that E and G are *not* independent.
   b. Given that E and G are independent events and $P(E) \neq 0$, $P(G) \neq 0$, prove that E and G are *not* mutually exclusive.

5. Let E be an event with $P(E) \neq 0$ and $P(E) \neq 1$. Are the events E and $\sim E$ independent?

6. Suppose that 120 students in a certain college took both mathematics and chemistry. Suppose also that 9 failed mathematics only, 10 failed chemistry only, and 6 failed both courses. A student is selected at random from the 120 students. Are the events "failed mathematics" and "failed chemistry" independent events?

7. In Problem 6, Exercises 3.2, are the events in parts a and b independent?

8. Three coins are tossed simultaneously. Let A be the event "both heads and tails turn up," and let B be the event "at most one coin falls tails." Show that A and B are independent events.

9. Suppose that two coins are tossed simultaneously. Letting A and B be the same as in Problem 8, are these events independent?

10. Consider the toss of two dice as in Problem 4, Section 3.1. Let A be the event "the red die shows a number ≤ 3," and B be the event "the green die shows a number ≤ 2." Show that A and B are independent events.

11. In the toss of two dice, let C be the event "the red die shows an even number," and D be the event "the green die shows an even number." Are C and D independent events?

12. A group of 100 students have elected to take 0, 1, or 2 of the mathematics, physics, and chemistry offerings. The diagram at the right shows the distribution of the students in these 3 courses: 40 chose mathematics only, 23 chose physics only, 3 chemistry only; 5 chose mathematics and physics, 5 mathematics and chemistry, 2 chemistry and physics, and 22 chose none of the three.

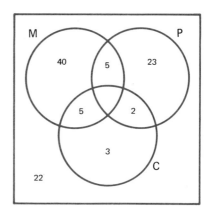

   a. Are the events "take mathematics" and "take chemistry" independent?
   b. Are the events "take mathematics" and "take physics" independent?
   c. Are the events "take chemistry" and "take physics" independent?

3.5 Repeated Independent Trials

One use of the concept of independent events is in connection with what is known as "independent trials" of an experiment. We have already had experience with independent trials, although we did not call them that.

Many of the experiments we have examined have been thought of as occurring in two steps. In some cases, the steps were quite different (for example, first toss a coin, and then throw a die). In others, the second step was, in a sense, simply a repetition of the first (for example, either toss a coin twice or draw twice from an urn with replacement). In fact, we made the remark in Example 3.4.1 that replacing the marble in the urn restores the contents to its original state. Here, we may think of the experiment as consisting of two separate experiments, each having the *same* set of outcomes and the *same* assigned probability function. The experiment "toss a fair coin twice" can be considered as two separate experiments each with

$$S = \{h, t\}$$

as a set of outcomes. Further,

$$f(h) = P(\{h\}) = \tfrac{1}{2}$$

and

$$f(t) = P(\{t\}) = \tfrac{1}{2}$$

for both experiments. We shall refer to this type of situation as two **trials** of the same experiment to avoid confusion with an experiment having two different steps.

Suppose that we make two trials of an experiment having a set of m outcomes. We can denote these outcomes as $s_1, s_2, \ldots, s_m$. Then

$$S = \{s_1, s_2, \ldots, s_m\}.$$

Let these outcomes have corresponding probabilities

$$f(s_1), f(s_2), \ldots, f(s_m),$$

with

$$\sum_{i=1}^{m} f(s_i) = 1.$$

The result of a pair of trials may be denoted by listing the outcomes in the order they were obtained; for example, (s_5, s_3) means that the outcome s_5 occurred on the first trial and s_3 on the second. The set of possible outcomes for the two-step experiment then consists of all m^2 possible pairs:

$$\{(s_1, s_1), (s_1, s_2), \ldots, (s_1, s_m),$$
$$(s_2, s_1), (s_2, s_2), \ldots, (s_2, s_m),$$
$$\cdots\cdots\cdots\cdots\cdots\cdots\cdots\cdots$$
$$(s_m, s_1), (s_m, s_2), \ldots, (s_m, s_m)\}.$$

This set is usually denoted by $S \times S$ (read "S cross S") and is called the **Cartesian product** of S with itself.

To describe a probability function for the events of $S \times S$, we need to determine

$$P((s_i, s_j))$$

for all values of i and j from 1 to m. This notation is to be interpreted as the probability that outcome s_i occurs on the first trial and outcome s_j occurs on the second trial. If the outcomes of the successive trials are *independent*, then we must have

$$P((s_i, s_j)) = P(s_i) \cdot P(s_j) = f(s_i) \cdot f(s_j).$$

Similarly, if s_i is the outcome on the first trial, s_j the outcome on the second trial, and s_k the outcome on the third trial, and if the trials are independent, then

$$P((s_i, s_j, s_k)) = f(s_i) \cdot f(s_j) \cdot f(s_k).$$

In general, we may refer to **repeated independent trials**.

A question remains. Does this assignment of probability really describe a probability function? First, since

$$0 \leq f(s_i) \leq 1$$

and

$$0 \leq f(s_j) \leq 1,$$

it follows that for each pair (s_i, s_j),

$$0 \leq P((s_i, s_j)) \leq 1.$$

It remains to check that the sum of these probabilities is 1. We shall make this verification for the case of two independent trials of an experiment having three outcomes.

Let $S = \{s_1, s_2, s_3\}$ be the set of outcomes we are considering, and let

$$f(s_1) = p_1, \qquad f(s_2) = p_2, \qquad f(s_3) = p_3$$

be the assigned probabilities. We know that

$$p_1 + p_2 + p_3 = 1,$$

and we may draw a tree diagram as in Figure 3.10.

Figure 3.10

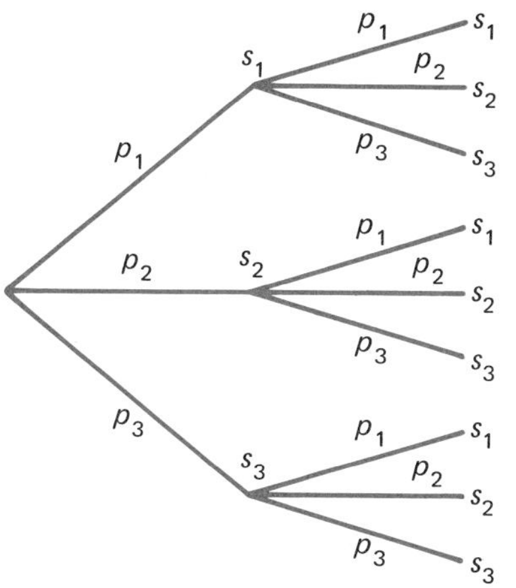

Outcomes	*Probabilities*
(s_1, s_1)	$p_1 \cdot p_1$
(s_1, s_2)	$p_1 \cdot p_2$
(s_1, s_3)	$p_1 \cdot p_2$
(s_2, s_1)	$p_2 \cdot p_1$
(s_2, s_2)	$p_2 \cdot p_2$
(s_2, s_3)	$p_2 \cdot p_3$
(s_3, s_1)	$p_3 \cdot p_1$
(s_3, s_2)	$p_3 \cdot p_2$
(s_3, s_3)	$p_3 \cdot p_3$

Adding the probabilities, we have

$$\begin{aligned} p_1 \cdot p_1 + p_1 \cdot p_2 + p_1 \cdot p_3 + p_2 \cdot p_1 + p_2 \cdot p_2 + p_2 \cdot p_3 + p_3 \cdot p_1 \\ + p_3 \cdot p_2 + p_3 \cdot p_3 \\ = p_1(p_1 + p_2 + p_3) + p_2(p_1 + p_2 + p_3) + p_3(p_1 + p_2 + p_3) \\ = \quad p_1 \quad + \quad p_2 \quad + \quad p_3 \\ = 1. \end{aligned}$$

The proof for the more general situation with n trials, each trial having m outcomes proceeds in a similar manner, but is more cumbersome.

Remark

If we reexamine Examples 3.3.1 and 3.4.1, we see that the discussion above is directly applicable. Each of these examples is an illustration of repeated, independent trials. In both examples each trial leads to exactly two outcomes. A special name is reserved for such cases. If there are exactly two outcomes for each of a series of independent trials of an experiment, we call these trials *Bernoulli trials* (named after Jacques Bernoulli, mentioned on page 3). In Chapter 7 we shall deal with Bernoulli trials in some detail.

EXERCISES 3.5

1. Suppose that the two-trial experiment of Example 3.3.1 is continued and a third trial is made. Construct the tree diagram, indicating probabilities of the branches. Verify that the sum of the probabilities of the eight outcomes is 1.

2. An urn contains 2 red and 2 blue marbles. Marbles are to be drawn one at a time with replacement. Let E be the event "at most one red" and G be the event "at least one of each color." Determine whether or not E and G are independent:
   a. if two draws are made (4 equally likely outcomes)
   b. if three draws are made (8 equally likely outcomes)

3. a. An X-ray picture will reveal the presence of a certain lung condition in afflicted patients with probability 0.8. If the X-ray is negative (fails to show the condition) a second X-ray is taken. Assuming that the results of the two X-rays are independent, what fraction of afflicted patients will be diagnosed by this technique? (In actual practice, of course, it is unlikely that the results would really be independent.)
   b. If a third, independent, X-ray is taken, what is the increase in the probability of diagnosis?
   c. A fourth X-ray?

4. Philip is faced with the task of answering three true-false questions whose answers he does not know. Let the probability that he guesses each individual question correctly be p. What is the probability that he guessed at least one question correctly?

5. If four cards are drawn, with replacement, from a shuffled deck, what is the probability that all the cards show a number less than 4?

6. Assume that male and female births are equally likely and assume further that the sex of the first born does not change the probability of the sex of the second born. Let H be the event "the family has children of both sexes" and A the event "there is at most one girl." Are the events A and H independent in families with:*
   a. two children?   b. three children?

7. For the 2- and 3-children families described in Problem 6, show that whether the complements $\sim H$ and $\sim A$ are independent, agrees with whether H and A are independent.

8. In an urn are 5 blue and 5 green marbles. Three draws are to be made at random one at a time with replacement. Paraphrase the events A and H described in Problem 6b and state whether or not they are independent.

*Compare Example (1.c), page 107, and Example (3.d), page 115, in William Feller's *An Introduction to Probability Theory and Its Applications*, Vol. I, 2d ed. (New York: John Wiley & Sons, Inc., 1957).

9. A series of games is being played in which the opponents are evenly matched. The winner of each game gets one point. The first player to accumulate $n = 3$ points wins a $100 prize. The contest is interrupted after opponent A has won 2 points and opponent B has won 1 point. How should the prize be divided? (This is a variation of the gambling problem presented to Pascal to which we referred in Section 1.1.)

★ 10. Solve Problem 9 if $n = 4$, assuming that the game is interrupted at the same point.

★ 11. Suppose that Able has 4 chips and Baker 2 chips. Suppose also that a fair coin is tossed. If it falls heads, Able takes a chip from Baker; if it falls tails, Baker takes a chip from Able. They decide to play four times or until one loses all his chips. Draw a tree diagram for this game. Let X represent the number of chips Baker has when the game ends. Find:

a. $P(X = 0)$ b. $P(X = 1)$ c. $P(X = 2)$
d. $P(X = 3)$ e. $P(X = 4)$ f. $P(X = 5)$
g. $P(X = 6)$ h. $P((X = 0) \mid$ heads on first toss$)$

★ 12. For Problem 11, what is the probability that Baker ends up with no more than two chips? (Even though the coin is fair, the fact that Able has more chips in the beginning gives him an advantage.)

★ 3.6 Bayes' Formula

In this section, we shall develop a formula which will enable us to proceed mechanically in computing certain conditional probabilities in complicated situations.

Recall that in Section 3.2, given $P(H)$, $P(T)$, $P(R \mid H)$, and $P(R \mid T)$, we found

$$P(R) = P(H \cap R) + P(T \cap R)$$

from

$$P(H \cap R) = P(R \cap H) = P(H) \cdot P(R \mid H)$$

and

$$P(T \cap R) = P(R \cap T) = P(T) \cdot P(R \mid T)$$

by applying the formula on page 50. Then in Example 3.2.2, we used $P(R)$ to find

$$P(H \mid R) = \frac{P(H \cap R)}{P(R)} \quad \text{and} \quad P(T \mid R) = \frac{P(T \cap R)}{P(R)}.$$

By substituting in these formulas the expressions given above, we obtain

$$P(H \mid R) = \frac{P(H) \cdot P(R \mid H)}{P(H) \cdot P(R \mid H) + P(T) \cdot P(R \mid T)}$$

and

$$P(T \mid R) = \frac{P(T) \cdot P(R \mid T)}{P(H) \cdot P(R \mid H) + P(T) \cdot P(R \mid T)}.$$

In this example the events H and T have the properties $H \cap T = \emptyset$, $H \cup T = S$, $P(H) \neq 0$, and $P(T) \neq 0$. It turns out that if $G_1, G_2, \ldots, G_k$ are k events of S such that

$$P(G_i) \neq 0 \quad \text{and} \quad G_i \cap G_j = \emptyset \quad \text{for } i \neq j,$$

and

$$G_1 \cup G_2 \cup \ldots \cup G_k = S,$$

then if E is any event of S with $P(E) \neq 0$, we have

$$P(G_1 \mid E) = \frac{P(G_1 \cap E)}{P(G_1 \cap E) + P(G_2 \cap E) + \cdots + P(G_k \cap E)}$$

$$= \frac{P(G_1) \cdot P(E \mid G_1)}{P(G_1) \cdot P(E \mid G_1) + \cdots + P(G_k) \cdot P(E \mid G_k)}.$$

Similar formulas hold for $P(G_2 \mid E), \ldots, P(G_k \mid E)$. (Notice that the denominator is $P(E)$.)

In general, we have:

BAYES' FORMULA

$$P(G_i \mid E) = \frac{P(G_i) \cdot P(E \mid G_i)}{\sum_{j=1}^{k} P(G_j) \cdot P(E \mid G_j)}$$

This is a generalization of a proposition that appeared in *The Philosophical Transactions* (1763) of the Royal Society by Rev. Thomas Bayes, F.R.S. What is now identified as Bayes' Theorem or Bayes' Formula is the generalized statement given above.

In essence, the proof for the general case follows the procedure indicated in our special case, $k = 2$. However, we shall omit the proof of Bayes' Formula, being content to present three more examples.

Example 3.6.1 A manufacturer has two machines, I and II, which produce the item he manufactures. Machine I is slower, but more reliable; it produces only 25% of the items with 1% defectives. Machine II produces the remainder of the items with 3% defectives. An item from the production line is selected at random. Let I be the event "the item came from Machine I," and II be the event "the item came from Machine II."

The *a priori* probability that it came from Machine I is 0.25. Suppose that the item selected is defective. With this additional information, we can find the probability that the item came from Machine I. Let us use D for the event "the item is defective." Then:

$$P(I \mid D) = \frac{P(I \cap D)}{P(D)}$$

$$= \frac{P(I) \cdot P(D \mid I)}{P(I) \cdot P(D \mid I) + P(II) \cdot P(D \mid II)}$$

$$= \frac{(0.25)(0.01)}{(0.25)(0.01) + (0.75)(0.03)} = \frac{0.0025}{0.0250} = 0.10$$

Example 3.6.2 We shall now work Problem 10c, Exercises 3.3, using Bayes' Formula, with H for "healthy," I for "ill," A for "positive," and N for "negative." The original data give $P(H) = 0.995$, $P(I) = 0.005$, $P(A \mid H) = 0.01$, and $P(A \mid I) = 0.94$.

$$P(H \mid A) = \frac{P(H) \cdot P(A \mid H)}{P(H) \cdot P(A \mid H) + P(I) \cdot P(A \mid I)}$$
$$= \frac{(0.995)(0.01)}{(0.995)(0.01) + (0.005)(0.94)} \doteq 0.68$$

A comparison of Example 3.6.2 with the solution given for Problem 10c, Exercises 3.3, shows that we actually made the same calculations here as we did earlier. The advantages of using Bayes' formula are that we did not need the tree diagram and did not need to write out the complete set of outcomes and their probabilities. We proceeded mechanically, once it was clear that $H \cap I = \emptyset$, $H \cup I = S$, and $P(H) \neq 0$, $P(I) \neq 0$. These advantages are greater if there are many outcomes.

Example 3.6.3 Suppose that a theoretical experiment has the following tree diagram:

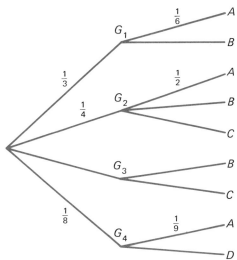

If we are interested only in $P(A)$ and in the conditional probabilities

$$P(G_1 \mid A), \quad P(G_2 \mid A), \quad \text{and} \quad P(G_4 \mid A),$$

then only the information shown on the diagram is needed. Using Bayes' formula, we have:

$$P(G_1 \mid A) = \frac{P(G_1) \cdot P(A \mid G_1)}{P(G_1) \cdot P(A \mid G_1) + P(G_2) \cdot P(A \mid G_2) + P(G_4) \cdot P(A \mid G_4)}$$
$$= \frac{\frac{1}{3} \cdot \frac{1}{6}}{\frac{1}{3} \cdot \frac{1}{6} + \frac{1}{4} \cdot \frac{1}{2} + \frac{1}{8} \cdot \frac{1}{9}} = \frac{\frac{1}{18}}{\frac{7}{36}} = \frac{2}{7} = \frac{4}{14}$$

$$P(G_2 \mid A) = \frac{\frac{1}{4} \cdot \frac{1}{2}}{\frac{7}{36}} = \frac{9}{14} \qquad P(G_4 \mid A) = \frac{\frac{1}{8} \cdot \frac{1}{9}}{\frac{7}{36}} = \frac{1}{14} \qquad P(A) = \frac{7}{36}$$

Check: $P(G_1 \mid A) + P(G_2 \mid A) + P(G_4 \mid A) = 1$

A major use of Bayes' formula is in situations for which it is difficult to determine accurate *a priori* probabilities. What may be done is to assign tentative, approximate, *a priori* probabilities to various events. After some partial evidence is received, these estimates can be revised. The revised probabilities are called *a posteriori* probabilities. These new probabilities may then be used as *a priori* probabilities in connection with further experimentations. These procedures, of course, are not unlike other mathematical approximations in which original estimates are systematically refined. The following example, while not a "real world" situation, illustrates the idea.

Example 3.6.4 Suppose that there are four countries in the Land of Oz. At home, the men of each country wear ties of a special color. Munchkins wear blue, Quadlings red, Gillikins purple, and Winkies yellow. When visiting the Emerald City, however, some men wear green ties. 20% of the Munchkins, 30% of the Quadlings, 40% of the Gillikins, and 50% of Winkies change their tie color. As a visitor approaches, the Guardian of Emerald City (knowing that Munchkins travel more) assigns *a priori* probabilities as follows:

$$P(M) = \tfrac{1}{2}, \quad P(Q) = P(G) = P(W) = \tfrac{1}{6}$$

Combining these assignments with the information given above, we have the following tree diagram:

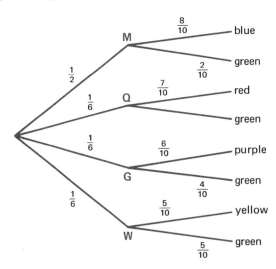

If the visitor is wearing a green tie, the Guardian would revise the probabilities by making the following computations:

$$P(\text{green}) = P(M) \cdot P(\text{green} \mid M) + P(Q) \cdot P(\text{green} \mid Q)$$
$$+ P(G) \cdot P(\text{green} \mid G) + P(W) \cdot P(\text{green} \mid W)$$
$$= \tfrac{1}{2} \cdot \tfrac{2}{10} + \tfrac{1}{6} \cdot \tfrac{3}{10} + \tfrac{1}{6} \cdot \tfrac{4}{10} + \tfrac{1}{6} \cdot \tfrac{5}{10} = \tfrac{18}{60}$$

$$P(M \mid \text{green}) = \frac{P(M) \cdot P(\text{green} \mid M)}{P(\text{green})} = \frac{\frac{1}{2} \cdot \frac{2}{10}}{\frac{18}{60}} = \frac{1}{3}$$

Similarly,

$$P(Q \mid \text{green}) = \frac{\frac{1}{6} \cdot \frac{3}{10}}{\frac{18}{60}} \qquad P(G \mid \text{green}) = \frac{\frac{1}{6} \cdot \frac{4}{10}}{\frac{18}{60}} \qquad P(W \mid \text{green}) = \frac{\frac{1}{6} \cdot \frac{5}{10}}{\frac{18}{60}}$$

$$= \tfrac{1}{6} \qquad\qquad\qquad = \tfrac{2}{9} \qquad\qquad\qquad = \tfrac{5}{18}$$

Check: $\frac{1}{3} + \frac{1}{6} + \frac{2}{9} + \frac{5}{18} = 1$

★ EXERCISES 3.6

1. From Example 3.6.1, we know that $P(I \mid D) = 0.10$. Therefore $P(II \mid D) = 1 - 0.10 = 0.90$. Verify this result by computing $P(II \mid D)$ using Bayes' formula.

2. One of three men, E, F, G, is chosen at random to shoot at a target. The probability that E will hit the target is $\frac{1}{3}$, that F will hit it is $\frac{1}{4}$, and that G will hit it is $\frac{1}{2}$. If you observe that the target is hit, what is the probability that the shot was made by E? by F? by G?

3. This one is a bit tricky. A lady has three sons (Adam, Barry, Cosmo) and six grandchildren. Adam has two sons, Barry one son, one daughter, and Cosmo has two daughters. While talking to you over the telephone, the lady states that one of her sons has left his children in her care for the day. In a few minutes you hear a boy's voice in the background. What is the probability that the children are Adam's?
   This is a version of the classic example of the need for care in using intuition in probability problems. The original problem is associated with the name of the French mathematician Joseph Bertrand (1822–1900).

4. A die is chosen at random from 3 dice, one red, one blue, and one green. The red and blue dice are fair, but the green die is loaded in such a way that the probability of each face turning up is proportional to the number of dots showing. Given that a 5 turns up, what is the probability that the green die was chosen?

5. An urn contains 5 red marbles and 3 blue marbles. One marble is drawn at random. This marble is replaced and 2 more marbles of the same color as that just drawn are added to the urn. A second random drawing is made. Find the probabilities:
   a. $P(\text{second red} \mid \text{first red})$
   b. $P(\text{first red} \mid \text{second red})$

6. Show that the result of Problem 5 is general. Suppose that an urn contains r red and b blue marbles. Whichever color is drawn first, it is replaced and t marbles of the same color are added. A new drawing is made. Show that

$$P(\text{second red} \mid \text{first red}) = P(\text{first red} \mid \text{second red}).$$

7. For the urn of Problem 6, find $P(\text{second red})$. (If $t = 0$, the drawing is "with replacement"; if $t = -1$, the drawing is "without replacement.") Verify that
$$P(\text{second red}) = P(\text{first red}).$$
(See Examples 3.4.1 and 3.4.2.)

Chapter Summary

1. If E and G are arbitrary events from a set of outcomes S and if $P(G) \neq 0$, then the *conditional probability* of E, given that G has occurred, is
$$P(E \mid G) = \frac{P(E \cap G)}{P(G)}.$$

2. The sum of conditional probabilities from any branch of a tree diagram is 1.

3. If the probability of success in an experiment is p, the probability of failure is $1 - p$.

4. Two events, E and G are called *independent events* if and only if
$$P(E \cap G) = P(E) \cdot P(G).$$

5. When an experiment is repeated n times, the situation is referred to as "n trials of the same experiment." These are also called *repeated independent trials*.

Chapter Review

1. For an experiment consisting of tossing a coin followed by throwing a die, a set of possible outcomes has 12 members:
$$\{(h, 1), (h, 2), (h, 3), \ldots, (t, 5), (t, 6)\}.$$
What is the reduced set given that the die shows a "4"?

2. If A and B are events from a set of possible outcomes, we have
$$P(A \cap B) = P(A) \cdot P(B \mid A).$$
Give (and explain) special cases for this relation when;
 a. A and B are independent events
 b. A and B are mutually exclusive events

3. A and B are events from a set of possible outcomes. If $P(A) = 0.6$, $P(B) = 0.4$, and $P(B \mid A) = 0.5$, find $P(A \mid B)$.

4. An urn contains 6 red and 2 blue marbles. A marble is drawn. Instead of replacing the marble that is drawn, a marble of the opposite color is added to the urn.
   a. Draw a tree diagram for this experiment and give the probabilities for each of the outcomes in
   $$\{(red, red), (red, blue), (blue, red), (blue, blue)\}$$
   b. Describe a fair game for a pair of drawings using the information from your tree diagram.
   c. What is the probability that the first marble drawn is red given the second is blue?
   d. Suppose that, after two drawings, P(red on third draw) = P(blue on third draw). What were the first two draws?

5. A defective light bulb was put together with two good bulbs by mistake. Each of the bulbs is then tested one at a time.
   a. Use a tree diagram to assign probabilities of finding the defective bulb after each trial.
   b. What is the probability that the defective bulb was found in one of the first two tests?

6. A fair coin is tossed until either two heads appear or two tails appear.
   a. Draw a tree diagram for the outcomes of this experiment.
   b. What is the greatest number of tosses that has to be made before stopping?
   c. Assign probabilities for each possible outcome.
   d. What is the probability that the first toss is tails given the last toss is heads?

7. There are two urns marked H and T. Urn H contains 2 red marbles and 1 blue marble. Urn T contains 1 red and 2 blue marbles. A coin is to be tossed. If it lands heads, a marble is drawn from Urn H. If it lands tails, a marble is drawn from Urn T. Find the following probabilities.
   a. P(heads, red) b. P(tails) c. P(red)
   d. P(blue) e. P(heads $\mid$ red)

8. In an experiment, each of two coins were tossed 10,000 times with the following results:

	Heads	Tails
Coin I	5,084	4,916
Coin II	8,347	1,653

   These two coins are to be tossed simultaneously. Explain which of the following games is more nearly fair:
   a. One wins if both coins fall heads; the other wins if both coins fall tails; otherwise, no one wins.
   b. One wins if both coins match; other wins otherwise.

9. Suppose that 120 students took both mathematics and chemistry. 15 failed mathematics; 16 failed chemistry. If the events "failed mathematics" and "failed chemistry" turned out to be independent events, how many failed both courses?

10. A certain type of pea occurs in two varieties, yellow and green. Three quarters of each generation is green, regardless of the color of the parent. A yellow pea of the third generation is selected at random. What is the probability that it came from a first generation green pea and a second generation yellow pea?

Chapter 4

Numerical Data

Eventually, we shall be applying our work with probability to the study of sets of numerical data. As mentioned in Section 1.4, it is essential to organize such data in an orderly way. We shall now explore some of the techniques that may be used. Primarily we are interested in

(a) organizing numerical data in tables,
(b) representing data in graphs, and
(c) using certain descriptive measures (numbers) to convey summarized information about a collection of data.

4.1 Frequency Distributions

Let us begin with a rather simple illustration. A ten-point quiz was administered to a class of size 50. When the scores were arranged alphabetically by student's name and read off, in that order, the scores were:

6, 4, 3, 7, 5,	3, 3, 8, 6, 6,	7, 2, 5, 7, 7,
9, 5, 4, 6, 6,	9, 5, 8, 5, 3,	8, 8, 7, 0, 8,
8, 5, 7, 3, 6,	7, 5, 7, 5, 5,	5, 8, 7, 5, 3,
2, 8, 7, 10, 9.		

Notice that for ease in reading, these numbers have been arranged in blocks of five. So there is already some sort of organization. We can further organize the data by rearranging them in order of increasing size:

0, 2, 2, 3, 3,	3, 3, 3, 3, 4,	4, 5, 5, 5, 5,
5, 5, 5, 5, 5,	5, 5, 6, 6, 6,	6, 6, 6, 7, 7,
7, 7, 7, 7, 7,	7, 7, 7, 8, 8,	8, 8, 8, 8, 8,
8, 9, 9, 9, 10.		

When this is done, we get a clearer picture of the scores obtained. For example, we can see a tendency for a "bunching" of the scores 5, 7, 8, and we notice relatively fewer students getting scores at either end, 0 or 10.

By organizing a collection of numbers according to size (in either ascending or descending order), we obtain a *rank-order listing*. For the fifty scores listed above, we did not gain a great deal of clarity, but we did make it easier to draw some conclusions. For example, if a score of 5 was necessary to pass the quiz, we can see from the rank-order listing that 11 students failed to pass.

We can refine this procedure by constructing a *tally* count (Figure 4.1) and, from this, a **frequency table** (Figure 4.2), which eliminates the necessity of listing the same score several times.

Figure 4.1

Score	Tally
0	/
1	
2	//
3	//// /
4	//
5	//// //// /
6	//// /
7	//// ////
8	//// ///
9	///
10	/

Figure 4.2

Score	Frequency (number of students)
0	1
1	0
2	2
3	6
4	2
5	11
6	6
7	10
8	8
9	3
10	1

$n = 50 =$ total frequency

Note that at the bottom of Figure 4.2, we use the letter n to represent the total number of students, that is, the **total frequency**. Although a comparison between frequencies of scores can be read directly from the table, the ratio of each frequency to the total gives a much easier comparison. By this **relative frequency**, we can visualize clearly the proportion of the class obtaining a given score. In Figure 4.3 we show the relative frequency both as a fraction and as a decimal.

Figure 4.3

Score	Frequency	Relative frequency
0	1	$\frac{1}{50} = 0.02$
1	0	$\frac{0}{50} = 0$
2	2	$\frac{2}{50} = 0.04$
3	6	$\frac{6}{50} = 0.12$
4	2	$\frac{2}{50} = 0.04$
5	11	$\frac{11}{50} = 0.22$
6	6	$\frac{6}{50} = 0.12$
7	10	$\frac{10}{50} = 0.20$
8	8	$\frac{8}{50} = 0.16$
9	3	$\frac{3}{50} = 0.06$
10	1	$\frac{1}{50} = 0.02$
	$n = 50$	Total $= 1.00$

For many purposes it is convenient to reverse the frequency table, placing the highest (largest) score on top, as in Figure 4.4. To prepare for a more general situation, we replace "score" by x, and "frequency," or "tally count," by t.

You can see that Figure 4.4 displays a function. For each x there is a corresponding entry t. To emphasize this, we shall often write $t(x)$ at the head of our frequency column. Since the table shows how the frequencies are distributed, we often speak of it as a **frequency distribution**.

Figure 4.4

x	t, or $t(x)$	$\frac{t}{n}$
10	1	$\frac{1}{50}$
9	3	$\frac{3}{50}$
8	8	$\frac{8}{50}$
7	10	$\frac{10}{50}$
6	6	$\frac{6}{50}$
5	11	$\frac{11}{50}$
4	2	$\frac{2}{50}$
3	6	$\frac{6}{50}$
2	2	$\frac{2}{50}$
1	0	0
0	1	$\frac{1}{50}$
	$n = 50$	$\frac{50}{50} = 1$

Remark

Many texts use f as the symbol for frequency instead of t. However, we wish to reserve f for probability functions.

We may present the information contained in the frequency distribution by means of graphs. There are several ways to do this. One possible form of graph is the *dot-frequency graph*, one dot representing each of the 50 scores, as in Figure 4.5.

Figure 4.5 *Dot-frequency graph*

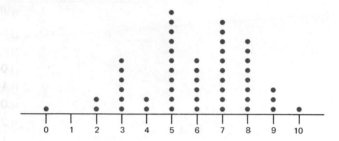

Here we see a piling of dots above each score to indicate its frequency of occurrence. This representation of the data is effective when the frequencies are reasonably low so that their counts are readily obtained. When the frequencies are so high that the number of dots is more confusing than enlightening, the dots may be replaced by segments having lengths proportional to the frequencies. The resulting *line-segment graph* for the data in the above dot-frequency diagram is shown in Figure 4.6.

Figure 4.6 *Line-segment graph*

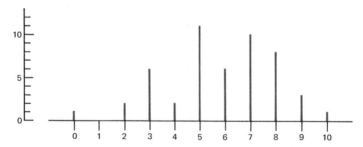

To make the segments stand out prominently, they can be thickened into bars. Thus the data are represented in a *bar graph* as in Figure 4.7.

Figure 4.7 *Bar graph*

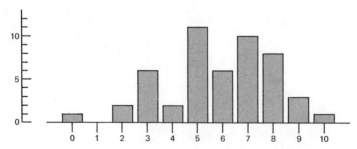

There are times when it is more useful to run the bars next to each other. For our data, the score of 6, for example, would be represented by a bar whose base extends from 5.5 to 6.5. The resulting graph is called a **histogram**. In Section 4.2 we shall see situations where histograms occur more naturally.

Figure 4.8 *Histogram*

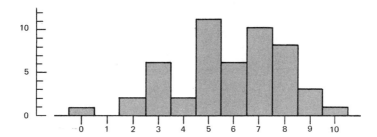

Example 4.1.1 If two dice are thrown, a score may be obtained by adding the numbers on the uppermost faces. The record of 30 such throws on one experiment showed the following results:

Line number	(Score) x_i	(Frequency) $t(x_i)$
1	12	1
2	11	2
3	10	2
4	9	4
5	8	4
6	7	6
7	6	3
8	5	4
9	4	2
10	3	0
11	2	2
		$n = 30$

The left-hand column in this table is not really necessary. We introduce it in order to explain the subscripts on the letters x_i. The symbol x_8 stands for the score entry in line 8. Hence
$$x_8 = 5$$
and correspondingly
$$t(x_8) = t(5) = 4.$$
Obviously, we could have numbered the lines in the reverse order. In the future we shall simply use x_i and $t(x_i)$ without listing the "line numbers."

A histogram for this distribution is given on the next page.

This is the histogram for the distribution of Example 4.1.1:

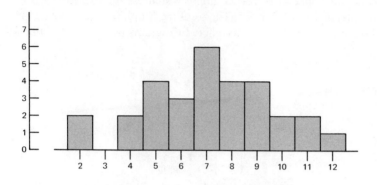

EXERCISES 4.1

1. In 1914, the English physicist, Henry G. Moseley (1887–1915) observed that changes occurred in wavelengths of X-rays as they were passed through crystals of various chemical elements. Some chemicals and their measures of changes are listed below. Give a rank-order listing (in ascending order) of these data.

aluminum (Al)	10.41	palladium (Pd)	39.06
calcium (Ca)	16.41	potassium (K)	15.55
chlorine (Cl)	13.82	ruthenium (Ru)	37.32
chromium (Cr)	19.87	silicon (Si)	11.29
cobalt (Co)	22.48	silver (Ag)	39.83
columbium (Cb)	34.72	titanium (Ti)	18.14
copper (Cu)	24.20	vanadium (V)	19.00
iron (Fe)	21.62	yttrium (Y)	32.95
manganese (Mn)	20.75	zinc (Zn)	25.07
molybdenum (Mo)	35.59	zirconium (Zr)	33.83
nickel (Ni)	23.34		

2. Below are given the areas of the 50 states to the nearest thousand square miles. Give a rank-order listing (in descending order) of these data:

AL	52	HI	6	MA	8	NM	122	SD	77
AK	586	ID	84	MI	58	NY	50	TN	42
AZ	114	IL	56	MN	84	NC	53	TX	267
AR	53	IN	36	MS	48	ND	71	UT	85
CA	159	IA	56	MO	70	OH	41	VT	10
CO	104	KS	82	MT	147	OK	70	VA	41
CT	5	KY	40	NE	77	OR	97	WA	68
DE	2	LA	49	NV	111	PA	45	WV	24
FL	59	ME	33	NH	9	RI	1	WI	56
GA	59	MD	11	NJ	8	SC	31	WY	98

3. Ten coins were tossed 25 times and the number of heads obtained on each toss was recorded (below). Make a tally count and a dot-frequency graph for these data.

   3, 6, 4, 7, 3,   3, 4, 2, 5, 4,   3, 5, 2, 5, 5,
   2, 5, 4, 6, 1,   4, 7, 4, 6, 3.

4. The number of days the 20 employees in a small business were absent during a year are listed below. Make a tally count and a dot-frequency graph for these data.

   1, 0, 0, 0, 2,   0, 3, 4, 0, 2,
   10, 0, 2, 5, 3,   0, 0, 0, 2, 4.

5. From a production line, batteries were selected at random and tested to determine the life of each (the number of hours each battery lasts). The numbers of hours were:

   186, 189, 182, 191, 183,   188, 176, 188, 189, 186,
   188, 191, 187, 181, 189,   186, 188, 181, 185, 183,
   189, 182, 186, 177, 188,   187, 188, 188, 189, 184,
   186, 180, 186, 189, 185,   187, 185, 183, 183, 188,
   190, 185, 188, 189, 188,   185, 186, 176.

   a. Make a tally count (in descending order) of the data.
   b. Construct a table showing the frequencies and the relative frequencies of the life of the batteries.
   c. Construct a histogram for the data.
   d. Between which two numbers does the base of the bar for 191 hours extend in the histogram?

In each of Problems 6–10:
a. *Construct a table showing frequencies and relative frequencies for the data.*
b. *Construct a histogram for the data.*

6. 7, 9, 10, 5, 7,   9, 8, 7, 11, 8,
   7, 9, 9, 7, 6,   10, 6, 9, 3, 4.

7. 17, 17, 14, 21, 17,   17, 21, 12, 15, 19,
   21, 14, 16, 19, 18,   20, 15, 17, 17, 21.

8. 80, 72, 70, 73, 74,   74, 75, 72, 73, 79,
   72, 73, 71, 75, 77,   72, 74, 71, 75, 78.

9. 90, 85, 90, 95, 100,   85, 80, 85, 95, 90,
   100, 85, 90, 85, 95,   80, 85, 90, 85, 90.

10. 30, 32, 33, 31, 35,   33, 33, 32, 35, 31,
    36, 32, 35, 34, 31,   32, 33, 33, 36, 30.

4.2 Grouped Data

The simple frequency tables of Section 4.1 are of somewhat limited usefulness. If we are faced with a great number of individual scores or measurements, even a frequency table is too cumbersome. One possible procedure is to group the scores as shown in the following example.

Example 4.2.1 One hundred forty-four college freshmen were given a "true-false" spelling test with 100 possible points:

$$0, 1, 2, \ldots, 100.$$

The results show that 56 different scores were made, ranging from 24 to 96. To make a frequency table with 56 lines would not be very helpful. Instead, the scores can be grouped in intervals (a width of 5 was selected), as shown in the following table:

Score interval	Midpoint of interval (x_i)	Frequency of interval ($t(x_i)$)
92.5–97.5	95	2
87.5–92.5	90	4
82.5–87.5	85	10
77.5–82.5	80	13
72.5 88.5	75	21
67.5–72.5	70	26
62.5–67.5	65	18
57.5–62.5	60	15
52.5–57.5	55	12
47.5–52.5	50	8
42.5–47.5	45	3
37.5–42.5	40	3
32.5–37.5	35	4
27.5–32.5	30	4
22.5–27.5	25	1
		$n = 144$

Here the interval endpoints as well as the midpoints are given. (For example, the interval endpoints of the lowest interval are 22.5 and 27.5, and the midpoint of this interval is 25. The interval endpoints of the highest interval are 92.5 and 97.5, and the midpoint of this interval is 95.) For many purposes it is not necessary to record the endpoints. By knowing the midpoints, x_i, we see that each interval is of width 5, centered at x_i.

Another situation is illustrated in the next example.

Example 4.2.2 The lengths of the drives of 100 professional golfers on a particular hole were measured to the nearest yard. Grouping the data into intervals of width 3 leads to the grouped frequency distribution shown here. (The interval with midpoint 266, for example, extends from 264.5 to 267.5.)

x_i	$t(x_i)$
278	3
275	2
272	1
269	6
266	15
263	17
260	20
257	10
254	9
251	5
248	6
245	3
242	2
239	1
$n = 100$	

There is an important difference between the data of Example 4.2.1 and those of Example 4.2.2. In the true-false test situation, the recorded scores were separate and distinct, that is, *discrete*. If two individuals both record a score of 70, they both have exactly that score. In the golf example, the data as recorded have already been "grouped." Two distances, both recorded as 260 yards might have been 260.385 and 259.576 $\left(\text{or even } 260 + \frac{\pi}{10} \text{ and } 260 - \frac{\sqrt{2}}{7}\right)$. The recorded measurements arose from a *continuous* interval of real numbers. We sometimes distinguish these cases by referring to *discrete data* (Example 4.2.1) and to *continuous data* (Example 4.2.2).

Once it has been decided to group data for reporting purposes, the individual scores or measurements have lost their identity. One way to think about the 26 spelling test scores between 67.5 and 72.5 is to assume that they are, somehow, spread evenly over the corresponding interval. (Thus, discrete grouped data are treated as though the distribution were continuous.) It is then natural to use a histogram for graphical representation. The distribution of Example 4.2.1 is shown in the histogram of Figure 4.9.

Figure 4.9 *Histogram*

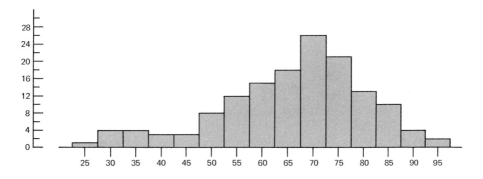

Each bar of a histogram is also called a *cell* of the histogram. If we locate the midpoint of the top of each cell and connect all consecutive midpoints by straight line segments, the resulting figure is a **frequency polygon.** In Figure 4.10, the line segments have been extended half an interval beyond the actual data at each end in order to "close" the polygon.

Figure 4.10   *Frequency polygon*

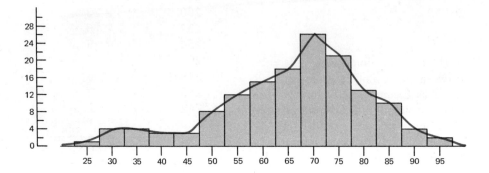

Sometimes it is convenient to sketch a smooth curve which approximates a histogram (or frequency polygon). Such a curve, a **frequency curve,** may be "fitted" by eye, by mechanical instruments, or by advanced mathematical methods. For our present discussion, we simply indicate a possible approximating curve in Figure 4.11.

Figure 4.11   *Frequency curve*

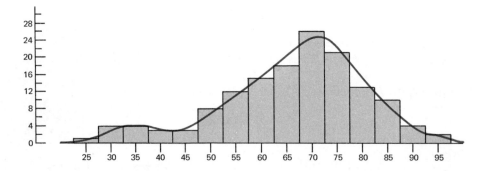

Remark

It is an observable fact that many different frequency distributions which occur in practice can be approximated by curves which have generally the same shape. These curves are bell-shaped, rising towards the center and tapering off at both ends. (See Sections 5.4 and 8.1.)

EXERCISES 4.2

- **1.** For the distribution of Example 4.2.2:
  - **a.** Construct a histogram.
  - **b.** Draw the corresponding frequency polygon.
  - **c.** Sketch an appropriate frequency curve.

- **2.** Twenty-five batteries were picked at random from an assembly line for testing. The life of each in number of hours was as follows:

    144, 162, 197, 173, 183, 129, 209, 190, 117, 160, 179, 177, 154, 132, 151, 159, 175, 154, 148, 166, 184, 157, 162, 150, 136.

  - **a.** Group the data into intervals of width 10 with the midpoint of the interval at 120, 130, ... (Assume that the right-hand endpoint belongs to the interval and the left-hand endpoint belongs to the next lower interval. That is, 135 belongs to the interval 125 to 135, whereas 125 belongs to the interval 115 to 125.)
  - **b.** Construct a table of relative frequencies for the grouped data in part a.
  - **c.** Draw a histogram for these data.

- **3.** For 60 selected cities, the number of days (in a given year) with precipitation (rain, snow, sleet, and so on) of 0.01 inches or more was:

124	121	124	100	110	34	105	112	92	101
82	58	103	35	121	167	117	56	112	121
79	62	126	110	44	87	112	115	103	125
128	96	86	117	133	153	148	112	163	135
114	116	116	82	162	115	99	152	116	91
98	117	147	119	97	148	119	108	47	123

  - **a.** Display a table of tallies for the data grouped into intervals of 8 (days) beginning with midpoints at 36.5, 44.5, and so on.
  - **b.** For the same data, give a table of tallies in intervals of 10 with the top interval being 161–170 (midpoint, 165.5).
  - **c.** Draw a frequency polygon for the grouped data in part b.
  - **d.** Superimpose a frequency curve on the frequency polygon obtained in part c.

In Problems 4–6:
a. *Group the data according to the specified intervals and make a frequency table. (Include right-hand endpoint in each interval.)*
b. *Draw a histogram for the data and superimpose a frequency polygon on it.*

- **4.** Width of interval: 6   Midpoint of first interval: 3
    52, 49, 18, 47, 16,   44, 29, 40, 47, 36,   21, 30, 42, 24, 10,
    17, 34, 9, 23, 45,   27, 36, 20, 25, 6,   14, 39, 2, 46, 8.

5. Width of interval: 5   Midpoint of first interval: 7.5
12, 27, 22, 7, 19,   13, 18, 21, 16, 11,   17, 15, 18, 23, 8,
26, 14, 17, 23, 6,   11, 22, 29, 10, 15,   24, 26, 21, 17, 13.

6. Width of interval: 10   Midpoint of first interval: 125
121, 135, 164, 176, 129,   142, 125, 157, 144, 133,   129, 152, 146, 131, 122,
132, 121, 143, 138, 128,   131, 172, 125, 162, 148,   136, 122, 143, 166, 157.

4.3 The Mean

We are all familiar with summarizing information about a collection of numbers in terms of "averages"; for example:

"The class average is 72."
"Jones's batting average is 0.281."
"The average price per share of stock traded is up 4¢."
"The average number of children per couple is 2.43."

These are all everyday illustrations. Although it is possible to define the word "average" in several ways, here we shall concentrate on the one that is most important for our purposes. This is the ordinary (*arithmetical*) mean.

For a list of n numbers, the **mean** is computed by adding the numbers and dividing by n; thus:

$$\bar{x} = \frac{x_1 + x_2 + x_3 + \cdots + x_n}{n}$$

Using the summation symbol (see Appendix), we write this as:

$$\bar{x} = \frac{\sum_{i=1}^{n} x_i}{n}$$

Example 4.3.1 For the list of quiz scores used in Section 4.1, we have:

$$\bar{x} = \frac{0 + 2 + 2 + 3 + 3 + \cdots + 9 + 9 + 9 + 10}{50}$$

$$= \frac{292}{50} = 5.84$$

If the data have been arranged in a frequency table, the number of addends may be reduced. Instead of adding each individual item, we multiply each item value by its frequency, and then add. In symbols, we have

$$\bar{x} = \frac{\sum_{i=1}^{m} x_i \cdot t(x_i)}{n}$$

where m is the number of distinct values.

Example 4.3.2 Using the data of Figure 4.4 to find the mean, we have:

x_i	$t(x_i)$	$x_i \cdot t(x_i)$
10	1	10
9	3	27
8	8	64
7	10	70
6	6	36
5	11	55
4	2	8
3	6	18
2	2	4
1	0	0
0	1	0
	$n = 50$	292

$$\bar{x} = \frac{\sum_{i=1}^{m} x_i \cdot t(x_i)}{n}$$

$$= \frac{\sum_{i=1}^{11} x_i \cdot t(x_i)}{50}$$

$$= \frac{292}{50} = 5.84$$

Since
$$\frac{\sum_{i=1}^{m} x_i \cdot t(x_i)}{n} = \sum_{i=1}^{m} x_i \cdot \frac{t(x_i)}{n},$$

we may calculate $\bar{x}$ by multiplying each value by its *relative frequency* and then adding these products. This method has the disadvantage of requiring many divisions. If, however, the relative frequencies are already known, the computations are not difficult. The major advantage to us is the comparison this method affords with the material that we shall be studying in Section 5.5.

Example 4.3.3 Using relative frequencies, the computation of $\bar{x}$ for the data of Example 4.3.2 is shown below. (See Figure 4.3.)

x_i	$\dfrac{t(x_i)}{n}$	$x_i \cdot \dfrac{t(x_i)}{n}$
10	0.02	0.20
9	0.06	0.54
8	0.16	1.28
7	0.20	1.40
6	0.12	0.72
5	0.22	1.10
4	0.04	0.16
3	0.12	0.36
2	0.04	0.08
1	0	0
0	0.02	0
	1.00	5.84

$$\bar{x} = \sum x_i \cdot \frac{t(x_i)}{n}$$

$$= 5.84$$

(Note: We have omitted the index, i, from the $\sum$ notation on the understanding that it is clear what terms are involved in the summation.)

FORMULAS FOR THE MEAN

$$\bar{x} = \frac{\sum_{i=1}^{n} x_i}{n} = \frac{\sum_{i=1}^{m} x_i \cdot t(x_i)}{n} = \sum_{i=1}^{m} x_i \cdot \frac{t(x_i)}{n}$$

where n is the number of items, m is the number of distinct values, and $t(x_i)$ is the frequency of x_i.

Remark

It can be noticed that $\sum_{i=1}^{m} t(x_i) = n$ and $\sum_{i=1}^{m} \frac{t(x_i)}{n} = 1$.

This observation is again useful in comparing $\bar{x}$ with the development of the idea of "expected value" in Section 5.5.

When data are ungrouped, the computation of $\bar{x}$ is straight-forward: simply find the sum and divide by the number of items. In working with grouped data, individual items are not available. How are we to get the sum of the items within each *interval?*. Clearly, the procedures for obtaining $\bar{x}$ must be modified.

Instead of adding separate items within the entire interval, we multiply the midpoint of each interval by the frequency of the interval. Justification of this procedure follows from our assumption that the scores in an interval are spread evenly over that interval. The average of the items in any one interval then equals the middle value. With this assumption, the computation proceeds as for ungrouped data.

Example 4.3.4 For the 144 test scores of Example 4.2.1, the calculation of the mean score is shown below.

x_i	$t(x_i)$	$x_i \cdot t(x_i)$
95	2	190
90	4	360
85	10	850
80	13	1040
75	21	1575
70	26	1820
65	18	1170
60	15	900
55	12	660
50	8	400
45	3	135
40	3	120
35	4	140
30	4	120
25	1	25
$n = 144$		9505

$\bar{x} = \frac{9505}{144} \doteq 66.01$

The rather unwieldy computation that may arise during the calculation of $\bar{x}$ can often be simplified by the use of a property of $\bar{x}$. Suppose that we let

$$y = ax + b;$$

then each value x_i of x gives a particular value y_i of y: $y_i = ax_i + b$. Since $t(y_i) = t(x_i)$, we have:

$$\bar{y} = \frac{\sum y_i \cdot t(y_i)}{n} = \frac{1}{n}\sum (ax_i + b) \cdot t(x_i)$$

$$= \frac{1}{n}\sum ax_i \cdot t(x_i) + \frac{1}{n}\sum b \cdot t(x_i) \quad \text{(See Appendix)}$$

$$= a \cdot \frac{\sum x_i \cdot t(x_i)}{n} + \frac{b}{n}\sum t(x_i)$$

$$= a\bar{x} + b, \text{ since } \sum t(x_i) = n \quad \text{(See Remark opposite)}$$

We shall now show how we can apply this idea to the calculations of Example 4.3.4. By some appropriate choices of the constants a and b, we obtain a y_i related to each x_i in such a way that computation of the mean is simplified. Since the interval width is 5, we take $a = \frac{1}{5}$. Since 65 is a value near the middle of the distribution, we take $b = -\frac{65}{5} = -13$. Thus, we replace each x_i by

$$y_i = \tfrac{1}{5}x_i - 13.$$

We may then compute $\bar{y}$ and find

$$\bar{x} = 5\bar{y} + 65.$$

x_i	y_i	$t(x_i) = t(y_i)$	$y_i \cdot t(y_i)$
95	6	2	12
90	5	4	20
85	4	10	40
80	3	13	39
75	2	21	42
70	1	26	26
65	0	18	0
60	−1	15	−15
55	−2	12	−24
50	−3	8	−24
45	−4	3	−12
40	−5	3	−15
35	−6	4	−24
30	−7	4	−28
25	−8	1	−8
		$n = 144$	29

$$\bar{y} = \tfrac{29}{144} \doteq 0.2014$$

$$\bar{x} \doteq 5(0.2014) + 65 \doteq 66.01$$

EXERCISES 4.3

1. Find the mean of the distribution in Problem 3, Exercises 4.1.
2. Find the mean of the distribution in Problem 4, Exercises 4.1.
3. Find the mean of the distribution in Problem 5, Exercises 4.1.
4. Find the mean of the distribution in Problem 7, Exercises 4.1.
5. From the distribution in Problem 5, Exercises 4.1, make up a "parallel" distribution by subtracting 185 from each entry. Find the mean of the related distribution and compare this result with the mean obtained in Problem 3 above.
6. From the distribution in Problem 7, Exercises 4.1, make up a "parallel" distribution by subtracting 16 from each entry. Find the mean of the related distribution and compare this result with the mean obtained in Problem 4 above.
7. Find the mean of the distribution in Example 4.2.2 from a table of the relative frequencies.
8. Find the mean of the distribution in Problem 2, Exercises 4.2, using a table of relative frequencies.
9. From the table obtained in Problem 3b, Exercises 4.2, find the mean of the distribution.
10. Find the mean of the distribution in Problem 5, Exercises 4.2.
11. Occasionally we receive pieces of information about the mean of a distribution. Suppose that we know that the mean score of 50 male freshmen on an arithmetic test is 74 and the mean score of 30 female freshmen is 68. If we use x_i for the male score and y_i for the female score:
    a. What is the value of $\sum x_i$?
    b. What is the value of $\sum y_i$?
    c. What is the mean score of the 80 freshmen?
12. We can use the idea of Problem 11 to generalize. If we know that the mean of n scores is $\bar{x}$, then we can find the total of these n scores. Likewise, we can find the total of m scores having a mean of $\bar{y}$.
    a. What is the mean of the $n + m$ scores?
    b. Under what conditions is it true that the combined mean is equal to $\dfrac{\bar{x} + \bar{y}}{2}$?
13. The mean of n scores is $\bar{x}$; the mean of m scores is $\bar{y}$; the mean of k scores is $\bar{z}$. What is the mean of the $n + m + k$ scores?

4.4 The Variance and the Standard Deviation

While the mean is familiar to us as a measure for presenting a "summary" of information about a distribution, the measure to be introduced in this section is less familiar. We seek to develop a number associated with a distribution of data which expresses the degree to which the numbers are spread, or scattered, or dispersed.

There are several ways we could proceed. An obvious measure of the spread of a distribution is the **range,** the difference between the largest and the smallest numbers. The fact that the range of a distribution is not a satisfactory measure is rather easily illustrated.

Example 4.4.1 Consider the three lists of numbers:

I: 1, 1, 1, 9, 9, 9   mean, 5; range, 8.
II: 1, 3, 5, 5, 7, 9   mean, 5; range, 8.
III: 1, 5, 5, 5, 5, 9   mean, 5; range, 8.

Clearly the range does not tell us that the numbers of II are more closely "bunched" near the mean than those of I. List III shows the numbers even more closely bunched near 5.

Since Example 4.4.1 shows that the range is an unsatisfactory measure, let us examine how each score deviates from the mean. Using the data of Example 4.4.1, we compute the quantity $x_i - \bar{x}$ for each number of the three lists:

I		II		III	
x_i	$x_i - \bar{x}$	x_i	$x_i - \bar{x}$	x_i	$x_i - \bar{x}$
9	4	9	4	9	4
9	4	7	2	5	0
9	4	5	0	5	0
1	−4	5	0	5	0
1	−4	3	−2	5	0
1	−4	1	−4	1	−4
	0		0		0

It is tempting to average these differences as a possible measure. In each case, however, $\sum(x_i - \bar{x}) = 0$, and so

$$\frac{\sum(x_i - \bar{x})}{n} = 0.$$

This result is general. In any distribution:

$$\frac{\sum(x_i - \bar{x})}{n} = \frac{\sum x_i}{n} - \frac{\sum \bar{x}}{n} = \frac{\sum x_i}{n} - \frac{n\bar{x}}{n}$$

$$= \bar{x} - \bar{x} = 0$$

(Note: In the first line of this computation, we made use of the fact that $\bar{x}$ is a constant and $\sum_{i=1}^{n} \bar{x} = n\bar{x}$; see Appendix.)

To avoid the unsatisfactory situation of having the result always 0, we could consider the average (mean) of the absolute values $|x_i - \bar{x}|$. In our three lists this technique yields:

I: $\dfrac{\sum |x_i - \bar{x}|}{n} = \dfrac{24}{6} = 4$ II: $\dfrac{\sum |x_i - \bar{x}|}{n} = \dfrac{12}{6} = 2$

III: $\dfrac{\sum |x_i - \bar{x}|}{n} = \dfrac{8}{6} \doteq 1.33$

These calculations reveal the fact that the numbers of list I are more scattered than those of list II and that the numbers of list III are the least scattered.

For some purposes, $\dfrac{\sum |x_i - \bar{x}|}{n}$ is quite satisfactory as a measure of "scatter." It turns out, however, that another measure is even more useful.

Remark

"It turns out ..." is a useful phrase in writing mathematics texts. It may be used in many ways. In the present connection it means that it is difficult to motivate what is to follow. Fortunately, Chapters 6–11 will reveal the reason we choose to adopt the measure which we shall now introduce.

Suppose that instead of finding the absolute value of the difference of each number from the mean, we square the quantities $x_i - \bar{x}$ and then find the average of these squared deviations. Again we illustrate with our three lists.

I

x_i	$x_i - \bar{x}$	$(x_i - \bar{x})^2$
9	4	16
9	4	16
9	4	16
1	−4	16
1	−4	16
1	−4	16
		96

II

x_i	$x_i - \bar{x}$	$(x_i - \bar{x})^2$
9	4	16
7	2	4
5	0	0
5	0	0
3	−2	4
1	−4	16
		40

III

x_i	$x_i - \bar{x}$	$(x_i - \bar{x})^2$
9	4	16
5	0	0
5	0	0
5	0	0
5	0	0
1	−4	16
		32

Finding the means of the last columns of each list gives us:

$$\text{I:} \quad \frac{\sum(x_i - \bar{x})^2}{n} = \frac{96}{6} = 16$$

$$\text{II:} \quad \frac{\sum(x_i - \bar{x})^2}{n} = \frac{40}{6} = 6.67$$

$$\text{III:} \quad \frac{\sum(x_i - \bar{x})^2}{n} = \frac{32}{6} = 5.33$$

The quantity
$$\frac{\sum(x_i - \bar{x})^2}{n}$$

is called the **variance** of the distribution and is denoted by s_x^2.

For a long list of numbers, the computation of s_x^2 becomes cumbersome, particularly if $\bar{x}$ has a value such as that for our 50 test scores, 5.84 (see Example 4.3.1). Certain alternate methods may be used, but the one that interests us for the purposes of this text is developed as follows. If we expand $(x_i - \bar{x})^2$, we have:

$$s_x^2 = \frac{\sum(x_i - \bar{x})^2}{n} = \frac{\sum(x_i^2 - 2x_i\bar{x} + \bar{x}^2)}{n}$$

$$= \frac{\sum x_i^2}{n} - 2\bar{x}\frac{\sum x_i}{n} + \frac{n\bar{x}^2}{n} \quad \text{(since } \bar{x} \text{ is a } constant\text{)}$$

$$= \frac{\sum x_i^2}{n} - 2(\bar{x})^2 + \bar{x}^2 = \frac{\sum x_i^2}{n} - \bar{x}^2$$

Thus, the variance equals the *mean of the squares minus the square of the mean*.

Example 4.4.2 To compute s_x^2 for our three lists of numbers, we may use the formula

$$s_x^2 = \frac{\sum x_i^2}{n} - \bar{x}^2.$$

I		II		III	
x_i	x_i^2	x_i	x_i^2	x_i	x_i^2
9	81	9	81	9	81
9	81	7	49	5	25
9	81	5	25	5	25
1	1	5	25	5	25
1	1	3	9	5	25
1	1	1	1	1	1
	246		190		182

$$s_x^2 = \frac{246}{6} - 5^2 \qquad s_x^2 = \frac{190}{6} - 5^2 \qquad s_x^2 = \frac{182}{6} - 5^2$$

$$= 41 - 25 \qquad \doteq 31.67 - 25 \qquad \doteq 30.33 - 25$$

$$= 16 \qquad \qquad = 6.67 \qquad \qquad = 5.33$$

There is one difficulty with s_x^2 as a measure of spread. Suppose that our lists I, II, III had been obtained by measuring certain lengths in inches. Then s_x^2 would have square inches as its units. It is often more useful to take the nonnegative square root of s_x^2,

$$s_x = \sqrt{s_x^2},$$

which is called the **standard deviation** of the distribution of numbers.

Example 4.4.3 For our three lists, we have the following standard deviations:

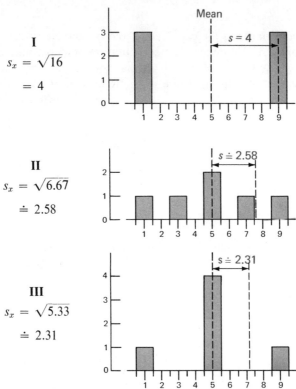

I
$s_x = \sqrt{16}$
$= 4$

II
$s_x = \sqrt{6.67}$
$\doteq 2.58$

III
$s_x = \sqrt{5.33}$
$\doteq 2.31$

(Table R at the back of the book may be used to find the square roots.)

Just as we did in computing the mean, we can take advantage of a frequency distribution to shorten some of the calculations. Thus, for frequency distributions, it is clear that we have

$$s_x^2 = \frac{\sum_{i=1}^{m}(x_i - \bar{x})^2 \cdot t(x_i)}{n} \quad \text{or} \quad s_x^2 = \frac{\sum_{i=1}^{m} x_i^2 \cdot t(x_i)}{n} - \bar{x}^2,$$

where m is the number of distinct values. For example, for list I, we have

$$s_x^2 = \frac{16(3) + 16(3)}{6} = \frac{96}{6} = 16$$

or

$$s_x^2 = \frac{81(3) + 1(3)}{6} - 5^2 = \frac{246}{6} - 25 = 16.$$

FORMULAS FOR THE VARIANCE

$$s_x^2 = \frac{\sum_{i=1}^{n}(x_i - \bar{x})^2}{n} = \frac{\sum_{i=1}^{n} x_i^2}{n} - \bar{x}^2$$

$$= \frac{\sum_{i=1}^{m}(x_i - \bar{x})^2 \cdot t(x_i)}{n} = \frac{\sum_{i=1}^{m} x_i^2 \cdot t(x_i)}{n} - \bar{x}^2,$$

where n is the number of items, m is the number of distinct values, and $t(x_i)$ is the frequency of x_i.

FORMULA FOR THE STANDARD DEVIATION

$$s_x = \sqrt{s_x^2}$$

Finding the variance and the standard deviation of a frequency distribution or of a grouped frequency distribution is a straight-forward matter, although the arithmetic may be annoying. In the calculations for a grouped frequency distribution, x_i may be taken to be the midpoint of the corresponding interval. This procedure agrees exactly with our practice in obtaining $\bar{x}$ for a grouped frequency distribution.

Occasionally we encounter lists of numbers that cannot be grouped conveniently. We illustrate below the computation of the standard deviation for one such list as an example of how the actual work may be organized.

Example 4.4.4 Find the mean and standard deviation for the normal monthly average temperature of San Francisco, given its distribution:

January	50.7	April	55.7	July	58.8	October	61.4
February	53.0	May	57.4	August	59.4	November	57.4
March	54.7	June	59.1	September	62.0	December	52.5

x_i	x_i^2
50.7	2570.49
53.0	2809.00
54.7	2992.09
55.7	3102.49
57.4	3294.76
59.1	3492.81
58.8	3457.44
59.4	3528.36
62.0	3844.00
61.4	3769.96
57.4	3294.76
52.5	2756.25
682.1	38912.41

$$\bar{x} = \frac{682.1}{12} \doteq 56.8$$

$$s_x^2 = \frac{38912.41}{12} - (56.8)^2$$

$$\doteq 16.46$$

$$s_x \doteq \sqrt{16.46} \doteq 4.06$$

EXERCISES 4.4

1. Find s_x^2 for the following data, using two methods.

    1, 3, 4, 5, 6, 7, 8, 8.

2. Find s_x^2 for the following data, using any method.

    10, 11, 11, 12, 12, 12, 12, 13, 13, 13.

3. Find s_x^2 and $\sqrt{s_x^2}$ for the frequency distribution of 50 test scores in Section 4.1 (see Figure 4.4), recalling that $\bar{x} = 5.84$ (Example 4.3.2).

4. Find s_x^2 and $\sqrt{s_x^2}$ for the frequency distribution of Problem 3, Exercises 4.1. (See answer to Problem 1, Exercises 4.3.)

5. The normal monthly average temperature for Kansas City is given below. Compare the mean and standard deviation of this distribution with those of Example 4.4.4.

    31.7, 35.8, 43.3, 55.7, 65.6, 75.9, 81.5, 79.8, 71.3, 60.2, 44.6, 35.8.

6. Find the standard deviation of the distribution of Problem 5, Exercises 4.1. (See answer to Problem 3, Exercises 4.3.)

7. From the table obtained in Problem 3b, Exercises 4.2, find the standard deviation of the distribution. (See answer to Problem 9, Exercises 4.3.)

8. For the following distributions, verify that the 6 numbers x_i have $\bar{x} = 5$ and $s_x^2 \doteq 6.67$ and that the 9 numbers y_i have $\bar{y} = 5$ and $s_y^2 \doteq 5.33$.

x_i	$t(x_i)$		y_i	$t(y_i)$
9	1		9	1
7	1		7	2
5	2		5	3
3	1		3	2
1	1		1	1
	$m = 6$			$n = 9$

Let z have the frequency distribution obtained by summing the corresponding frequencies of x and y. How is s_z^2 related to s_x^2 and s_y^2? (See Problem 12, Exercises 4.3.)

★ 9. Let y have the frequency distribution in Problem 8 and w have the frequency distribution given below. Further, let z have the frequency distribution obtained by summing the corresponding frequencies of w and y.

w_i	$t(w_i)$
9	1
7	2
5	3
3	4
1	5
	$k = 15$

a. Find s_z^2.

b. Show that $\bar{z} = \dfrac{k\bar{w} + n\bar{y}}{k + n}$.

c. Compare s_z^2 with $\dfrac{ks_w^2 + ns_y^2}{k + n}$. Explain why these are not equal.

★ COMPUTER INVESTIGATIONS

If you have access to an electronic computer, try to carry out the following investigations.

1. Write a computer program for finding the mean of a set of numbers.
2. Write a computer program for finding the standard deviation of a set of numbers.
3. Make any necessary changes in your programs for Problems 1 and 2 in order to find the mean and the standard deviation of a frequency distribution.
4. Test the programs of Problem 3 above with Problem 1, Exercises 4.3, and Problem 4, Exercises 4.4.
5. Test the programs of Problem 3 above with Problem 3, Exercises 4.3, and Problem 6, Exercises 4.4.

For which parts of the following problems is an electronic computer most useful?

Find an alphabetical listing of states of the United States with the District of Columbia and assign identifying numbers 1–51 to them.

6. Find a listing of population by states, round off the values to a convenient number of digits, and find:
   a. the mean population
   b. the state having the population nearest to the mean
   c. the variance and standard deviation of the distribution
7. From the listing used in Exercise 6, find:
   a. the state having the largest population
   b. the state having the smallest population
   c. the state having the median population
8. Find a listing of area (in square miles) by states, round off the values to a convenient number of digits, and find:
   a. the mean area
   b. the state having the area nearest to the mean
   c. the variance and standard deviation of the distribution
9. From the listing used in Exercise 8, find:
   a. the state having the largest area
   b. the state having the smallest area
   c. the state having the median area

10. Using the lists in Exercises 6 and 8, find:
    a. the average number of persons per square mile in each state
    b. the state having the greatest number of persons per square mile
    c. the state having the least number of persons per square mile
    d. the state having the median number of persons per square mile

4.5 What Do You Think?

The mean and the standard deviation of a distribution of numbers provide summarized information about the distribution. In later chapters we shall give a more detailed discussion of the significance and usefulness of the standard deviation. At this point, it is sufficient that we understand that the standard deviation is a measure of variability from the mean.

Suppose that we receive reports of the results of an arithmetic test given to five different groups of 100 sixth-grade pupils.

Group	$\bar{x}$	s_x
I	50	5
II	50	15
III	60	10
IV	70	20
V	70	2

Without knowing any further information, we can only **guess** the answers to the questions in Exercises 4.5a. What do you think?

EXERCISES 4.5a

● 1. Which group (I, II, III, IV, V, above) would you guess:
    a. shows the highest individual score?
    b. shows the lowest individual score?

● 2. a. Which group is the most homogeneous (or uniform) in its performance?
    b. Which distribution has the largest range?

● 3. Suppose that a score of 40 or more is "passing" and a score of 90 or more will receive special commendation. Which group
    a. has the fewest "passing" scores?
    b. the most "commendations"?

● 4. Which of the five groups of 100 students most closely resembles the entire 500 on test performance?

We have used the word "average" to denote the arithmetic mean of a distribution of numbers. It is common practice to use the word "average" in several different senses. One such use is made in reference to the score, or measure, which occurs with the greatest frequency. This is more properly called the **mode.**

Example 4.5.1 If we accept as true the statement "the average American male wears a size 9 shoe," we can be reasonably sure that this refers to the most frequently occurring shoe size.

The mode is an important notion when one has data which cannot be ordered in the ordinary sense, for example, hair color.

For grouped data we cannot pick out a particular number as a mode. We may, however, specify the interval having the greatest frequency. This is called the *modal interval*. In frequency polygons, histograms, or frequency curves, the mode (or modal interval) is associated with the highest point of the graph.

A third measure is sometimes used to summarize information about a distribution. This is the **median,** which is the score, or measure, that is the halfway point in a rank-order listing of numbers. Certain annoying questions about the median arise if we consider frequency tables or grouped data. These questions, fortunately, are not of major concern to us in this text.

Some of the problems in Exercises 4.5b deal with small collections of numbers. For small collections, of course, there is no need to summarize the data; all the information can be absorbed at a glance.

EXERCISES 4.5b

- 1. Find the mean, median, and mode of:

    1, 4, 4, 4, 4, 5, 10, 10, 10, 12, 13

- 2. If there is an odd number of distinct scores, the median is clearly the middle number (in order of size). What do you think should be the median of:

    10, 10, 12, 14, 20, 20?

- 3. Some data reveal two (or more) modes or modal intervals. Can you think of some practical case where this might occur, even when a large amount of data is obtained?

- 4. The median is often used in connection with reporting "average income." Can you think of a reason why? (Hint: Why is the use of the mean misleading?)

Sometimes measures called **percentiles** are used. For example, the 25th percentile is the number below which 25% of the scores fall.

In Problem 2, if the median is taken as 13, then 13 may be called the 50th percentile of that set of data.

- **5. a.** Between which two percentiles does the "middle 50%" of the distribution lie?
  - **b.** The 25th percentile is often called the first *quartile;* the 75th percentile is called the third quartile. What do you think is meant by the "semi-interquartile range"? (The semi-interquartile range is often used as another measure of spread.)
  - **c.** In using the term percentile, is it ever possible to designate one number in a distribution as the 100th percentile?
- **6. a.** Find the mode of the distribution of Example 4.3.2.
  - **b.** Find the modal interval for the data of Example 4.3.4.
- ★ **7.** In the grouped frequency distribution of Example 4.3.4 we do not have the individual scores. What is a reasonable guess as to the median?

Chapter Summary

1. It is helpful to organize a collection of numbers by constructing a *frequency table*. Such an organization is called a *frequency distribution* and may be represented by a graph called a *histogram*.
2. **a.** It is often convenient to *group* data before constructing a frequency table or a histogram.
   b. From a histogram, a *frequency polygon* may be constructed or a *frequency curve* may be sketched.
3. The *mean* of a distribution may be computed thus:

$$\bar{x} = \frac{\sum_{i=1}^{n} x_i}{n} = \sum_{i=1}^{m} \frac{x_i \cdot t(x_i)}{n} = \sum_{i=1}^{m} x_i \cdot \frac{t(x_i)}{n},$$

where n = number of items, m = number of distinct values, and $t(x_i)$ is the frequency of x_i.

4. **a.** The *variance* of a distribution may be computed thus:

$$s_x^2 = \frac{\sum_{i=1}^{n} (x_i - \bar{x})^2}{n} = \frac{\sum_{i=1}^{n} x_i^2}{n} - \bar{x}^2$$

$$= \frac{\sum_{i=1}^{m} (x_i - \bar{x})^2 \cdot t(x_i)}{n} = \sum_{i=1}^{m} \frac{x_i^2 \cdot t(x_i)}{n} - \bar{x}^2,$$

where n = number of items, m = number of distinct values, and $t(x_i)$ is the frequency of x_i.

b. The *standard deviation* of a distribution is found as

$$s_x = \sqrt{s_x^2}.$$

Chapter Review

1. The table below shows the number of days in a particular year that Washington, D.C., had precipitation of 0.01 inch (trace) or more.

Jan.	11	Apr.	10	July	9	Oct.	7
Feb.	9	May	11	Aug.	9	Nov.	8
Mar.	12	June	9	Sept.	8	Dec.	9

   a. Construct a table listing the frequency corresponding to the number of days having at least a trace of precipitation.
   b. Construct a histogram for this frequency distribution. Draw in the corresponding frequency polygon.
   c. Find the mean and variance for this distribution.

2. A quiz was given to two physics classes with the following results.

Class X		Class Y	
Score	Frequency	Score	Frequency
10	1	10	4
9	1	9	6
8	2	8	8
7	5	7	6
6	9	6	4
5	2	5	2
	$n = 20$		$n = 30$

   a. Find the mean for each of the classes.
   b. Suppose the results for the two classes were combined as a "Class Z," consisting of all the students who took this quiz. Construct a frequency table for Class Z and find its mean.
   c. Let the mean, $\bar{x}$, of Class X be 6.7, the mean, $\bar{y}$, of Class Y be 7.8. Find the mean of Class Z without using the corresponding table in part b.

3. a. Group the 25 scores listed below into intervals of 10 with the top interval from 85 to 95, and construct the frequency table for these data.

93, 89, 84, 80, 77, 72, 72, 70, 68, 66, 66, 64,
63, 63, 60, 60, 57, 52, 51, 50, 49, 46, 44, 38, 36.

b. Find the mean, variance, and standard deviation for the frequency distribution in part a.

Chapter 5

Random Variables and Probability Functions

Variables and functions play a basic role in mathematics, and we shall assume that the reader has some familiarity with (real) variables and (real-valued) functions. The idea of a graph of such a function will be particularly useful in this chapter.

The phrase "probability function" has been applied (Sections 2.3 and 2.5) to the assignment of probabilities $f(s_i)$ to the elements of a set of outcomes, $S = \{s_1, s_2, \ldots, s_n\}$. We recall that

$$0 \leq f(s_i) \leq 1 \quad \text{and} \quad \sum f(s_i) = 1.$$

A probability function has as its set of "inputs" (the domain) the set of outcomes, S, and as its set of "outputs" (range) a subset of the interval [0, 1] of the real numbers. An unusual feature of the probability functions in the examples we have been using is that frequently the elements in S have not been numbers. Rather the elements of S have often been described verbally, as "the coin falls heads," "a red marble is drawn," "the item is defective," and so on.

In our earlier work, we have been more accustomed to dealing with functions which map real numbers into real numbers. There are several advantages in dealing with functions from the real numbers to the real numbers, since we may graph such functions and we may analyze them by using mathematical methods such as algebraic techniques. Thus, in this chapter we shall develop ways of expressing outcomes of experiments as real numbers.

5.1 Random Variables

We have already seen examples of experiments in which it is rather natural to associate real numbers with the outcomes. (See Examples 3.2.3 and 3.3.2, Problem 5 of Exercises 3.2, and Problem 2 of Exercises 3.3 as illustrations.)

As we have already seen (Example 3.3.1), if a coin is tossed twice, one possible set of (equally likely) outcomes is

$$\{hh, ht, th, tt\}.$$

By "ht" we mean, of course, "heads on the first toss, tails on the second." Another possible set of outcomes (not equally likely) is obtained by considering the *number* of heads that may occur. This set could be written as

$$\{2, 1, 0\},$$

where "1," for example, represents the statement "we obtain exactly one head on the two tosses."

We may now proceed a bit more formally. Let the *variable* X represent the possible number of heads that occur when a coin is tossed twice. The actual value that X takes on depends on the result of an experiment. The outcome of an experiment is, by the nature of experimentation, uncertain. In order to emphasize this, we refer to X as a **random variable.** For two tosses of a coin, the possible values of X are 2, 1, 0. More generally:

A variable whose values are determined by the outcomes of an experiment is called a *random variable*.

Remark 1

If we think carefully about the idea of a random variable X we see that the variable is itself a function. It maps each outcome of the experiment into a real number. For the coin example, X is the function described by:

$$\{hh\} \to 2$$
$$\{ht\} \to 1$$
$$\{th\} \to 1$$
$$\{tt\} \to 0$$

For the purposes of this text, we need not stress this aspect of a random variable. We shall require only an understanding of the discussion and definition of the preceding paragraph.

Remark 2

It is common practice to designate a random variable by a capital letter. This is perhaps unfortunate, since we have been using capital letters for sets (events). In what follows, we shall try to reserve the latter part of the alphabet (X, Y, Z) for random variables. If the context is not clear, we shall specifically refer to "the random variable X."

Example 5.1.1 A die is to be tossed. Let X be the random variable denoting the number on the upper face of the die. The possible values of X are 1, 2, 3, 4, 5, 6.

Example 5.1.2 An urn contains 4 red and 2 blue marbles. Three marbles are to be drawn one at a time with replacement. Let X represent the number of red marbles occurring. Then X is a random variable whose possible values are 0, 1, 2, 3. Let Y be the random variable denoting the number of blue marbles drawn. Again the possible values of Y are 0, 1, 2, 3.

Example 5.1.3 Using the urn of Example 5.1.2, we now assume that the three draws are to be made *without* replacement. Using X and Y as before, we see that the possible values of X are 1, 2, 3, and the possible values of Y are 0, 1, 2.

Example 5.1.4 Two dice, one red and one green, are to be tossed. One random variable we may use for this experiment is the sum of the two faces. If we let Z denote this sum, the possible values of Z are 2, 3, 4, 5, 6, 7, 8, 9, 10, 11, 12.

Example 5.1.4 may be interpreted a bit differently. Suppose that we toss the two dice, but think of this as two experiments. If we let X represent the number appearing on the red die and Y the number appearing on the green die, then X and Y are two random variables. Using Z as before, we have

$$Z = X + Y.$$

In general, the sum of two (or more) random variables is a random variable.

Similarly, if three coins are tossed, we could let X_1, X_2, and X_3 represent the number of heads appearing on the first, second, and third coin, respectively. Then

$$Z = X_1 + X_2 + X_3$$

is a new random variable representing the total number of heads obtained. If n coins are tossed, we could consider

$$Z = \sum_{i=1}^{n} X_i,$$

where X_i is the number of heads appearing on the ith coin. Each X_i has possible values 0 and 1, and Z has the possible values 0, 1, 2, ..., n.

Our examples thus far have dealt with random variables whose possible values are nonnegative integers. In general, a random variable may assume any real number as a value.

Example 5.1.5 Consider a game played as follows. A card is drawn from a standard deck. The player wins $3 if an ace is drawn. He receives $2 for a king, $0.50 for either a queen or a jack. If a 10, 9, or 8 is selected, he pays $0.50. For the 7, 6, 5, 4, 3, or 2, he pays $0.75. Let X be a random variable representing his possible gain (positive values of X) or loss (negative values of X) in dollars. Then X has the values 3, 2, $\frac{1}{2}$, $-\frac{1}{2}$, $-\frac{3}{4}$.

EXERCISES 5.1

1. Describe an appropriate random variable and its possible values for the experiment of tossing four coins.
2. On each of three slips of paper is a numeral "1," "2," or "3." The slips are to be drawn at random one at a time without replacement. A "match" is obtained if the slip numbered k is selected on the kth draw. Let X be a random variable representing the number of matches obtained. What are the possible values of X?

In Problems 3–6, let X be a random variable whose value corresponds to the number showing on (the uppermost face of) a tossed die. Several possible games are described. For each game, a player is awarded points in agreement with the rules given.

In each case, let Z be a random variable corresponding to the number of points awarded for a single toss. List the possible values of Z and indicate the relation between Z and X.

3. Double the number on the face of the die.
4. Subtract $3\frac{1}{2}$ from the number shown on the face of the die.
5. Square the number shown on the face of the die.
6. Subtract $3\frac{1}{2}$ from the number shown on the face of the die and then square the difference.
7. In another game two dice are tossed. Points are awarded by multiplying the numbers shown on the individual dice. Let Z indicate the points awarded on a single toss of the two dice. List the possible values of Z. (Hint: There are 18 such values.)
8. We may consider a set of 36 equally likely outcomes of the experiment of tossing two dice. How many of these correspond to each value of Z of Problem 7?
9. During the day, the stock market quotation for J. B. J. Bigdome Company fluctuated between $67\frac{1}{2}$ and $68\frac{3}{4}$.
   a. If X is a random variable denoting the *change* from one quotation to another that day, what are the possible values of X? (Note: Stocks can be quoted by eighths of a dollar. The quotation, "$67\frac{1}{2}$" means $67.50. Hint: X has 21 possible values.)
   b. Would the possible values of X be changed if it is known at what price the stock was quoted when the market opened? Explain.
   c. If the quotation for the stock opened at 67 and remained unchanged throughout the day, what are the possible values for X?
10. The cost of living index is 100 times the percentage that the cost in a given year is of that in the base year, taken to be 100.0, reported in tenths. Suppose that the index ranges from 100.0 in 1958 to 131.3 in 1970. If X is the random variable representing this index, then possible values of X are 100.0, 100.1, 100.2, ..., 131.3 for this period.
    a. How many possible values for X are there during this period?

b. Suppose, instead, that the cost of living is taken to be 100.0 in 1970. Then the index for 1958 is less than 100 (but not by 31.3). If Y is the random variable representing the new index (rounded to tenths), what are the possible values of Y?

c. How many possible values for Y are there for this period?

★ 11. A game is played on a board as shown on the diagram below.

Instructions

2* advance 4 squares
4* go back to "Start"
5* go back 2 squares
7* go back 4 squares
8* advance 1 square

A player begins at square 1 and proceeds along a spiral path according to the numbering. He spins the spinner at the right, takes the number of steps indicated, and follows the directions on the square he lands on. He finishes when he gets to or passes square 9. If X denotes the number of steps a player takes on any spin (positive for forward, negative for backward), what are the possible values of X?

★ 12. In Example 5.1.5, suppose that two cards are drawn. If X is a random variable representing the player's possible gain or loss, what are the possible values for X?

5.2 Probability Functions and Their Graphs

In this section we shall be concerned with the probability that the random variable X will take on a given value from a finite set of values. We need an agreement on notation. In place of the phrase "the probability that the random variable X takes on the value x," we shall often write $P(X = x)$. That is, we shall use a lowercase letter to represent a *particular value* of the random variable X (compare Example 3.2.3). [Note that in earlier courses, it is customary to use a lowercase x as a variable. Here it is the capital X that stands for the variable. This is simply a convention in current use.]

We shall also have occasion to write $f(x)$ in place of $P(X = x)$. Thus,

$$f(x)$$

is the *probability that the random variable X takes on the value x.*

Example 5.2.1 A single marble is to be drawn from an urn containing 2 red marbles and 1 blue marble. Let X be the random variable representing the number of red marbles drawn. The possible values of X are 0 and 1. It is not difficult to see that the appropriate probabilities are

$$P(X = 1) = \tfrac{2}{3} \quad \text{and} \quad P(X = 0) = \tfrac{1}{3}.$$

The foregoing example is an illustration of an important special type of random variable:

If X takes on exactly two possible values (usually 1 and 0), then X is called a *Bernoulli random variable*.[*]

For the experiment of tossing a coin twice, we start with a set of four equally likely outcomes
$$S = \{hh, ht, th, tt\}.$$
Let the random variable X denote the number of heads that appear on one trial of the experiment. Each outcome of S corresponds to one value of X:

Outcome of S	hh	ht	th	tt
Corresponding value of X	2	1	1	0

Since we know the probabilities of the outcomes of S, we may immediately assign probabilities to the values of X:

$$f(2) = P(X = 2) = P(\{hh\}) = \tfrac{1}{4}$$
$$f(1) = P(X = 1) = P(\{ht, th\}) = \tfrac{1}{2}$$
$$f(0) = P(X = 0) = P(\{tt\}) = \tfrac{1}{4}$$

In tabular form, we have:

Number of heads, x	2	1	0
Probability, f(x)	$\tfrac{1}{4}$	$\tfrac{1}{2}$	$\tfrac{1}{4}$

In this way, we have constructed a **probability function for the random variable X.** As we mentioned earlier (page 109), one advantage of introducing the concept of a random variable is that we can graph its probability function. In the two-toss experiment, we have the graph shown in Figure 5.1.

Figure 5.1

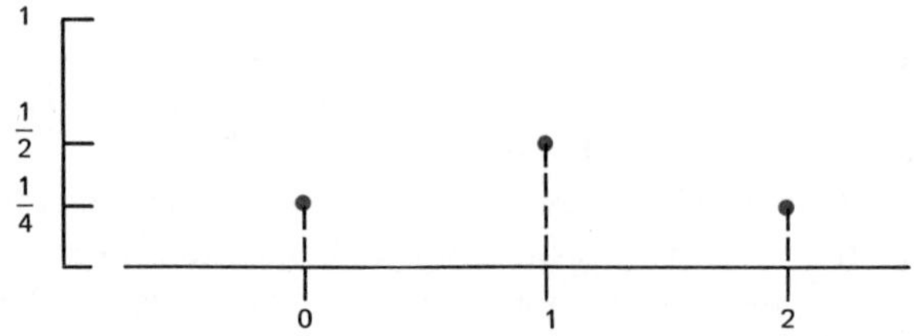

A particularly simple graph occurs when the random variable associated with an experiment is a Bernoulli random variable. Let X be a random variable which takes on only two possible values 0 and 1. If we assign probability
$$f(1) = P(X = 1) = p,$$
then
$$f(0) = P(X = 0) = 1 - p.$$
Using q in place of $1 - p$, we have the graph of the probability function shown in Figure 5.2.

[*]Named after Jacques Bernoulli (1654–1705), who was mentioned in Section 1.1.

Figure 5.2

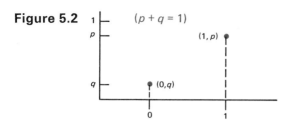

Example 5.2.2 A marble is to be drawn from an urn containing 3 red and 2 blue marbles. Let X represent the number of red marbles drawn. Then X is a random variable with probability function given by the table:

x, the value of X	0	1
$f(x)$, probability	$\frac{2}{5}$	$\frac{3}{5}$

A graph of this function is shown at the right.

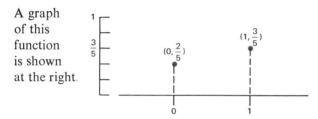

Example 5.2.3 The probability function for X, the random variable representing the number shown after one throw of a fair die, is given by:

x, the value of X	1	2	3	4	5	6
$f(x)$, probability of x	$\frac{1}{6}$	$\frac{1}{6}$	$\frac{1}{6}$	$\frac{1}{6}$	$\frac{1}{6}$	$\frac{1}{6}$

A graph of this function is shown at the right.

Example 5.2.4 Let Z represent the sum of the faces when two fair dice are thrown. (See Problem 4, Exercises 2.4.)

x, the value of X	2	3	4	5	6	7	8	9	10	11	12
$f(x)$	$\frac{1}{36}$	$\frac{2}{36}$	$\frac{3}{36}$	$\frac{4}{36}$	$\frac{5}{36}$	$\frac{6}{36}$	$\frac{5}{36}$	$\frac{4}{36}$	$\frac{3}{36}$	$\frac{2}{36}$	$\frac{1}{36}$

A graph of this function is shown at the right.

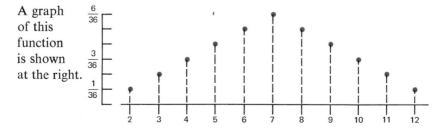

Example 5.2.5 A number is to be selected at random from the set

$$\{1, 2, 3, 4, 5\}.$$

Let Y be a random variable which takes on the value 1 if the number selected is a prime. $Y = 0$ if the number selected is not a prime.

Since there are five equally likely outcomes and three lead to $Y = 1$, we have:

y, the value of Y	0	1
$f(y)$, probability	$\frac{2}{5}$	$\frac{3}{5}$

Notice that in each case, the graph of the probability function has consisted of a set of isolated points, $(x, f(x))$. We might think of the situation as follows. Imagine that a dump truck travelling along the X-axis is carrying a full load of probability. When the truck is at a value of X, say x_1, for which $P(X = x_1) > 0$, the truck "dumps" some probability. The "height" $f(x)$ of the point $(x, f(x))$ is proportional to the probability that has been dumped at x. In Example 5.2.5, $\frac{2}{5}$ of the load is dumped at "0" and $\frac{3}{5}$ of the load at "1." The truck "distributes" the probability at isolated points along the X-axis.

If we compare Examples 5.2.2 and 5.2.5, we see that although X and Y are different random variables (they arise from different experiments), the distribution of probability is the same. Any probability question about the random variable X can be answered equally well by studying random variable Y. For this reason, we shall speak of a **probability distribution** without necessarily referring to the particular random variable involved.

For any Bernoulli random variable, the probability function is completely determined when the value of p is known. In Examples 5.2.2 and 5.2.5, $p = \frac{3}{5}$. We call this distribution the *Bernoulli distribution with* $p = \frac{3}{5}$.

Example 5.2.6 A fair coin is tossed once. If X represents the number of heads appearing, then X has a Bernoulli distribution with $p = \frac{1}{2}$.

If we are given a random variable and a probability function, we are in a position to answer questions about the probability distribution. For example, using X as in Example 5.2.4, we find the following probabilities:

$P(X = 6) = \frac{5}{36}$

$P(X < 6) = P(X = 2) + P(X = 3) + P(X = 4) + P(X = 5)$
$= \frac{1}{36} + \frac{2}{36} + \frac{3}{36} + \frac{4}{36} = \frac{10}{36} = \frac{5}{18}$

$$P(X \le 6) = P(X < 6) + P(X = 6) = \tfrac{15}{36} = \tfrac{5}{12}$$
$$P(X > 10) = P(X = 11) + P(X = 12) = \tfrac{3}{36} = \tfrac{1}{12}$$
$$P(6 \le X \le 8) = P(X = 6) + P(X = 7) + P(X = 8)$$
$$= \tfrac{5}{36} + \tfrac{6}{36} + \tfrac{5}{36} = \tfrac{16}{36} = \tfrac{4}{9}$$

We may interpret $P(X < 6)$, for example, as the amount of probability that is distributed to the left of the point $X = 6$. In particular, "$P(6 \le X \le 8)$" indicates the probability over the *interval* [6, 8].

For an arbitrary random variable, it is reasonable to ask:

"What is $P(4.5 \le X \le 6.98)$?"

For our dice example,

$$P(4.5 \le X \le 6.98) = P(X = 5) + P(X = 6) = \tfrac{9}{36} = \tfrac{1}{4}.$$

Suppose that we were asked to find $P(X > 14)$. We look at our distribution and find that no probability has been distributed to the right of 14. Our answer then is "$P(X > 14) = 0$."

In order to make our geometric interpretation complete, it is convenient to think of X as being allowed to take on *any real value*. A complete description of the Bernoulli distribution of Example 5.2.2 for instance, is:

$$f(x) = \begin{cases} \tfrac{2}{5} & \text{if } x = 0 \\ \tfrac{3}{5} & \text{if } x = 1 \\ 0 & \text{if } x = \text{any other value} \end{cases}$$

With this interpretation, our graph now appears as in Figure 5.3. The graph

Figure 5.3

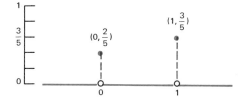

consists of all the points of the X-axis except 0 and 1, plus the two isolated points $(0, \tfrac{2}{5})$ and $(1, \tfrac{3}{5})$. As we move from left to right, the graph of $f(x)$ is broken only by two "jumps" at 0 and at 1. These jumps are called *discontinuities* of the function.

If the graph of a probability distribution is of this form—points along the X-axis plus a finite number of isolated points $(x, f(x))$—we call this distribution a *discrete* probability distribution. The corresponding probability function is called a *discrete* probability function, and the random variable is called a *discrete* random variable.

In Section 5.3 we shall examine some random variables that are not discrete, but continuous. We shall discover, eventually, that certain discrete distributions may be approximated by continuous ones (or conversely). This is much the same idea that we discussed in Chapters 1 and 4, using a sequence of discrete photographs to approximate a continuous action.

Remark 1

If we wish to display the probability function of a discrete random variable by means of a table, we usually list only those values x of X for which $P(X = x) > 0$. It is convenient to use the form:

x
$f(x)$

Remark 2

If we have occasion to consider more than one random variable or probability function at a time, we shall use distinctive capital letters (X, Y, Z, and so on) and distinctive functional notations ($f(x), g(y), h(z)$, and so on). We agree that any functional symbol, unless otherwise indicated, shall designate a *probability* function. We note, in particular, that

$$\sum f(x) = 1, \quad \sum g(y) = 1, \quad \sum h(z) = 1, \quad \text{and so on.}$$

EXERCISES 5.2

- 1. A fair coin is to be tossed. Let X be the random variable representing the number of heads that appear. Show the probability distribution of X by means of a table and a graph if the coin is to be tossed:
    - **a.** once **b.** twice **c.** three times

Describe and graph the appropriate probability distribution for the random variable of:

- 2. Problem 3, Exercises 5.1
- 3. Problem 4, Exercises 5.1
- 4. Problem 5, Exercises 5.1
- 5. Problem 6, Exercises 5.1
- 6. Problem 7, Exercises 5.1. (Use the results of Problem 8, Exercises 5.1.)

- 7. Let X be the random variable of Example 3.2.3. Describe and graph its probability distribution.
- 8. Let X be the random variable of Example 3.3.2. Describe and graph its probability distribution.

Graph the Bernoulli distribution for these different values of p, when $P(X = 1) = p$ and $P(X = 0) = 1 - p = q$.

- 9. $p = 0$
- 10. $p = 0.1$
- 11. $p = 0.25$
- 12. $p = 0.50$
- 13. $p = 0.75$
- 14. $p = 1$

★ 15. The words "When," "in," "the," "course," "of," "human," "events" are written, one on each of seven slips of paper. A slip is drawn from a hat. The number of points a player gets is: 1 point for every consonant and -1 (he loses one) for every vowel in the word. Let X be the random variable representing the number of points a player gets in the drawing. Show the probability distribution of X by a table and a graph.

★ 16. Letters are written on little cards, one letter on a card, with exactly enough letters to make up the message, "When, in the course of human events." The cards are shuffled and a card is picked at random. A player gets 4 points if he picks a card with the letter f, h, v, or w; 3 points for c or m; 1 point otherwise. Let Y be the random variable representing the number of points a player gets. Graph the probability distribution of Y.

★ 17. Refer to the game described in Problem 11, Exercises 5.1. Let the random variable X denote the number of the square a player lands on. Then, by the tree diagram below, it can be seen that the possible values of X after the first spin are 1, 3, and 6. (Note: The number at each vertex denotes the square the player lands on; the number along each path denotes the number he spins.)

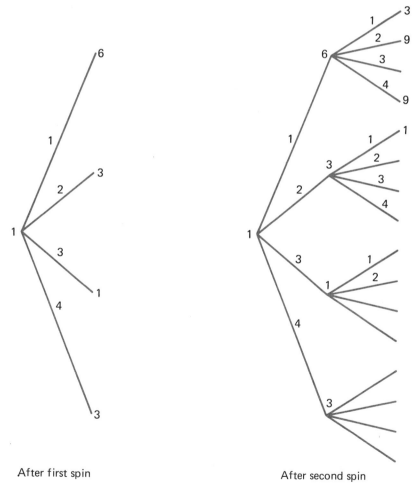

After first spin   After second spin

  a. Complete the tree diagram for the second spin.
  b. What are the possible values of X after the second spin?
  c. What is the probability of $X = 9$ after the first spin?

★ 18. For the game described in Problem 17, let Y be the random variable denoting the number of the square after the first spin and Z after the second spin.
  a. Find $P(Z = 9)$.
  b. What is $P(Z = 6 \mid Y = 6)$?
  c. Display the probability distribution of Z by a table.

5.3 Continuous Probability Distributions

In the last section, we discussed several examples of discrete random variables and discrete probability distributions. The graphs of the distributions consisted of all points on the X-axis together with jumps, or discontinuities, at a finite number of isolated points.

For a thorough understanding of continuous distributions, we need more advanced mathematical tools than we have assumed are available to us. We shall, then, try to present informal arguments, avoiding definitions and proofs.

The analogy with a dump truck which was given in Section 5.2 can be modified. Suppose that we have a random variable X whose probability function is $f(x) = 0$ except for some interval $[a, b]$. We could ask our driver to spread the probability over the interval rather than to deposit it in separate piles. There are, of course, many ways in which he could do this. Any discussion of the general problem involves difficulties. Let us examine two special cases—two different types of instructions we could give the driver—for spreading the probability, say $[0, 10]$.

Case 1. Instructions: Spread (or distribute) the probability evenly, uniformly over the interval. (This is comparable to spreading asphalt in repaving a highway.)

When the job is done, the result would look like Figure 5.4.

Figure 5.4

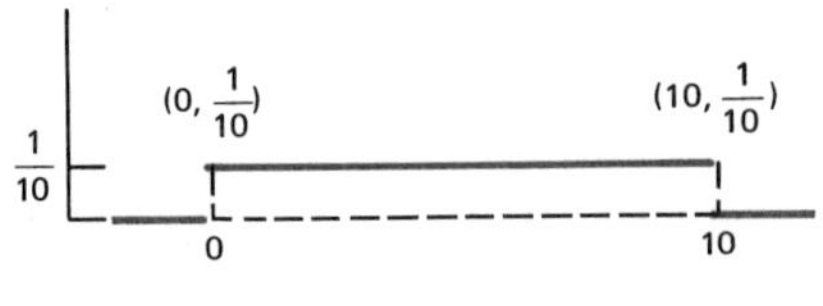

Since one load is spread over the interval of length $b - a$, the height at each point of $[a, b] = [0, 10]$ must be

$$\frac{1}{b-a} = \frac{1}{10}.$$

The total *area* above the X-axis is 1. The resulting distribution is called the *uniform* probability distribution over the interval $[a, b]$. Our precise instructions are: distribute the probability in accordance with the rule:

$$f(x) = \begin{cases} 0 & \text{if } x < 0 \\ \frac{1}{10} & \text{if } 0 \leq x \leq 10 \\ 0 & \text{if } x > 0 \end{cases}$$

Case 2. Instructions: Start spreading probability at $X = 0$, gradually increasing the amount distributed until, at $X = 10$, the truck is empty.

The resulting distribution (called the *triangular distribution*) is shown graphically in Figure 5.5a.

Figure 5.5a

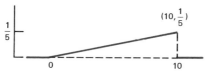

If $f(10) = h$, then, since the area of the triangle must be 1, we have

$$\tfrac{1}{2} \cdot 10 \cdot h = 1,$$

or

$$h = \tfrac{1}{5}.$$

So $f(10)$ must equal $\tfrac{1}{5}$. The equation of the line joining $(0, 0)$ and $(10, \tfrac{1}{5})$ is given by

$$f(x) = \tfrac{1}{50}x.$$

Therefore our detailed instructions are: distribute the probability in accordance with the rule:

$$f(x) = \begin{cases} 0 & \text{if } x < 0 \\ \tfrac{1}{50}x & \text{if } 0 \leq x \leq 10 \\ 0 & \text{if } x > 0 \end{cases}$$

This does not seem too complicated. We must, however, be careful in our interpretation of $f(x)$.

For a discrete probability distribution, $f(x)$ represents the height of the graph corresponding to $X = x$. In addition, $f(x) = P(X = x)$.

For a continuous probability distribution, $f(x)$ still represents the height of the graph corresponding to $X = x$. In Case 2, above, $f(5) = \tfrac{1}{10}$, the height of the graph for $X = 5$. Suppose that we assume, as before, that $f(x) = P(X = x)$. Then $P(X = 5) = f(5) = \tfrac{1}{10}$. Since $f(x)$ is increasing in the triangular distribution, $f(x) > \tfrac{1}{10}$ for $x > 5$.

Select any ten distinct points, x, in the interval $5 < x \leq 10$. For each of these, $P(X = x) > \tfrac{1}{10}$; the sum of all ten probabilities is clearly greater than 1. This tells us that we can no longer depend on the height of the distribution to represent the probability.

Referring back to our analogy of the dump truck, we can agree that the amount of probability the truck driver spreads is proportional to the area under the curve in an interval. Since the entire area is 1, the *area over a given interval*, in fact, measures the probability. Hence, for a continuous distribution over the interval $[a, b]$, we agree that:

1. The total *area* bounded by $X = a$, $X = b$, the X-axis, and the curve whose *height* is given by $f(x)$ is 1.
2. $P(X = x) = 0$ for every real number x.

This second point is disturbing. If every individual value of X has zero probability, what sorts of things have positive probability? In other words, for a continuous random variable, what are the *events*? Again, a complete answer to this question is quite complicated. A partial answer, sufficient for our purposes, is illustrated by the following question. How much probability has been spread over the *interval* [0, 3]?

For Case 1, the uniform distribution, $\frac{3}{10}$ of the probability has been distributed over this subinterval. $\frac{3}{10}$ of the total area lies between $X = 0$ and $X = 3$.

For Case 2, the triangular distribution, the calculation in Figure 5.5b shows that $\frac{9}{100}$ of the probability has been distributed over [0, 3]. $\frac{9}{100}$ of the total area lies between $X = 0$ and $X = 3$.

Figure 5.5b

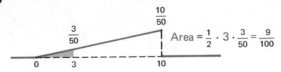

Our illustrations lead us to assert that for a continuous probability distribution over the interval [a, b]:

3. *Intervals* (and subintervals) have probabilities. Intervals are events.
4. $P(c \leq X \leq d)$ equals the *area* bounded by $X = c$, $X = d$, the X-axis, and the curve whose height is given by $f(x)$. $f(x)$ gives the *density* of the probability at point $X = x$.

Example 5.3.1 Consider the experiment "pick a real number at random from the interval [3, 15]." Let X be a random variable having a uniform probability distributed over the interval. Then $f(x) = \frac{1}{12}$ for $3 \leq x \leq 15$ and $f(x) = 0$ for all other values. The area above any subinterval is found by multiplying $\frac{1}{12}$ times the length of the subinterval.

a. Illustrations below show calculations for different subintervals, some not including either endpoint and some including one of the endpoints. For an interval including both endpoints, of course the probability is 1.

$$P(5 \leq X \leq 8) = \tfrac{1}{12}(8 - 5) = \tfrac{3}{12}$$
$$P(X \leq 8) = P(3 \leq X \leq 8) = \tfrac{1}{12}(8 - 3) = \tfrac{5}{12}$$
$$P(X \leq 5) = P(3 \leq X \leq 5) = \tfrac{1}{12}(5 - 3) = \tfrac{2}{12}$$

(We observe that $P(5 \leq X \leq 8) = P(X \leq 8) - P(X \leq 5)$.)

$$P(X > 8) = 1 - P(X \leq 8) = 1 - \tfrac{5}{12} = \tfrac{7}{12}$$
$$= P(8 \leq X \leq 15)$$

Since $P(X = 8) = 0$, $P(X > 8) = P(X \geq 8)$, and so on. (See Problem 3 of Exercises 5.3.)

b. Suppose now that we are given the probability and want to determine the subinterval given one of its boundaries. If $P(X \leq x) = 0.50$, then since

$$\tfrac{1}{12}(x - 3) = \tfrac{1}{2},$$

we have $x = 9$.

Example 5.3.2 Given X having the probability distribution of Case 2 (see Figure 5.5), to find $P(4 \leq X \leq 8)$, we compute the area of the trapezoid bounded by (4, 0), (4, $\frac{4}{50}$), (8, $\frac{8}{50}$), (8, 0). The area of the trapezoid is $\frac{1}{2}(8-4)[\frac{4}{50}+\frac{8}{50}]$, or $\frac{24}{50}$, or $\frac{12}{25}$.

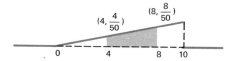

Our major interest in continuous probability distributions is in one special case which will be discussed in Section 5.4. Before proceeding to that discussion, it is appropriate to add four remarks.

Remark 1

We have not given a formal definition of a continuous distribution. The general situation can be quite complicated. It is sufficient for our purposes that a distribution have no jumps in the interval under consideration.

Remark 2

Another technically difficult matter is the definition of events in reference to a continuous random variable. We are content to restate our earlier assertion: intervals and subintervals are events and therefore are assigned probabilities. The union of two or more intervals is also an event.

Remark 3

For the uniform and triangular distributions discussed in this section, it is not difficult to find the probability associated with a given interval. We need only to find the area of rectangles, triangles, and trapezoids. For more complicated distributions, we would need many techniques from that branch of mathematics called analysis, which includes the study of the calculus.

Remark 4

We have seen examples of discrete random variables having a finite number of values with positive probability. It is possible for a random variable to assume an infinite number of values with positive probability and still be discrete. This type of random variable falls somewhat between the finite and continuous cases. Fortunately, we have no real need to discuss this so-called countably infinite case.

One illustration as to how such a random variable arises might be illuminating. Imagine an urn containing an infinite number of slips of paper. Half of them are marked "1," $\frac{1}{4}$ are marked "2," $\frac{1}{8}$ are marked "3," ..., $\frac{1}{2^n}$ are marked "n." Our experiment is "choose one slip at random." X represents the number drawn. We would agree that

$$P(X=1) = \frac{1}{2}, \quad P(X=2) = \frac{1}{4}, \quad \ldots, \quad P(X=n) = \frac{1}{2^n}.$$

To verify that this is, in fact, a probability function, we add the probabilities

$$\frac{1}{2} + \frac{1}{4} + \frac{1}{8} + \cdots + \frac{1}{2^n} + \cdots$$

The sum of this infinite series (a geometric progression) as defined in algebra is exactly 1.

EXERCISES 5.3

● 1. Let X have a uniform distribution over the interval $[2, 12]$. Find:
   a. $P(X \leq 5)$
   b. $P(X < 5)$
   c. $P(X \geq 5)$
   d. $P(4 \leq X \leq 8)$
   e. $P(2 \leq X \leq 3 \text{ or } 9 \leq X \leq 11)$
   f. $P(4 \leq X \leq 9 \text{ or } 6 \leq X \leq 10)$

● 2. For the distribution of Problem 1 determine x so that
   a. $P(X \leq x) = 0.50$
   b. $P(X \leq x) = 0.25$
   c. $P(3 \leq X \leq x) = 0.50$

● 3. Prove: For any continuous distribution over an interval $[a, b]$,

$$P(X \leq c) = P(X < c) \quad \text{if} \quad a \leq c \leq b.$$

(Hint: Write the event $X \leq c$ as the union of the two mutually exclusive events $\{X < c\}$ and $\{X = c\}$.)

● 4. For the triangular distribution of Case 2, shown in Figure 5.5a, find:
   a. $P(X \leq 4)$
   b. $P(X \geq 8)$ (Hint: You may use the result of Example 5.3.2.)
   c. $P(2 \leq X \leq 6)$
   d. x such that $P(X \leq x) = 0.50$

● 5. Write the probability function for the uniform distribution over the interval. (The experiment may be considered as being "choose a real number at random from the interval.")
   a. $[0, 1]$ b. $[-5, 5]$ c. $[a, b]$

★ 6. Write the probability function for the triangular distribution over the interval.
   a. $[0, 1]$ b. $[-5, 5]$ c. $[a, b]$

★ 7. Let X have a continuous distribution over the interval $[a, b]$. Suppose that $a < c < d < e < f < b$ as shown:

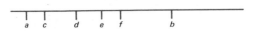

Explain why the following is true:

$P(c \leq X \leq e \text{ or } d \leq X \leq f)$
$= P(c \leq X \leq e) + P(d \leq X \leq f) - P(d \leq X \leq e).$

5.4 The Standard Normal Distribution

There is one particular continuous distribution that is of extreme importance. This distribution has the graph shown in Figure 5.6.

Figure 5.6

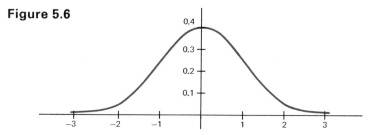

Letting Z be the random variable having this distribution, we point out the following special features:

1. $f(z) > 0$ for all real numbers z. That is, our imaginary truck driver must distribute his probability over the entire Z-axis.
2. As for any continuous probability, the probability that an experiment will lead to a value of Z falling in the interval $[a, b]$ is equal to the area bounded by the curve, the Z-axis, and the two vertical lines $Z = a$ and $Z = b$.
3. The curve is symmetric about the vertical line through 0.
4. The curve is "bell-shaped."
5. The probability function is given by the formula

$$f(z) = \frac{1}{\sqrt{2\pi}} e^{\frac{-z^2}{2}}$$

We include this formula for the sake of completeness. π has its usual meaning, the ratio of the circumference of a circle to the length of a diameter ($\pi \doteq 3.142$). e is another important mathematical constant, the base of the natural logarithms ($e \doteq 2.718$). For our purposes, we may think of the distribution as being defined by a table (Table N in the Appendix).

The distribution pictured in Figure 5.6 and having features 1–5 listed above is called the *standard normal distribution*, and Z is called the *standard normal random variable*.

The significance of the adjective "standard" will be discussed in Chapter 6. With reference to the word "normal," we can think of this distribution as being typical of many experiments and observations, particularly of natural occurrences. Frequently, graphs of data from such sources approximate this curve in general character.

We shall see many uses of the standard normal distribution in Chapters 8 through 11. For the moment we shall concentrate on using Table N which enables us to determine probabilities, given a standard normal random variable Z.

Our first observation is that the smallest value of Z which is listed in Table N is -3.00. The corresponding entry in the table is 0.0013. This is to be interpreted as

$$P(Z \leq -3.00) \doteq 0.0013.$$

Only 0.0013 (0.13%) of the probability has been distributed to the left of -3.00. Thus, it is extremely unlikely that we will observe a value of Z less than -3.00. [Recall that

$$P(Z < -3.00) = P(Z \leq -3.00),$$

since $P(Z = -3.00) = 0$.]

Secondly, the symmetry of the curve implies that the amount of probability to the right of 3.00 equals the amount of probability to the left of -3.00. This may be verified from the table, where we find that

$$P(Z \leq 3.00) \doteq 0.9987;$$

so

$$P(Z \geq 3.00) \doteq 1 - 0.9987 = 0.0013.$$

We may combine these results in the following way:

$$P(Z \leq -3.00 \text{ or } Z \geq 3.00) \doteq 0.0013 + 0.0013 = 0.0026$$

or

$$P(-3.00 \leq Z \leq 3.00) \doteq 1 - 0.0026 = 0.9974.$$

This means that about 99.74% of the probability lies between -3.00 and 3.00.

To express this efficiently, we can use the absolute value symbol and write our last conclusions as:

$$P(|Z| \geq 3.00) \doteq 0.0026$$

$$P(|Z| \leq 3.00) \doteq 0.9974$$

Our final general observation is that

$$P(a \leq Z \leq b) = P(Z \leq b) - P(Z \leq a).$$

This conclusion may be explained graphically. The area between the ordinates at b and at a may be found by subtracting the area to the left of the ordinate at a from the total area to the left of the ordinate at b. (See Figure 5.7.)

Figure 5.7

We may use Table N to determine probabilities connected with other integral values of Z. For example:

$$P(Z \leq -2.00) = P(Z \geq 2.00) \doteq 0.0227$$
$$P(Z \leq -1.00) = P(Z \geq 1.00) \doteq 0.1587$$
$$P(Z \leq 0) = P(Z \geq 0) = 0.5000$$

These results can be combined with those obtained earlier for $Z = -3.00$ and $Z = 3.00$ to find the following:

$$P(Z \leq -3.00) \doteq 0.0013$$
$$P(-3.00 \leq Z \leq -2.00) \doteq 0.0227 - 0.0013 = 0.0214$$
$$P(-2.00 \leq Z \leq -1.00) \doteq 0.1587 - 0.0227 = 0.1360$$
$$P(-1.00 \leq Z \leq 0) \doteq 0.5000 - 0.1587 = 0.3413$$
$$P(0 \leq Z \leq 1.00) \doteq 0.8413 - 0.5000 = 0.3413$$
$$P(1.00 \leq Z \leq 2.00) \doteq 0.9773 - 0.8413 = 0.1360$$
$$P(2.00 \leq Z \leq 3.00) \doteq 0.9987 - 0.9773 = 0.0214$$
$$P(Z \geq 3.00) \doteq 1 - 0.9987 = 0.0013$$
$$\overline{1.0000}$$

These results are shown on the graph of Figure 5.8.

Figure 5.8

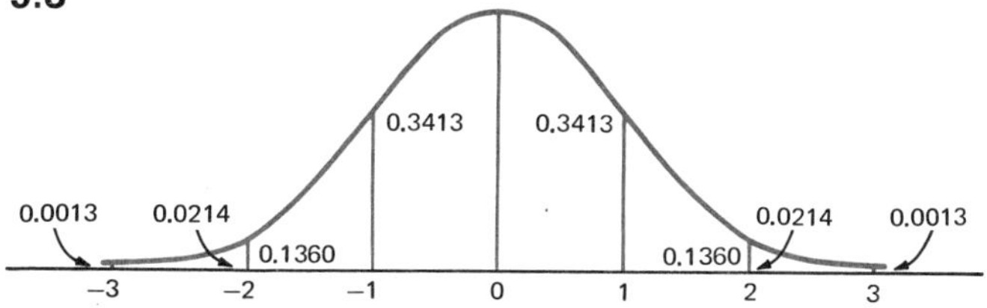

In statistics, one often refers to regions centered at 0 in which 90, 95, and 99 percent of the distribution falls (Chapters 8 through 11). For this reason, the following values of Z deserve particular attention:

$$P(|Z| \leq 1.64) \doteq 0.9495 - 0.0505 = 0.8990 \doteq 0.90$$
$$P(|Z| \leq 1.96) \doteq 0.9750 - 0.0250 = 0.95$$
$$P(|Z| \leq 2.58) \doteq 0.9951 - 0.0049 = 0.9902 \doteq 0.99$$

Another interesting fact is this:

$$P(|Z| \leq 1.00) \doteq 0.68$$

EXERCISES 5.4

The following exercises provide practice in using Table N. Assume that Z is a random variable having the standard distribution.

Find the following probabilities.

- 1. a. $P(Z \leq 0.50)$ b. $P(|Z| \leq 0.50)$
- 2. a. $P(Z \geq 0.50)$ b. $P(|Z| \geq 0.50)$
- 3. a. $P(Z \leq 1.32)$ b. $P(Z \leq 0.92)$
- 4. a. $P(-1.00 \leq Z \leq 1.00)$ b. $P(-2.00 \leq Z \leq 2.00)$
- 5. a. $P(0.92 \leq Z \leq 1.32)$ b. $P(-1.07 \leq Z \leq -0.36)$
- 6. a. $P(-1.25 \leq Z \leq 2.25)$ b. $P(-0.74 \leq Z \leq 1.30)$

Determine z, such that:

- 7. $P(Z \leq z) = 0.5000$
- 8. $P(Z \leq z) = 0.1020$
- 9. $P(Z \geq z) = 0.1020$
- 10. $P(|Z| \leq z) = 0.8664$
- 11. $P(-1.00 \leq Z \leq z) = 0.6408$
- 12. $P(-1.50 \leq Z \leq z) = 0.8664$

★ 13. Explain why
   a. $P(Z \geq z) = 1 - P(Z \leq z)$
   b. $P(|Z| \geq z) = 2[1 - P(Z \leq z)], z \geq 0$
   c. $P(|Z| \leq z) = 1 - 2[1 - P(Z \leq z)] = 2P(Z \leq z) - 1$

5.5 The Expected Value of a Random Variable

We have seen one advantage of introducing a random variable associated with the outcome of an experiment. Since a random variable takes on real numbers x as its values, we have been able to study some probability distributions for both discrete and continuous random variables.

Still another advantage of identifying a value of a random variable with an event is that we are able to apply mathematical techniques of algebra and geometry to functions of a real variable. A simple, but extremely important, illustration is the calculation of what will be called the *expected value* of a random variable. In the development of this idea in this section, it is helpful to keep in mind the development of the concept of the mean given in Chapter 4. We concentrate on a discrete random variable.

We note that the mean is a "weighted average." That is, if our data consist of n scores having m distinct values, the mean may be found (page 93) by multiplying each distinct score by its relative frequency and adding the results:

$$\bar{x} = x_1 \cdot \frac{t_1}{n} + x_2 \cdot \frac{t_2}{n} + \cdots + x_m \cdot \frac{t_m}{n}$$

$$= \sum x_i \cdot \frac{t_i}{n}$$

In a comparable way, we can calculate the weighted average of the values of a random variable. Instead of multiplying by a relative frequency, we multiply each value of a random variable by its probability:

The sum of the products of each value of a random variable and its probability is called the *expected value* of the random variable. If the random variable is denoted by X, the expected value is denoted by E(X). If X takes on the distinct values, $x_1, x_2, \ldots, x_n$, then

$$E(X) = \sum_{i=1}^{n} x_i \cdot f(x_i).$$

Example 5.5.1 Let X be the random variable representing the number of heads that appear when two fair coins are tossed. We already know the probability distribution of this random variable:

x	0	1	2
$f(x)$	$\frac{1}{4}$	$\frac{1}{2}$	$\frac{1}{4}$

Thus, the expected value of X is

$$E(X) = 0 \cdot \tfrac{1}{4} + 1 \cdot \tfrac{1}{2} + 2 \cdot \tfrac{1}{4} = 1.$$

Example 5.5.2 Let X represent the number appearing on the upper face when a fair die is thrown:

x	1	2	3	4	5	6
$f(x)$	$\frac{1}{6}$	$\frac{1}{6}$	$\frac{1}{6}$	$\frac{1}{6}$	$\frac{1}{6}$	$\frac{1}{6}$

$$E(X) = 1 \cdot \tfrac{1}{6} + 2 \cdot \tfrac{1}{6} + 3 \cdot \tfrac{1}{6} + 4 \cdot \tfrac{1}{6} + 5 \cdot \tfrac{1}{6} + 6 \cdot \tfrac{1}{6}$$
$$= \tfrac{1}{6}(1 + 2 + 3 + 4 + 5 + 6) = \tfrac{21}{6} = 3\tfrac{1}{2}$$

This last example illustrates that the term "expected value" must not be interpreted in any way except as the definition indicates. We do *not* really "expect" that the face of the die will show $3\tfrac{1}{2}$. But see Remark 1 at the end of this section.

Example 5.5.3 Let X be a Bernoulli random variable with probability distribution given by the following table:

x	0	1
$f(x)$	q	p

$(q = 1 - p)$

Then
$$E(X) = 0 \cdot q + 1 \cdot p = p.$$

The result of Example 5.5.3 is of sufficient importance to justify emphasis:

The expected value of a Bernoulli random variable with

$$P(X = 0) = 1 - p \quad \text{and} \quad P(X = 1) = p$$

is precisely p.

Example 5.5.4 An urn contains 5 red and 3 blue marbles. One marble is drawn at random. Let X represent the number of red marbles drawn. Then X is a Bernoulli variable with $P(X = 1) = \frac{5}{8}$ and $P(X = 0) = \frac{3}{8}$. The expected value of X is $\frac{5}{8}$.

It is possible for a random variable to have a negative expected value.

Example 5.5.5 Suppose that X is a random variable having the distribution:

x	-3	-2	0	1	4
$f(x)$	$\frac{1}{6}$	$\frac{1}{3}$	$\frac{1}{12}$	$\frac{1}{3}$	$\frac{1}{12}$

$$E(X) = (-3)(\tfrac{1}{6}) + (-2)(\tfrac{1}{3}) + 0(\tfrac{1}{12}) + 1(\tfrac{1}{3}) + 4(\tfrac{1}{12})$$
$$= \tfrac{1}{12}(-6 - 8 + 0 + 4 + 4) = -\tfrac{1}{2}$$

In playing a game, a player may be able to compute the expected value of his gain (positive expected value) or loss (negative expected value). A game is called **fair** for the player if the expected value is 0. (Compare this with the definition of a "fair" game on page 3.)

Example 5.5.6 Let X be the random variable representing the possible gain in the game discussed in Example 5.1.5: A card is drawn from a standard deck. The player wins $3 if an ace is drawn. He receives $2 for a king, $0.50 for either a queen or a jack. If a 10, 9, or 8 is selected, he pays $0.50. For the 7, 6, 5, 4, 3, or 2 he pays $0.75. This game is fair for the player since

$$E(X) = \tfrac{4}{52}[3 + 2 + 2(\tfrac{1}{2}) + 3(-\tfrac{1}{2}) + 6(-\tfrac{3}{4})]$$
$$= \tfrac{1}{13}(3 + 2 + 1 - \tfrac{3}{2} - \tfrac{9}{2}) = 0.$$

Remark 1

It is perhaps clear that there is a long-run interpretation of expected value. If a single fair die is tossed once, the expected value of $3\frac{1}{2}$ has no obvious meaning. If, however, we throw the die 100 times and add the different scores obtained, it is meaningful to say that we "expect" the total to be about 350. We shall take a more careful look at this interpretation in later chapters. At this point, it is worth recalling that 100 throws of a die may be regarded as 100 repeated, independent trials.

Remark 2

If X is a random variable having expected value $E(X)$, we should remember that $E(X)$ is a *constant*, not a variable.

Remark 3

Several of the problems in Exercises 5.5 are designed to yield further insight into the nature of the expected value of a random variable. In Chapter 6 we shall continue our study of $E(X)$.

EXERCISES 5.5

Find the expected value of the random variable described in:

- 1. Problem 3, Exercises 5.1
- 2. Problem 4, Exercises 5.1
- 3. Problem 5, Exercises 5.1
- 4. Problem 6, Exercises 5.1
- 5. Problem 5, Exercises 3.2
- 6. Example 3.2.3

- 7. Two fair dice are tossed. What is the expected value of Y if Y represents the sum of the numbers appearing on the upper faces of the dice? (See Problem 4c, Exercises 2.4a.)
- 8. Two fair dice are tossed. Find the expected value of Z if Z represents the product of the two numbers tossed. (See Problem 7, Exercises 5.1.)
  9. Let X represent the number of heads obtained on tosses of a coin. (See Problems 3 and 4, Exercises 2.2.) Find the expected value for:
     a. a single toss
     b. two tosses
     c. three tosses
     ★ d. four tosses
- ★ 10. From the results of Problem 9 above, what would you guess to be the expected value on n tosses of a fair coin?
- 11. An urn contains 5 red and 3 blue marbles. Draws are made, one marble at a time, with replacement.
     a. Find $E(X)$ if X denotes the number of red marbles drawn and two draws are made.

b. Find $E(X)$ if X denotes the number of red marbles drawn and three draws are made.

c. Find $E(Y)$ if Y denotes the number of blue marbles drawn for two draws with replacement.

● 12. Suppose that the theory of inheritance predicts that one fifth of the offspring of a certain breed of dogs will have floppy ears ($p = \frac{1}{5}, q = \frac{4}{5}$). In a litter of 3 puppies of this breed, if X represents the number of dogs with floppy ears, what is the expected value of X?

13. A container has 3 good items and 1 defective one. The items are drawn one at a time and tested until the defective item is found. (Problem 8, Exercises 3.3.) Find $E(X)$ if X represents the number of the draw on which the defective item is found.

14. Two urns are marked H and T respectively. In Urn H are two marbles, 1 red, 1 blue; in Urn T are three marbles, 1 red, 1 blue, 1 green. A coin is tossed. If the coin falls heads, a marble is drawn from Urn H; if tails, from Urn T.

a. If a red marble is drawn, you win \$2; blue, you lose \$3; green, you win \$3: Is this a fair game?

b. If you win \$2 on a red marble drawn and lose \$3 on blue, at least how much should green pay before you can consider the game fair?

15. Roulette is a well-known game in which a mechanical device randomly selects one of 38 equally likely numbers. If a player decides to play this game by choosing one particular number, he gains \$35 if his number is selected by the device. If any other number is selected, he loses \$1. Calculate the expected value of this game and notice that it is negative.

16. The 38 numbers in a roulette wheel are associated with colors: 18 red, 18 black, and 2 white. If a player chooses, say, red, he wins \$1 if a red number is selected; otherwise he loses \$1. Is the game fair for the player?

● 17. Suppose that X is a (finite) discrete random variable with $E(X)$ known. Let $Y = X - k$, k a constant, be a new random variable. Prove: $E(Y) = E(X) - k$. (Hint: $E(Y) = \sum_i y_i \cdot P(Y = y_i)$ by definition. How is y_i related to x_i? How is $P(Y = y_i)$ related to $P(X = x_i)$?)

● 18. $E(X - k)$ is sometimes called the *first moment* of the probability distribution about k. (Readers who have had experience with the law of leverage in elementary physics can recall the idea of a moment in connection with a seesaw. The product of a weight with its distance from the pivot point is its moment.) If we imagine that our probabilities are weights distributed on a lever arm, then

$$E(X - k) = (x_1 - k)f(x_1) + (x_2 - k)f(x_2) + \cdots + (x_n - k)f(x_n)$$

is the sum of the products of the weights each multiplied by the length of its "lever arm" to k.

If $E(X - k) = 0$, then the "lever" balances at k. k would be the center of mass (the balance point) of the system.

Use Problem 17 to prove that the center of mass of a finite discrete probability distribution is at $E(X)$.

19. Problem 18 applies to finite discrete probability distributions. Assume that the same physical interpretation can be given to the expected value of a continuous random variable. That is, that the distribution will balance about the line $X = E(X)$. Find $E(X)$ if:
   a. X is uniformly distributed over the interval $[a, b]$
   b. X has standard normal distribution
   ★ c. X has the triangular distribution over the interval $[0, 10]$. (See Example 5.3.2. Hint: The center of mass of a triangle is at the intersection of the medians.)

● 20. Two fair dice are tossed. If X is the random variable for the number obtained in the first die and Y is the corresponding random variable for the second, then $E(X) = E(Y) = 3\frac{1}{2}$. In Problem 8, $E(X \cdot Y)$ was found.
   a. Compare $E(X \cdot Y)$ with $E(X) \cdot E(Y)$.
   b. From the results of Problem 3 on page 131, compare $E(X^2)$ with

$$(E(X))^2 = E(X) \cdot E(X).$$

21. Can you think of some difference between $X \cdot Y$ and $X \cdot X$ that might explain the results of Problem 20?

5.6 What Do You Think?

The notion of the expected value of a random variable is of fundamental importance for decision making. In choosing among several courses of action in situations involving uncertainty, it is reasonable to choose that course which offers the largest expected value.

If the decision involves monetary worth, then the expected value can be expressed in dollars. Even in personal situations it is not uncommon to think in terms of some units of personal satisfaction. (The authors know of husbands who speak of earning "points" at home by buying flowers, wiping dishes, and so on!) It is perhaps easier to confine our attention to problems for which we can interpret expected value in terms of dollars. It is simplest if we think of a decision in terms of deciding whether to play or to refuse to play a game.

As noted on page 130, a game is called *fair* for a player if his expected value is 0. Taking a long-run view of expected value, playing a game with positive or negative expected value is not really gambling. Problems 15 and 16 of Exercises 5.5 make it clear that the operator of a roulette wheel does not "gamble." If enough people play often enough, he can count on an expected gain of about $0.05 for each dollar played. Anyone who decides to play against the operator must anticipate losing $0.05 for each dollar played —if he plays long enough. (As we have noted in Problems 11 and 12, Exercises 3.5, even in a game with a fair coin, where $E(X) = 0$, the player with the small stake has a disadvantage.) There may be additional moral, social, or legal reasons to refuse to play, but we emphasize the financial one.

It is interesting to think about the following two questions. From a strictly financial viewpoint, is it ever wise:

a. to refuse to play a game with positive expected value?

or

b. to decide to play a game with negative expected value?

The interesting facts of the real world are that people *do* make decisions which are in apparent conflict with the anticipated expected value. (Before working Exercises 5.6, it is worth some thought to see whether we can imagine such real world situations.) Economists have devised the concept of "utility" to explain why such decisions are made. Exercises 5.6 provide problems for discussion.

EXERCISES 5.6

- **1.** In Chapter 1 we mentioned a college student who owned an expensive sports car. Comment on whether he should pay premiums for collision insurance even though these payments reduce his budget for other expenses.
- **2.** Similar questions might be asked about life insurance, health insurance, and accident insurance. Comment on these.
- **3.** Another interesting aspect of automobile insurance is that many states *require* the owner of an automobile to carry liability insurance. Thus it appears that these states force us to play a game with negative expected value. Comment.
- **4.** Sometimes a choice is made between alternatives offering the same expected value. A farmer has the opportunity to use part of his land to plant either a staple crop A or a more expensive crop B which is difficult to grow. He estimates his profits (which depend on weather, market condition, and so on) in accordance with the following tables.

Crop A:	x, profit	$1000	$3000	$6000
	$f(x)$, probability	$\frac{1}{2}$	$\frac{1}{3}$	$\frac{1}{6}$

Crop B:	y, profit	$-$1000	0	$18,000
	$g(y)$, probability	$\frac{1}{2}$	$\frac{1}{3}$	$\frac{1}{6}$

  Which crop should he plant?
- **5.** Many people in foreign countries and in some states in this country participate in government-run lotteries. Is this participation wise? (Suppose, for example, that there are 1,000,000 tickets selling for $0.25, the prize being $100,000.)

- **6.** Another type of situation arises less often, but is interesting to consider. Suppose that you have the opportunity of paying $20,000 to play a game in which you have one chance in 10,000 of winning $500 million. What do you think?
- **7.** In Chapter 1 we considered a doctor faced with the decision of beginning a treatment for a patient, knowing that the treatment may produce harmful side effects. The doctor could proceed as follows. The treatment may be of major help (with 10 points to the patient), of minor help (2 points), or of no help. The side effects may be major (-5), minor (-2), or not occurring. From past experience he judges the probabilities of these events according to the following table:

		Help		
		major	minor	none
Side Effects	major	0.20	0.10	0.10
	minor	0.05	0.20	0.05
	none	0.05	0.10	0.15

The doctor decides to begin the treatment if the expected gain to the patient is 1 or more "points," since he judges the expense and inconvenience to the patient as -1. Does he begin the treatment?

- **8.** Work Problem 7 if the estimated probabilities are:

		Help		
		major	minor	none
Side Effects	major	0.10	0.30	0.20
	minor	0.10	0.05	0.10
	none	0.05	0	0.10

The mathematics of expected value has some unusual features. Sometimes our intuition leads us astray. Consider a game played with two dice (Problem 20, Exercises 5.5). Points are won by multiplying the numbers which appear on the individual dice. If we let Z denote the number of points won on an individual throw, we may compute $E(Z)$. Now the value of Z is obtained by multiplying the value of X (number on first die) by the value of Y (number on second die). That is, $Z = XY$. Now X and Y have the same probability distribution.

We might argue as follows. Why bother with two dice? If X and Y have the same distribution does it matter, in finding the expected value, whether we multiply the values of X and Y or simply use one die and square the number on the face? We would then compute $E(Z)$, where $Z = X \cdot X = X^2$. What do you think? Did you conclude from Problem 20, Exercises 5.5, that $E(XY) = E(X^2)$?

Chapter Summary

1. A variable whose values are determined by the outcomes of an experiment is called a *random variable*.
2. a. A *probability function for a random variable X* may be represented by a graph.
   b. The graph of a discrete probability distribution consists of all points on the X-axis plus a finite number of isolated points $(x, f(x))$.
   c. A random variable that takes on exactly two values is called a *Bernoulli random variable*.
3. a. For a continuous probability distribution, the area under its graph over a given interval measures the probability. The total area under the curve is 1.
   b. Intervals and unions of intervals are events and thus have probabilities.
4. a. The standard normal distribution is a continuous probability distribution that has the probability function

$$f(z) = \frac{1}{\sqrt{2\pi}} e^{\frac{-z^2}{2}}$$

   b. The graph is "bell-shaped" and is symmetric about the vertical line through 0.
5. If X takes on the distinct values $x_1, x_2, \ldots, x_n$, then the *expected value* of the random variable X is found by using this formula:

$$E(X) = \sum_{i=1}^{n} x_i \cdot f(x_i)$$

Chapter Review

1. A game is played on a toss of a coin. If the coin falls heads, the player wins $1; if tails, he loses $1. Let X be a random variable representing the amount the player wins (loses) after one play.
   a. Graph the probability function for X.
   b. Find the expected value of X.
2. Suppose, in the game of Problem 1, the player wins $1 on heads and wins $4 on tails. Let Y be a random variable representing the amount the player wins after one play. How much should he be charged for playing if the game is to be fair?

3. A marble is to be drawn from an urn containing 3 red and 2 blue marbles. If a red marble is drawn, the player loses $4. If a blue marble is drawn, the player wins $6. Let Y be a random variable representing the number of dollars the player wins.
   a. Construct a table giving the probability function of Y.
   b. Graph the probability function for Y.
   c. Find $E(Y)$.
   d. How much should be charged for playing if this game is to be fair?

4. An urn contains 3 red and 2 blue marbles. Two drawings with replacement are made. Each time, one marble is drawn. Let X be a random variable representing the number of red marbles drawn.
   a. Draw a tree diagram showing the outcomes and their probabilities.
   b. Show the probability distribution of X by means of a table.

5. Six slips of paper, numbered 1, 2, 3, 4, 5, 6, respectively, are placed in a container. Two drawings are made, without replacement, one slip at a time. Let X be a random variable corresponding to the number on the first slip and Z correspond to the second number drawn.
   a. Let $Y = X + Z$. Make a table to show the possible outcomes for Y.
   b. Find $E(Y)$.
   c. Since $E(Y) = E(X) + E(Z)$, find $E(Z)$.

6. Let W be a random variable corresponding to the sum of the numbers on a toss of a pair of dice except when the toss results in a "double." In the event of a "double," Y represents the difference of the numbers instead of the sum.
   a. Construct a table showing the possible values of W.
   b. Find $E(W)$.
   c. Why does $E(W)$ differ from $E(Y)$ of Problem 5b?

7. Suppose that a certain characteristic, say green eyes, has probability $\frac{1}{3}$ of appearing in a particular family. Graph the Bernoulli distribution for $p = \frac{1}{3}$.

8. Let X have a uniform (continuous) distribution over the interval [4, 16]. Find:
   a. $P(X \leq 7)$
   b. $P(X \geq 7)$
   c. $P(X > 7)$
   d. $P(6 < X < 10)$
   e. x so that $P(X < x) = \frac{1}{3}$
   f. x so that $P(X < x \text{ or } x < X) = \frac{1}{3}$

9. A triangular distribution over the interval [4, 16] has $P(X \leq 4) = P(X \geq 16) = 0$, increasing toward maximum height at $X = 16$.
   a. Find the height of the distribution at $X = 16$.
   b. Find the height at $X = 13$.
   c. Find $P(7 < X < 13)$.

10. The probabilities for $P(Z \leq z)$ in a standard normal distribution for certain values of z are given in the table at the right. Find:

z	$P(Z \leq z)$
−3	0.0013
−2	0.0228
−1	0.1587
0	0.5000

   a. $P(Z \leq -1)$
   b. $P(-3 \leq Z < -1)$
   c. $P(Z \geq 2)$
   d. $P(|Z| < 1)$
   e. $P(|Z| > 2)$
   f. $P(|Z - 1| \leq 2)$

11. Using the table in Problem 10, determine z such that:
   a. $P(z < Z < -1) = 0.1574$
   b. $P(|Z| \leq z) = 0.9544$
   c. $P(0 < Z < z) = 0.3413$

Chapter 6

Functions of Random Variables

In Problem 17, Exercises 5.5, we found that if X is a random variable with expected value $E(X)$, then

$$Z = X - k$$

for a constant k is also a random variable with expected value

$$E(Z) = E(X) - k.$$

This is a special case of a more general situation, where we have a random variable Z expressed in terms of a random variable X. Let us consider an arbitrary function, ϕ,* defined by

$$Z = \phi(X),$$

and ask whether, knowing $E(X)$, we can determine $E(Z)$. It happens that our primary interest is in rather simple functions, such as those defined by

$$\phi(X) = aX + b,$$

$$\phi(X) = X^2,$$

and

$$\phi(X) = (aX + b)^2,$$

where a and b are constants. We shall develop our results through examples and informal arguments, supplying proofs wherever possible.

---

* The symbol ϕ (phi) is a letter of the Greek alphabet corresponding to the English "f" or "ph."

We need to remind ourselves that, while X is a random variable, $E(X)$ is a *number*, a constant. To emphasize this, we shall make increasing use of a new symbol to denote $E(X)$, thus,

$$E(X) = \mu_X.*$$

The subscript is necessary if we are dealing with more than one random variable. Thus, if $Z = aX + b$,

$$E(Z) = \mu_Z = \mu_{aX+b}.$$

If there is no confusion, the subscript may be omitted.

6.1 $E(aX + b)$, $E(X^2)$

In considering $E(Z)$ where $Z = X - k$, we have already examined the special case of

$$E(aX + b),$$

where $a = 1$ and $b = -k$. For this, we have shown that

$$E(X - k) = E(X) - k.$$

Further, we have seen (Section 5.5) an analogy between the expected value of a random variable and the mean of a set of data, and in Section 4.3 we found that

$$\overline{ax + b} = a\bar{x} + b.$$

From this it is natural to guess that

$$E(aX + b) = aE(X) + b,$$

or using our alternate notation,

$$\mu_{aX+b} = a\mu_X + b.$$

We now illustrate the correctness of this result with some examples. The proof is given at the end of this section.

Example 6.1.1 Consider the random variables X, U, V, W, and Z having distributions given by the tables below. We find and compare the corresponding expected values:

x, values of X	-2	0	2	5
Probability	0.1	0.3	0.4	0.2

---

* Read "mu sub-X." The symbol μ is a letter of the Greek alphabet corresponding to the English "m."

$E(X) = 1.6$

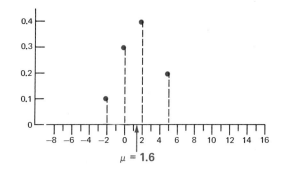

u, values of U	−4	−2	0	3
Probability	0.1	0.3	0.4	0.2

$E(U) = -0.4$
$ = 1.6 - 2$

We observe that
$U = X - 2$,
$E(U) = E(X) - 2$.

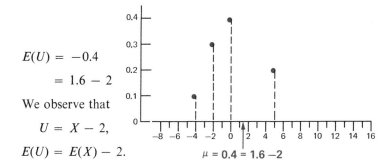

v, values of V	−6	0	6	15
Probability	0.1	0.3	0.4	0.2

$E(V) = 4.8$
$ = 3(1.6)$

We observe that
$V = 3X$,
$E(V) = 3 \cdot E(X)$.

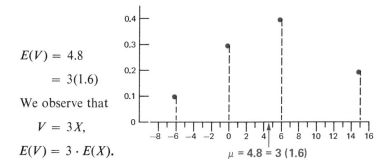

w, values of W	-3.6	-1.6	0.4	3.4
Probability	0.1	0.3	0.4	0.2

$E(W) = 0$

$\qquad = 1.6 - 1.6$

We observe that

$\qquad W = X - E(X),$

$\qquad E(W) = E(X) - E(X)$

$\qquad\quad = 0.$

z, values of Z	-8	-2	4	13
Probability	0.1	0.3	0.4	0.2

$E(Z) = 2.8$

$\qquad = 3(1.6) - 2$

We observe that

$\qquad Z = 3X - 2,$

$\qquad E(Z) = 3 \cdot E(X) - 2.$

Notice that the probabilities (and hence the heights of the graphs) are the same in all cases. Only the horizontal placement is changed.

Example 6.1.2 A game is played by tossing two coins. Points are awarded by the rule "add 1 to twice the number of heads appearing." Our set of outcomes is

$$\{hh, th, ht, tt\}.$$

Letting X represent the number of heads, we have:

x	0	1	2
$f(x)$	$\frac{1}{4}$	$\frac{1}{2}$	$\frac{1}{4}$

$E(X) = 1$

Now if Z represents the number of points awarded, Z takes on the values

$$2 \cdot 0 + 1 = 1, \qquad 2 \cdot 1 + 1 = 3, \qquad 2 \cdot 2 + 1 = 5.$$

Then:
$$P(Z = 1) = P(2X + 1 = 1) = P(X = 0) = f(0) = \tfrac{1}{4}$$
$$P(Z = 3) = P(2X + 1 = 3) = P(X = 1) = f(1) = \tfrac{1}{2}$$
$$P(Z = 5) = P(2X + 1 = 5) = P(X = 2) = f(2) = \tfrac{1}{4}$$

Finally, we note that
$$E(Z) = 1 \cdot \tfrac{1}{4} + 3 \cdot \tfrac{1}{2} + 5 \cdot \tfrac{1}{4} = 3 = 2 \cdot E(X) + 1.$$

Any function of X has the property that for each value of X there is precisely one value of Y. A *linear* function, defined by
$$\phi(X) = aX + b,$$
has the additional property that each value of Y arises from precisely *one* particular value of X. However, in the case of a quadratic function, defined by
$$\phi(X) = X^2,$$
two values of X may yield the same value of Y. We next consider the expected value of $Z = X^2$.

In Example 6.1.1 we began with a random variable X having its distribution defined by a table and we looked at several other closely (linearly) related random variables having corresponding distributions. In the next example, we shall use the same random variable X and compare its expected value with that of X^2.

Example 6.1.3 Letting $Z = X^2$, we see that as X takes on the values $-2, 0, 2, 5$, Z takes on the values $0, 4, 25$:

x, values of X	−2	0	2	5
z, values of Z	4	0	4	25

Using the probabilities from Example 6.1.1, we have:
$$P(Z = 0) = P(X^2 = 0) = P(X = 0) = 0.3$$
$$P(Z = 4) = P(X^2 = 4) = P(X = 2 \text{ or } X = -2)$$
$$= P(X = 2) + P(X = -2)$$
$$= 0.1 + 0.4 = 0.5$$
$$P(Z = 25) = P(X^2 = 25) = P(X = 5) = 0.2 \quad (\text{since } P(X = -5) = 0).$$

Hence,
$$\mu_{X^2} = \mu_Z = E(Z) = (0)(0.3) + (4)(0.5) + (25)(0.2) = 7.$$

Just as we observed in Problem 20, Exercises 5.5, we note again in Example 6.1.3 that
$$E(X^2) \neq [E(X)]^2.$$

Example 6.1.4 In a game involving tossing two coins, points are awarded by squaring the number of heads appearing. Using X as in Example 6.1.2 and letting $Z = X^2$, we have:

$$P(Z = 0) = P(X^2 = 0) = P(X = 0) = \tfrac{1}{4}$$
$$P(Z = 1) = P(X^2 = 1) = P(X = 1) = \tfrac{1}{2}$$
$$P(Z = 4) = P(X^2 = 4) = P(X = 2) = \tfrac{1}{4}$$

Hence,

$$\mu_{X^2} = \mu_Z = E(Z) = 0 \cdot \tfrac{1}{4} + 1 \cdot \tfrac{1}{2} + 4 \cdot \tfrac{1}{4} = \tfrac{3}{2}.$$

We have not been able to obtain a "nice" formula for $E(X^2)$ as we were able to do for $E(aX + b)$. However, we see that the computation of $E(X^2)$ is not difficult for finite probability distributions. The reason for the emphasis on $E(X^2)$ will become apparent in Section 6.2. In order to lead to that section, we may combine the two results we have already obtained. Let $U = aX + b$ and $Z = U^2$. That is, $Z = (aX + b)^2$. We illustrate the method of computing

$$E((aX + b)^2)$$

in the following example, where $a = 3$ and $b = -2$.

Example 6.1.5 Again we take X as in Example 6.1.1 and let $Z = 3X - 2$. Then:

x, values of X	−2	0	2	5
z, values of Z	−8	−2	4	13

Thus the values of Z^2 are 64, 4, 16, 169. The associated probabilities are:

$$P(Z^2 = 64) = P(Z = -8) = P(X = -2) = 0.1$$
$$P(Z^2 = 4) = P(Z = -2) = P(X = 0) = 0.3$$
$$P(Z^2 = 16) = P(Z = 4) = P(X = 2) = 0.4$$
$$P(Z^2 = 169) = P(Z = 13) = P(X = 5) = 0.2$$

Hence:

$$\mu_{(aX+b)^2} = E(Z^2)$$
$$= (64)(0.1) + (4)(0.3) + (16)(0.4) + (169)(0.2)$$
$$= 47.8$$

Now $E(Z) = 2.8$ (from Example 6.1.1), and it is clear that here again $E((3X - 2)^2) \neq [E(3X - 2)]^2$.

Although we have examined only certain functions of X, we are led, on the basis of our examples, to a useful definition.

Suppose that the random variable X takes on the values $x_1, x_2, \ldots, x_n$ with probabilities
$$P(X = x_i) = f(x_i).$$
If $Z = \phi(X)$, then Z is also a random variable, and $z_i = \phi(x_i)$. We make this definition:

If $Z = \phi(X)$, then $E(Z) = E(\phi(X)) = \sum_{i=1}^{n} \phi(x_i) \cdot f(x_i).$

We use this definition to prove the following theorem:

If X is a random variable with expected value $E(X)$, or μ_X, then
$$E(aX + b) = aE(X) + b,$$
or
$$\mu_{aX+b} = a\mu_X + b,$$
where a and b are arbitrary constants.

Proof: $\phi(X) = aX + b$, and so from the definition above:
$$E(aX + b) = \sum_i (ax_i + b) \cdot f(x_i)$$
$$= \sum_i ax_i \cdot f(x_i) + \sum_i b \cdot f(x_i)$$
$$= a \sum_i x_i \cdot f(x_i) + b \sum_i f(x_i) \quad \text{(see Appendix)}$$

But $\sum x_i \cdot f(x_i) = E(X)$ by definition (Section 5.5) and $\sum f(x_i) = 1$ (page 24). Therefore,
$$E(aX + b) = aE(X) + b = a\mu_X + b.$$

As immediate consequences, we have our earlier results (Problems 17 and 18, Exercises 5.5, and Example 6.1.1):
$$E(X - k) = E(X) - k$$
and
$$E(X - \mu_X) = 0.$$

EXERCISES 6.1

● **1.** Let X have the probability distribution shown at the right. Find:

x	−1	0	4
f(x)	$\frac{1}{3}$	$\frac{1}{6}$	$\frac{1}{2}$

   a. μ_X

   b. μ_Z, where $Z = 6X + 2$ **c.** μ_Z, where $Z = -6X + 2$

   d. μ_Z, where $Z = 6X - 2$ **e.** $E(X^2)$

2. Repeat Problem 1 if X has the probability distribution shown at the right.

x	-2	1	4
$f(x)$	$\frac{1}{2}$	$\frac{1}{4}$	$\frac{1}{4}$

● 3. Let X have the Bernoulli distribution with $P(X = 0) = \frac{1}{3}$ and $P(X = 1) = \frac{2}{3}$. Find
   a. μ_X
   b. $E(X - \frac{2}{3})$
   c. $E(X^2)$
   d. $E((X - \frac{2}{3})^2)$
   e. $E(X^2) - \mu_X^2$
   (Recall from Section 5.5 that in a Bernoulli distribution $\mu_X = P(X = 1)$.)

4. Repeat Problem 3, replacing $\frac{1}{3}$ by $\frac{2}{5}$ and $\frac{2}{3}$ by $\frac{3}{5}$.

● 5. Repeat Problem 3, replacing $\frac{2}{3}$ by p and $\frac{1}{3}$ by q.

● 6. For the experiment of tossing two fair coins, compare $E((X - \mu_X)^2)$ with $E(X^2) - \mu_X^2$, where X represents the number of heads appearing.

● 7. Repeat Problem 6 for the experiment of tossing three fair coins. (Recall Problem 9c, Exercises 5.5.)

● 8. A fair die is to be tossed. Let X represent the number appearing. We know that $E(X) = \frac{7}{2}$ (Example 5.5.2).
   a. Compute $E(X^2)$. (Recall Problem 3, Exercises 5.5).
   b. Compute $E((X - \frac{7}{2})^2)$.
   c. Show that $E((X - \mu_X)^2) = E(X^2) - \mu_X^2$.

● 9. It is possible to think of a constant c as a random variable by the following device. Let U be a random variable with $P(U = c) = 1$. Prove that $E(U) = c$.

● 10. Let random variables X and Z be related by $Z = 3X - 2$, and let the probability distribution of X be as shown in the following table:

x	-2	0	2	5
$f(x)$	0.1	0.3	0.4	0.2

z	-8			
x^2	4			
z^2			169	

$E(X) = (-2)(0.1) + \cdots = \underline{\ ?\ }$

$E(Z) = (-8)(0.1) + \cdots = \underline{\ ?\ }$

$E(X^2) = \underline{\ ?\ }$

$E(Z^2) = \underline{\ ?\ }$

Copy and complete the above table, and compute the information required to the right of the table.

● 11. It was proved that $E(aX + b) = aE(X) + b$.
   a. Since $(aX + b)^2 = (aX)^2 + 2(aX)(b) + b^2$
   $= a^2X^2 + 2abX + b^2$,
   make a conjecture about $E((aX + b)^2)$.
   b. In Problem 10, $Z = 3X - 2$; therefore $a = 3$ and $b = -2$. Verify by computation that here,
   $$E((aX + b)^2) = a^2 E(X^2) + 2ab E(X) + b^2.$$

  c. Prove that the conclusion in part b is valid assuming that if X and Y are two random variables, then $E(X + Y) = E(X) + E(Y)$. (This will be discussed further, Section 6.3.)

★ **12.** Let $Z = 4X - 3$ and let X be defined by

x	−3	1	4	7
f(x)	0.4	0.3	0.2	0.1

Calculate $E((4X - 3)^2)$ by making use of the results in Problem 11b.

6.2 Variance and Standard Deviation

Several of the problems of Exercises 6.1 called for the computation of $E((X - \mu_X)^2)$. This particular constant associated with a random variable X plays an important role in subsequent chapters and is given a special name, the *variance of X* (abbreviated "var(X)") and a special symbol, σ_X^2.*

Thus:

VARIANCE

If X is finite,
$$\sigma_X^2 = \text{var}(X) = E((X - \mu_X)^2).$$

$$E((X - \mu_X)^2) = \sum_i (x_i - \mu_X)^2 \cdot f(x_i).$$

We occasionally omit the subscript in σ_X^2 if we are considering only one random variable.

The analogy between σ_X^2 defined above and s_x^2 of Chapter 4 is evident. As in Chapter 4, we often have use for the (positive) square root of var(X), σ_X. This is called the *standard deviation of X*. Thus:

STANDARD DEVIATION

$$\sigma_X = \sqrt{\sigma_X^2}$$

  To describe a probability distribution completely, we need to know the probability function of X. However, useful information about a probability distribution is obtained if $E(X)$ is known—it locates the center of mass of the distribution (Problem 18, Exercises 5.5). Even more information is obtained if, in addition to $E(X)$, we also have a measure of spread, or variability, of the distribution. The choice of σ_X as such a measure is

---
* Read "sigma-squared sub-X." The symbol σ is a letter of the Greek alphabet corresponding to the English "s."

justified largely because it lends itself to mathematical treatment. This treatment in turn leads to several powerful results. Some of the results we shall explore in later chapters. As one indication of the usefulness of the standard deviation, we state, without proof, the following theorem due to the Russian mathematician, P. L. Chebyshev (1821–1894):

CHEBYSHEV'S INEQUALITY Let X be a random variable with $E(X) = \mu_X$ and variance σ_X^2. Then at most $\dfrac{1}{k^2}$ of the probability is distributed outside the interval from $\mu_X - k\sigma_X$ to $\mu_X + k\sigma_X$, where $k \geq 1$. That is,

$$P(|X - \mu_X| > k\sigma_X) \leq \frac{1}{k^2}$$

or, alternately,

$$P(|X - \mu_X| \leq k\sigma_X) \geq 1 - \frac{1}{k^2}.$$

For example, at most $\frac{1}{9}$ of any probability distribution falls outside the interval $\mu_X - 3\sigma_X$ to $\mu_X + 3\sigma_X$:

$$P(|X - \mu_X| > 3\sigma_X) \leq \tfrac{1}{9}$$

and

$$P(|X - \mu_X| \leq 3\sigma_X) \geq \tfrac{8}{9}$$

or

$$P(\mu_X - 3\sigma_X \leq X \leq \mu_X + 3\sigma_X) \geq \tfrac{8}{9}.$$

In other words, if we pick a value of the random variable arbitrarily, we can state the minimum probability that the *random* value falls within given limits. Moreover, it can be shown that if we consider specific distributions, we can obtain even more accurate probability statements. For us, the present significance of Chebyshev's Inequality is to emphasize the general importance of the standard deviation of a random variable.

Although we have already calculated $E((X - \mu_X)^2)$ for some earlier problems, we carry out the computations for one example in some detail.

Example 6.2.1 We use the random variable X of Example 6.1.1 and prepare a table as shown on the next page. The sum of the second column is

$$\sum f(x_i) = 1,$$

and the sum of the third column is

$$\mu = \sum x_i f(x_i) = 1.6.$$

Having obtained μ, we can then complete the table. Finally the sum of the last column is

$$\sum (x_i - \mu)^2 f(x_i) = \operatorname{var}(X) = 4.44.$$

Functions of Random Variables / 149

x	f(x)	xf(x)	x − μ	(x − μ)²	(x − μ)²f(x)
−2	0.1	−0.2	−3.6	12.96	1.296
0	0.3	0	−1.6	2.56	0.768
2	0.4	0.8	0.4	0.16	0.064
5	0.2	1.0	3.4	11.56	2.312
Sum	1	1.6 = μ			4.440

As a quick check, it can be verified that

$$\sum (x - \mu) f(x) = 0.$$

The standard deviation is

$$\sigma = \sqrt{4.44} \doteq 2.11.$$

In this case, all values of the variable lie within the interval

$$[\mu - 2\sigma, \mu + 2\sigma] \doteq [-2.6, 5.8],$$

although Chebyshev's Inequality only guarantees that at most $\frac{1}{4}$ will fall outside that interval.

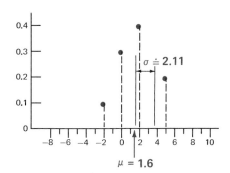

Several problems of Exercises 6.1 revealed the fact that

$$\text{var}(X) = E((X - \mu)^2) = E(X^2) - \mu^2.$$

This formula, which we shall prove at the end of this section, sometimes enables us to simplify the computation of var(X).

Example 6.2.2 Applying the formula "var$(X) = E(X^2) - \mu^2$" to the random variable of Example 6.2.1, we have the following:

x	f(x)	xf(x)	x²	x²f(x)
−2	0.1	−0.2	4	0.4
0	0.3	0	0	0
2	0.4	0.8	4	1.6
5	0.2	1.0	25	5.0
Sum	1	1.6		7.0

The sum of the last column is $E(X^2) = \sum x_i^2 f(x_i)$, and so

$$\text{var}(X) = E(X^2) - \mu^2 = 7 - (1.6)^2 = 4.44.$$

Example 6.2.3 Because of its importance for later work, we repeat the calculation of var(X) for the Bernoulli distribution (see Problem 5, Exercises 6.1):

x	$f(x)$	$xf(x)$	x^2	$x^2f(x)$
0	q	0	0	0
1	p	p	1	p
Sum	1	p		p

$$\text{var}(X) = E(X^2) - \mu^2 = p - p^2 = p(1-p) = pq,$$

where, of course, $p + q = 1$.

In Section 6.1 we considered functions of a finite random variable X. If $Z = \phi(X)$, we know how to compute $E(Z)$. It is natural to inquire about var(Z). In Section 6.4 we shall return to this question, confining ourselves to the special case $Z = aX + b$.

We shall now prove the formula:

$$E((X - \mu)^2) = E(X^2) - \mu^2.$$

Proof: $\phi(X) = (X - \mu)^2$, and so from the definition on page 145:

$$\begin{aligned}
E((X - \mu)^2) &= \sum_i (x_i - \mu)^2 \cdot f(x_i) \\
&= \sum_i (x_i^2 - 2\mu x_i + \mu^2) \cdot f(x_i) \\
&= \sum_i x_i^2 \cdot f(x_i) + \sum_i (-2\mu) x_i \cdot f(x_i) + \sum_i \mu^2 \cdot f(x_i) \\
&= \sum_i x_i^2 \cdot f(x_i) - 2\mu \sum_i x_i \cdot f(x_i) + \mu^2 \sum_i f(x_i) \\
&= \sum_i x_i^2 \cdot f(x_i) - 2\mu^2 + \mu^2 \\
&= E(X^2) - \mu^2
\end{aligned}$$

Remark 1

Our definitions of expected value and variance and our conclusions (theorems) have been stated in terms of finite, discrete random variables. It is possible to extend the definitions to continuous random variables and to prove the corresponding theorems. The definitions and proofs for the continuous case involve more advanced mathematics than we have at our disposal.

Remark 2

Our primary concern with continuous random variables is with the standard normal distribution (Section 5.4). We assert, without proof, that *the variance of the standard normal distribution is 1*. It follows that *the standard devia-*

tion is 1. Looking back at Figure 5.8, we can see that about 68% of the probability falls within the interval

$$[\mu - \sigma, \mu + \sigma]$$

while about 95% of it falls within the interval

$$[\mu - 2\sigma, \mu + 2\sigma].$$

In the latter case, Chebyshev's Inequality only guarantees that at most $\frac{1}{4}$ will fall outside that interval.

EXERCISES 6.2

1. Compute σ_X^2 and σ_X for the following distribution by:
   a. using the definition of var(X)
   b. using the formula, var(X) = $E(X^2) - \mu^2$

x	-3	1	2	6
$f(x)$	$\frac{3}{12}$	$\frac{1}{12}$	$\frac{4}{12}$	$\frac{4}{12}$

2. Repeat Problem 1 for the following distribution:

x	-2	0	1	3
$f(x)$	0.1	0.4	0.3	0.2

3. Find σ_X^2 for the random variable associated with the experiment "a fair die is thrown."

4. Find σ_X^2 for the random variable associated with the experiment of Problem 5, Exercises 2.4.

5. Find σ_X^2 for the random variable representing the sum of the faces if two fair dice are thrown. (For the distribution of X see the solution of Problem 4, Exercises 2.4.)

6. Find σ_X^2 for the random variable Z associated with the game in Problem 7, Exercises 5.1.

7. Let X be as in Problem 1 and suppose that $Z = 3X$. Find σ_Z^2.

8. Repeat Problem 7 with $Z = 3X + 2$.

9. In Problems 7 and 8 compare σ_Z^2 with σ_X^2.

10. We have seen that $E(X)$ may be negative. Is it possible for σ_X^2 to be negative?

11. We found that the "expected value" of a constant k is k. What is a reasonable meaning of the "variance" of a constant?

12. Find var(X) where X represents the number of heads appearing on:
    a. one toss of a fair coin
    b. two tosses of a fair coin
    c. three tosses of a fair coin

- 13. Examining the results of Problem 12, make a reasonable guess at the variance of X where X denotes the number of heads appearing on n tosses of a fair coin.
- 14. Find the variance for the random variable, X, described in Problem 5, Exercises 3.2: A fair coin is tossed until a tail appears or until three heads appear; X represents the number of heads that appear.
- 15. A container has 3 good items and 1 defective item. The items are drawn one at a time and tested until the defective item is found. X represents the number of the draw on which the defective item is found. (Problem 8, Exercises 3.3.) Find $\text{var}(X)$.
- 16. From Problem 5, if X is the random variable representing the sum of the faces if two dice are thrown, we find $\mu = 7$, $\sigma = \sqrt{\frac{35}{6}}$.
  - a. Prove, using Chebyshev's Inequality, that the probability that the sum is from 4 to 10 (inclusive) is at least $\frac{19}{54}$.
  - b. Verify by Problem 4b, Exercises 2.4, that the actual probability of obtaining a sum from 4 to 10 is greater than $\frac{19}{54}$.

6.3 Sums of Random Variables

Suppose that we have any two random variables X and Y, with $E(X) = \mu_X$ and $E(Y) = \mu_Y$. If $Z = X + Y$, then Z is also a random variable, and it is of interest to investigate $E(Z) = E(X + Y)$. It is tempting to **guess** what turns out, happily, to be a fact:

$$E(X + Y) = E(X) + E(Y),$$

or

$$\mu_{X+Y} = \mu_X + \mu_Y$$

We shall accept this result without proof. We shall also accept an extension to the sum of a finite number of random variables, given their expected values:

$$E(X_1 + X_2 + \cdots + X_n) = E(X_1) + E(X_2) + \cdots + E(X_n)$$

The proof for two independent variables is not difficult, but the proof for the general case would become complicated.

Example 6.3.1 Consider an urn containing 6 red and 2 blue marbles. Two marbles are drawn, one at a time *with* replacement. Let X represent the number of red marbles obtained on the first draw and Y the number obtained on the second draw. If Z represents the total number of red marbles drawn, then $Z = X + Y$. The possible values of X are 0 and 1, of Y are 0 and 1, and of Z are 0, 1, 2.

$$P(Z = 0) = P(X = 0 \text{ and } Y = 0) = P(X = 0) \cdot P(Y = 0)$$
$$= \tfrac{1}{4} \cdot \tfrac{1}{4} = \tfrac{1}{16}$$
$$P(Z = 1) = P(X = 0 \text{ and } Y = 1) + P(X = 1 \text{ and } Y = 0)$$
$$= P(X = 0) \cdot P(Y = 1) + P(X = 1) \cdot P(Y = 0)$$
$$= \tfrac{1}{4} \cdot \tfrac{3}{4} + \tfrac{3}{4} \cdot \tfrac{1}{4} = \tfrac{6}{16}$$
$$P(Z = 2) = P(X = 1 \text{ and } Y = 1) = P(X = 1) \cdot P(Y = 1)$$
$$= \tfrac{3}{4} \cdot \tfrac{3}{4} = \tfrac{9}{16}$$

Check: $\tfrac{1}{16} + \tfrac{6}{16} + \tfrac{9}{16} = 1$

$$E(Z) = 0 \cdot \tfrac{1}{16} + 1 \cdot \tfrac{6}{16} + 2 \cdot \tfrac{9}{16} = \tfrac{24}{16} = \tfrac{3}{2}.$$

Now X and Y are Bernoulli random variables with $p = \tfrac{3}{4}$, and so $E(X) = \tfrac{3}{4}$, $E(Y) = \tfrac{3}{4}$, and $\tfrac{3}{4} + \tfrac{3}{4} = \tfrac{3}{2}$. Thus, in this case,

$$E(X + Y) = E(X) + E(Y),$$

or

$$\mu_{X+Y} = \mu_X + \mu_Y.$$

Example 6.3.2 In the preceding example we made use of our previous argument that the events connected with the second draw are independent of the events connected with the first draw. If we consider the same urn but make the two draws *without* replacements, the situation no longer involves independence:

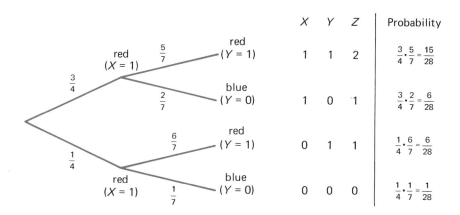

$$P(X = 0) = \tfrac{1}{4}, \quad P(X = 1) = \tfrac{3}{4}; \quad E(X) = \tfrac{3}{4}$$
$$P(Y = 0) = \tfrac{6}{28} + \tfrac{1}{28} = \tfrac{1}{4}, \quad P(Y = 1) = \tfrac{15}{28} + \tfrac{6}{28} = \tfrac{3}{4}; \quad E(Y) = \tfrac{3}{4}$$

X and Y have precisely the same probability distributions as in Example 6.3.1. The probability distribution of Z, however, is quite different. If we compute $E(Z)$, we find

$$E(Z) = 2 \cdot \tfrac{15}{28} + 1 \cdot \tfrac{6}{28} + 1 \cdot \tfrac{6}{28} + 0 \cdot \tfrac{1}{28} = \tfrac{42}{28} = \tfrac{3}{2}.$$

Thus, even though the distribution of Z differs from the Z of Example 6.3.1, we still have
$$E(X + Y) = E(X) + E(Y),$$
or
$$\mu_{X+Y} = \mu_X + \mu_Y.$$

Example 6.3.3 Example 6.3.1 is a special case of independent Bernoulli trials. This concept was introduced in the Remark in Section 3.5. Let X_1 and X_2 be the random variables associated with two independent Bernoulli trials with
$$P(X_1 = 1) = P(X_2 = 1) = p.$$
Then by the method of Example 6.3.1, we can show that
$$E(X_1 + X_2) = E(X_1) + E(X_2) = p + p = 2p.$$
For n independent Bernoulli trials, we have
$$E(X_1 + X_2 + \cdots + X_n) = np.$$

Having guessed correctly that $E(X + Y) = E(X) + E(Y)$, we quite naturally ask whether $\text{var}(X + Y) = \text{var}(X) + \text{var}(Y)$. To test this, we compute the variance of the random variables of Examples 6.3.1 and 6.3.2 in Examples 6.3.4 and 6.3.5.

Example 6.3.4 For the variables of Example 6.3.1, we have:

$\text{var}(X) = pq = \frac{3}{4} \cdot \frac{1}{4} = \frac{3}{16}$ (See Example 6.2.3.)

$\text{var}(Y) = \frac{3}{16}$

$\text{var}(Z) = \text{var}(X + Y) = E(Z^2) - \mu_Z^2$

$\qquad = (0^2 \cdot \frac{1}{16} + 1^2 \cdot \frac{6}{16} + 2^2 \cdot \frac{9}{16}) - (\frac{3}{2})^2$

$\qquad = \frac{42}{16} - \frac{9}{4} = \frac{3}{8}$

It is indeed true, for this example, that
$$\text{var}(X + Y) = \text{var}(X) + \text{var}(Y),$$
or
$$\sigma_{X+Y}^2 = \sigma_X^2 + \sigma_Y^2.$$

Example 6.3.5 For the variables of Example 6.3.2, we have:

$\text{var}(X) = \text{var}(Y) = 0^2 \cdot \frac{1}{4} + 1^2 \cdot \frac{3}{4} - (\frac{3}{4})^2$

$\qquad = \frac{3}{4} - (\frac{3}{4})^2 = \frac{3}{16}$

$\text{var}(X) + \text{var}(Y) = \frac{3}{16} + \frac{3}{16} = \frac{3}{8}$

On the other hand:
$$\text{var}(Z) = \text{var}(X + Y) = E(Z^2) - \mu_Z^2$$
$$= (0^2 \cdot \tfrac{1}{28} + 1^2 \cdot \tfrac{12}{28} + 2^2 \cdot \tfrac{15}{28}) - (\tfrac{3}{2})^2$$
$$= \tfrac{72}{28} - \tfrac{9}{4} = \tfrac{9}{28}$$

Thus, for this example,
$$\text{var}(X + Y) \neq \text{var}(X) + \text{var}(Y).$$

Example 6.3.5 is sufficient to show that the variance of the sum does not equal the sum of the variance in every case. Since the formula does hold in Example 6.3.4, we are tempted to suspect that the notion of independence may be important.

Two events, A and B, are independent if $P(A \cap B) = P(A) \cdot P(B)$ (Section 3.4). An extension of this definition to two random variables is:

Let X and Y be random variables having values $x_1, x_2, \ldots, x_n$ and $y_1, y_2, \ldots, y_m$, respectively, and probability functions $f(x)$ and $g(y)$. Then X and Y are *independent* if, for all i and j,
$$P(X = x_i \text{ and } Y = y_j) = P(X = x_i) \cdot P(Y = y_j)$$
$$= f(x_i) \cdot g(y_j).$$

It might be well to recall the discussion in Section 3.4 which pointed out that the formal definition of independence does not always agree with our intuition.

We next prove the following theorem:

If X and Y are independent random variables having probability functions $f(x)$ and $g(y)$, respectively, then
$$\text{var}(X + Y) = \text{var}(X) + \text{var}(Y).$$

Proof: We have assumed that $\mu_{X+Y} = \mu_X + \mu_Y$.

$\text{var}(X + Y) = E([(X + Y) - \mu_{X+Y}]^2)$	by definition
$= E((X + Y - \mu_X - \mu_Y)^2)$	by substitution
$= E([(X - \mu_X) + (Y - \mu_Y)]^2)$	by regrouping
$= E((X - \mu_X)^2 + (Y - \mu_Y)^2 + 2(X - \mu_X)(Y - \mu_Y))$	
$= E((X - \mu_X)^2) + E((Y - \mu_Y)^2) + 2E((X - \mu_X)(Y - \mu_Y))$	

The last step is justified by the fact that the expected value of the sum of several random variables equals the sum of the expected values.

Now $E((X - \mu_X)^2) = \text{var}(X)$ and $E((Y - \mu_Y)^2) = \text{var}(Y)$, and so we have
$$\text{var}(X + Y) = \text{var}(X) + \text{var}(Y) + 2E((X - \mu_X)(Y - \mu_Y)).$$

We need to show that the last term at the right, $E((X - \mu_X)(Y - \mu_Y))$, equals 0. We have:

$$\begin{aligned} E((X - \mu_X)(Y - \mu_Y)) &= E(XY - \mu_X Y - \mu_Y X + \mu_X \mu_Y) \\ &= E(XY) - \mu_X E(Y) - \mu_Y E(X) + \mu_X \mu_Y \\ &= E(XY) - \mu_X \mu_Y - \mu_Y \mu_X + \mu_X \mu_Y \\ &= E(XY) - \mu_X \mu_Y \end{aligned}$$

So our attention is focused on $E(XY)$. If it can be shown that this is exactly $\mu_X \mu_Y$, then the expected value of the product of the differences would be zero.

$$\begin{aligned} E(XY) &= \sum_i \sum_j x_i y_j P(X = x_i \text{ and } Y = y_j) \\ &= \sum_i \sum_j x_i y_j f(x_i) g(y_j) \quad \text{(Notice that here we use our assumption of independence.)} \\ &= \sum_i \sum_j x_i f(x_i) \cdot y_j g(y_j) \\ &= \sum_i x_i f(x_i) \cdot \sum_j y_j g(y_j) \quad \text{(See Appendix.)} \\ &= \mu_X \mu_Y \end{aligned}$$

Hence
$$E((X - \mu_X)(Y - \mu_Y)) = 0,$$
and
$$\text{var}(X + Y) = \text{var}(X) + \text{var}(Y).$$

Remark 1

Once again our results have been stated only for finite discrete random variables. They hold true for continuous random variables also.

Remark 2

The converse of the above theorem is not true. It is possible that

$$\text{var}(X + Y) = \text{var}(X) + \text{var}(Y)$$

even though X and Y are not independent random variables. (See Problem 4 in the following set of exercises.)

Example 6.3.6 In Example 6.3.3 we found that $E(X_1 + X_2) = 2p$ if X_1 and X_2 were independent Bernoulli variables with

$$P(X_1 = 1) = P(X_2 = 1) = p.$$

We also know that $\text{var}(X_i) = pq$ from Example 6.2.3. We conclude that

$$\text{var}(X_1 + X_2) = pq + pq = 2pq.$$

For n independent Bernoulli trials,

$$\text{var}(X_1 + X_2 + \cdots + X_n) = npq.$$

EXERCISES 6.3

- 1. A number is selected at random from the set

$$\{1, 2, 3, 4, 5, 6, 7, 8\}.$$

  Let $X = 2$ if the number selected is even, $X = 0$ otherwise. Let $Y = 3$ if the number selected is divisible by 3, $Y = 0$ otherwise.
  a. Find the probability distributions of X and of Y.
  b. Compute μ_X, σ_X^2, μ_Y, and σ_Y^2.
  c. Find the probability distribution of $Z = X + Y$.
  d. Compute μ_Z and σ_Z^2.
  e. Are the random variables X and Y independent? (Does this agree with your intuition?)

- 2. Repeat Problem 1, using X as before but let $Y = 3$ if the number selected is prime, $Y = 0$ otherwise. (Note that 1 is not a prime number.)

- 3. Five marbles are drawn, one at a time, with replacement from an urn containing 3 red marbles and 1 blue marble. Find the expected value and the variance of the number of:
  a. red marbles obtained
  b. blue marbles obtained

- 4. This example shows that the converse of the theorem of this section is false. Let the distribution of X be given as:

x	-2	0	2
$f(x)$	$\frac{1}{8}$	$\frac{3}{4}$	$\frac{1}{8}$

  a. Find the probability distribution of $Y = X^2$.
  b. Let $Z = X + Y$. Find the probability distribution of Z and show that

  $$P(Z = 0) \neq P(X = 0) \cdot P(Y = 0).$$

  (Therefore X and Y are not independent, confirming the really obvious fact that the values of X^2 "depend" on the values of X.)
  c. Show that $\text{var}(Z) = \text{var}(X + Y) = \text{var}(X) + \text{var}(Y)$.

5. Suppose that an urn contains 3 red marbles and 1 blue marble and that two draws are made without replacement. Let X represent the number of red marbles obtained on the first draw, Y the number obtained on the second draw, and Z the total number of red marbles drawn.
  a. Find $E(X)$, $E(Y)$, and $E(Z)$.
  b. Find $\text{var}(X)$, $\text{var}(Y)$, and $\text{var}(Z)$.

6.4 A Standardized Random Variable

As we mentioned in Section 6.2, we now examine the variance of $aX + b$. In order to do this, we first recall the following.

1. If X is a random variable having $f(x)$ as its probability function, and with expected value μ_X, then
   a. $E(aX + b) = a\mu_X + b = \mu_{aX+b}$ (Section 6.1)
   b. $\text{var}(X) = E(X^2) - \mu_X^2 = \sigma_X^2$ (Section 6.2)

2. The expected value of the sum of two or more random variables is the sum of the expected values. (Section 6.3)

3. The expected value of a constant c is equal to c. (Problem 9, Exercises 6.1)

The theorem we wish to prove is:

If X is a random variable with expected value μ_X and variance σ_X, then
$$\text{var}(aX + b) = a^2 \text{var}(X) = a^2\sigma_X^2.$$

Proof:
$$\begin{aligned}
\text{var}(aX + b) &= E((aX + b)^2) - \mu_{aX+b}^2 && \text{by 1b above} \\
&= E(a^2X^2 + 2abX + b^2) \\
&\quad - (a\mu_X + b)^2 && \text{by 1a and algebra} \\
&= E(a^2X^2) + E(2abX) + E(b^2) \\
&\quad - (a^2\mu_X^2 + 2ab\mu_X + b^2) && \text{by 2 and algebra} \\
&= a^2E(X^2) + 2abE(X) + b^2 \\
&\quad - (a^2\mu_X^2 + 2ab\mu_X + b^2) && \text{by 1a and 3} \\
&= a^2E(X^2) + 2ab\mu_X + b^2 \\
&\quad - a^2\mu_X^2 - 2ab\mu_X - b^2 && \text{definition of } \mu_X \\
&= a^2E(X^2) - a^2\mu_X^2 && \text{simplifying} \\
&= a^2[E(X^2) - \mu_X^2] && \text{factoring} \\
&= a^2\sigma_X^2 && \text{by 1b}
\end{aligned}$$

Since $\text{var}(aX + b) = \sigma_{aX+b}^2$, it follows as a consequence, that
$$\sigma_{aX+b} = |a|\,\sigma_X.$$

One use of this theorem is to simplify the calculation of σ^2 in certain cases.

Example 6.4.1 Suppose that we wished to find μ_X and σ_X for the distribution:

x	−35	25	85	175
f(x)	0.1	0.3	0.4	0.2

We could, of course, proceed directly, using the definitions of μ_X and σ_X. Suppose, instead, that we first subtract 25 from each value x of X, obtaining

$$-60, \quad 0, \quad 60, \quad 150.$$

Dividing each of these by 30, we have

$$-2, \quad 0, \quad 2, \quad 5.$$

Consider now a new random variable Z having these values z, with distribution corresponding to that of X:

z	−2	0	2	5
g(z)	0.1	0.3	0.4	0.2

Recalling the changes made, we see that the relation between the two random variables may be expressed as

$$Z = \frac{X - 25}{30}, \quad \text{or} \quad X = 30Z + 25.$$

It simplifies the arithmetic to deal with the smaller numbers in the second table. Here, we find:

$$\mu_Z = 1.6 \text{ (Example 6.1.1)} \quad \text{and} \quad \sigma_Z \doteq 2.11 \text{ (Example 6.2.1)}$$

Now, $X = 30Z + 25$. So

$$\mu_X = E(X) = E(30Z + 25) = 30(1.6) + 25 = 73$$

and

$$\sigma_X = \sigma_{30Z+25} \doteq 30(2.11) = 63.3.$$

Remark 1

The technique of Example 6.4.1 may be applied to finding the mean and standard deviation of a distribution of numbers (as discussed in Chapter 4).

Remark 2

Our proof was for finite, discrete random variables. The result also holds for continuous random variables.

Remark 3

The equation $Z = aX + b$ may be interpreted as a "change of variable" or a "transformation." Each value x of X is multiplied by a, and b is added to each product. For such a transformation, we note the simple relations existing between the expected value and the variance of the new random variable and the constants μ and σ of the original random variable:

$$\mu_{aX+b} = a\mu_X + b \quad \text{and} \quad \sigma_{aX+b} = |a|\sigma_X.$$

A special choice of the constants a and b yields a result which will be of importance in Chapter 8. Suppose that we choose

$$a = \frac{1}{\sigma} \quad \text{and} \quad b = -\mu.$$

Then

$$Z = \frac{1}{\sigma} X - \frac{\mu}{\sigma} = \frac{X - \mu}{\sigma}.$$

For this

$$E(Z) = \frac{1}{\sigma} E(X) - \frac{\mu}{\sigma} = \frac{1}{\sigma} \mu - \frac{\mu}{\sigma} = 0$$

and

$$\text{var}(Z) = \frac{1}{\sigma^2} \text{var}(X) = \frac{1}{\sigma^2} \cdot \sigma^2 = 1.$$

If X is any random variable with expected value μ_X and variance σ_X^2, then

$$Z = \frac{X - \mu_X}{\sigma_X}$$

is a new random variable with expected value 0 and variance 1. Since $\sqrt{1} = 1$, the standard deviation of Z is also 1. Z is called a **standardized random variable**, and the corresponding probability distribution is called a **standardized distribution**.

Example 6.4.2 Applying the above ideas, we can standardize the distribution of the random variable X of Example 6.2.1 as follows.

x	-2	0	2	5	$\mu_X = 1.6$
$f(x)$	0.1	0.3	0.4	0.2	$\sigma_X \doteq 2.11$

$$Z = \frac{X - \mu_X}{\sigma_X} \doteq \frac{X - 1.6}{2.11}$$

z	-1.7	0.76	0.2	1.6	$\mu = 0$
$g(z)$	0.1	0.3	0.4	0.2	$\sigma \doteq 1$

Compare the graphs of X and Z as shown on the next page.

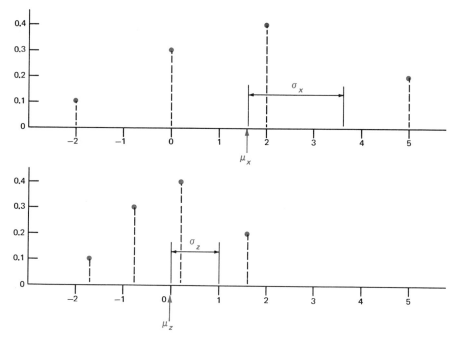

We now have an explanation for the adjective "standard" applied to the standard normal distribution of Section 5.4. We have pointed out (see Remark 2, Section 6.2) that the "standard normal distribution" does have $\mu = 0$ and $\sigma = 1$.

We remark here that in standardizing, a Bernoulli distribution remains Bernoulli, a triangular distribution remains triangular, and a uniform distribution remains uniform.

Example 6.4.3 To standardize the Bernoulli distribution

x	0	1
f(x)	q	p

we first recall that $\mu_X = p$, $\sigma_X = \sqrt{pq}$. We then let $Z = \dfrac{X - p}{\sqrt{pq}}$.

Then, since $1 - p = q$, we have the adjoining table.

z	$\dfrac{-p}{\sqrt{pq}}$	$\dfrac{q}{\sqrt{pq}}$
g(z)	q	p

Check: $\mu_Z = \dfrac{-p}{\sqrt{pq}} \cdot q + \dfrac{q}{\sqrt{pq}} \cdot p = 0$

$\sigma_Z^2 = E(Z^2) - \mu_Z^2 = \left[\left(\dfrac{-p}{\sqrt{pq}} \right)^2 \cdot q + \left(\dfrac{q}{\sqrt{pq}} \right)^2 \cdot p \right] - 0^2 = p + q = 1.$

EXERCISES 6.4

1. For $p = \frac{1}{2}$, find μ and σ and graph:
   a. the Bernoulli distribution
   b. the standardized Bernoulli distribution
2. Repeat Problem 1 for $p = \frac{3}{5}$.
3. Construct a table and a graph for the probability distribution of the standardized random variable corresponding to:

x	0	1	2	3
f(x)	$\frac{1}{8}$	$\frac{3}{8}$	$\frac{3}{8}$	$\frac{1}{8}$

4. Repeat Problem 3 for this table:

x	−1	0	1	3
f(x)	0.2	0.2	0.3	0.3

5. Compute μ_X and σ_X for the distribution shown in the table by first replacing X with $Z = aX + b$ using convenient values of a and b.

x	−305	−105	95	195	395
f(x)	0.15	0.25	0.10	0.20	0.30

6. If X has $\mu = 10$ and $\sigma = 3$, find μ_Y, σ_Y^2, and σ_Y when:
   a. $Y = X + 4$
   b. $Y = -2X + 3$
   c. $Y = \frac{X}{4} + 1$
   d. $Y = \frac{X - 10}{3}$

7. Graph μ_X and the interval $[\mu_X - \sigma_X, \mu + \sigma_X]$ for X given in Problem 6. Then graph μ_Y and the interval $[\mu_Y - \sigma_Y, \mu_Y + \sigma_Y]$ for parts a–d.

★ 6.5 What Do You Think?

In the course of the proof of the theorem of Section 6.3, we found that
$$\text{var}(X + Y) = \text{var}(X) + \text{var}(Y) + 2E((X - \mu_X)(Y - \mu_Y)).$$
We have seen that if X and Y are independent random variables, then
$$E((X - \mu_X)(Y - \mu_Y)) = 0.$$
However, another term is used in this connection:

If $E((X - \mu_X)(Y - \mu_Y)) = 0$, the random variables X and Y are said to be *uncorrelated*.

Thus, if X and Y are *independent* random variables, they are uncorrelated. However, we have also discovered that uncorrelated variables are not necessarily independent. (See Problem 4, Exercises 6.3.)

On the other hand, if $E((X - \mu_X)(Y - \mu_Y)) \neq 0$, then the random variables X and Y, are said to be **correlated**. If we know the value of this quantity, then we know by how much the variance of $X + Y$ differs from the sum of the variances of X and Y. Since this quantity has interesting properties of its own, it has been given a special name.

$E((X - \mu_X)(Y - \mu_Y))$ is called the *covariance* of random variables X and Y and is denoted by

$$\text{cov}(X, Y).$$

During the proof in Section 6.3, we obtained the formula

$$\text{cov}(X, Y) = E(XY) - \mu_X \mu_Y.$$

Thus, there are two ways to determine the value of $\text{cov}(X, Y)$:

We assume that the distributions of X and Y are known and that μ_X, μ_Y, $\text{var}(X)$, and $\text{var}(Y)$ are also known.

1. Find the probability distribution of $X + Y$. Compute $E(X + Y)$ and $\text{var}(X + Y)$. Then (page 155):

$$\text{cov}(X, Y) = \tfrac{1}{2}[\text{var}(X + Y) - \text{var}(X) - \text{var}(Y)]$$

2. Find the probability distribution of XY. Find $E(XY)$ and use the formula:

$$\text{cov}(X, Y) = E(XY) - \mu_X \mu_Y.$$

Which of these do you think is the easier method for most cases?

★ EXERCISES 6.5a

1. Let X have the distribution given by

x	1	2
f(x)	$\tfrac{1}{2}$	$\tfrac{1}{2}$

   Find $\text{cov}(X, Y)$ by both methods, where:
   a. $Y = 2X$
   b. $Y = -2X$
   c. $Y = X^2$

2. Suppose that we know that $\mu_X = 1$, $\mu_Y = 5$, and that $W = XY$ has the distribution

w	1	2	5	20
h(w)	$\frac{1}{4}$	$\frac{1}{4}$	$\frac{1}{4}$	$\frac{1}{4}$

Find cov(X, Y).

Even these simple examples illustrate that using

$$\text{cov}(X, Y) = E(XY) - \mu_X \mu_Y$$

offers some advantage. If X and Y take on many values, the computation of $E(XY)$ is easier than the computation of var$(X + Y)$, since we avoid much squaring. In any event, cov(X, Y) may be considered apart from its connection with var$(X + Y)$. (The name **covariance** reminds us of the connection.)

Earlier we have seen that

$$E(aX) = aE(X) \qquad \text{(Section 6.1)}$$

and

$$\text{var}(aX) = a^2 \text{var}(X). \qquad \text{(Section 6.4)}$$

It seems possible that cov(aX, bY) may be found if we know cov(X, Y). We also know that $E(X - \mu_X) = 0$, and that var$(X - \mu_X) = $ var(X). It is natural to ask whether knowing cov(X, Y), we can determine cov(aX, bY) and cov$(X - \mu_X, Y - \mu_Y)$ in some simple way. What do you think?

★ EXERCISES 6.5b

1. If X and Y are random variables with cov(X, Y) known, prove:
   a. cov$(X, Y) = $ cov(Y, X)
   b. cov$(aX, bY) = ab$ cov(X, Y)
   c. cov$(X - \mu_X, Y - \mu_Y) = $ cov(X, Y)

2. Find cov$(a(X - \mu_X), b(Y - \mu_Y))$ in terms of a, b, and cov(X, Y).

3. a. Let X be any random variable with μ and σ known. Find cov(X, X).
   b. Use the above result to find cov(X, Y) for the variables of Problem 1a, Exercises 6.5a.

The numerical value of cov(X, Y) is a measure of how the variables X and Y are "co-related." Unfortunately it is not a good measure, as the following discussion shows.

Suppose that the values of X and Y are obtained from some physical experiment. Perhaps the values of X represent height in inches and the values of Y, the weight in pounds of a group of men. It is reasonable to guess that height and weight are correlated; generally taller men tend to be heavier than short ones. Cov(X, Y) may be computed. However, we might have chosen different units of measurement. Suppose that we had recorded the weights in ounces; call this variable Y'. Then, $\text{cov}(X, Y') = 16 \text{cov}(X, Y)$. This is not satisfactory; surely the measure of degree of relationship should not depend on the units used.

Do you think it possible to "rig" a measure of co-relationship which would be more stable than $\text{cov}(X, Y)$? Consider the following exercises.

★ EXERCISES 6.5c

1. If X is to be replaced by $Z = X + b$, we have seen that $b = -\mu_X$ is often a "nice" choice. If $X - \mu_X$ is to be replaced by $Z = a(X - \mu_X)$, what is a nice choice of a?

2. X and Y will usually have quite different expected values and variance. In order to compare two distributions, it is convenient if they have the *same* expected value and the *same* variance. Suggest functions $X' = \phi(X)$ and $Y' = \psi(X)$* so that X' and Y' have the desired properties.

3. Prove:
$$\text{cov}\left(\frac{X - \mu_X}{\sigma_X}, \frac{Y - \mu_Y}{\sigma_Y}\right) = \frac{1}{\sigma_X \sigma_Y} \text{cov}(X, Y)$$

Correlation coefficient. "$\frac{1}{\sigma_X \sigma_Y} \text{cov}(X, Y)$" is called the "*correlation coefficient*" of the variables X and Y and is denoted by the symbol $\rho(X, Y)$.†

4. Compute $\rho(X, Y)$ for the variables of Exercises 6.5a:
   a. Problem 1a
   b. Problem 1b
   c. Problem 1c
   d. Problem 2

5. Prove: If $Y = aX + b$, $a \neq 0$, then
$$\rho(X, Y) = 1 \quad \text{if} \quad a > 0$$
and
$$\rho(X, Y) = -1 \quad \text{if} \quad a < 0.$$

Problem 5 shows that if Y is a linear function of X, then $\rho(X, Y)$ takes on the values $+1$ (see Problem 4a above) or -1 (see Problem 4b above). It can be shown that for *any* random variables
$$-1 \leq \rho(X, Y) \leq 1.$$

---
* The symbol ψ (psi) is a letter of the Greek alphabet transliterated in English as "psy."
† The symbol ρ (rho) is a letter of the Greek alphabet corresponding to the English "r".

The correlation coefficient is frequently used in statistical reports. If the points (x_i, y_i) are graphed, then $\rho = 1$ implies that these points lie along a straight line with positive slope. Similarly, $\rho = -1$ implies a straight line with negative slope. These situations are shown schematically in Figures 6.1a and 6.1b.

Figure 6.1

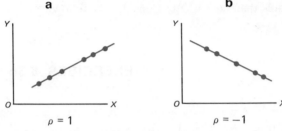

For these situations, since the functional relationship is determined (it is linear), we know precisely the value of y_i given a particular value of x_i. For $\rho = 1$, a larger value of Y accompanies a larger value of X; for $\rho = -1$, a smaller value of Y accompanies a larger value of X.

With experimental data, one would not expect to find graphs of data to be as "well-behaved" as either of those in Figure 6.1a or Figure 6.1b. In particular situations, if we graph pairs of values, we often find the points to be scattered in an elliptical pattern.

Example 6.5.1 The accompanying table gives data for several observed weather conditions in 38 cities of the United States. If a city is chosen at random, we may consider the following random variables:

T: average annual temperature (degrees Fahrenheit);
R: average annual precipitation—rain, snow, and so on (in inches);
P: average amount of annual air pollution—suspended particulate matter (in micrograms per cubic meter).

Four our purpose, actual data from the *Statistical Abstract of the United States*, 1970, 91st annual edition (Washington, D.C.: U.S. Bureau of the Census) have been rounded off to make our computations easier.

The temperature-precipitation graph shows a characteristic pattern mentioned above. (The points are scattered in an elliptical pattern.) There seems to be a tendency for a positive correlation: higher temperatures are associated with greater amounts of precipitation.

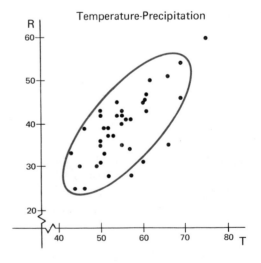

City	T	R	P
Hartford, CT	50	43	67
Wilmington, DE	54	45	125
District of Columbia	57	41	92
Miami, FL	75	60	49
Atlanta, GA	61	47	90
Chicago, IL	51	33	124
Indianapolis, IN	52	39	128
Des Moines, IA	49	30	100
Wichita, KS	57	28	60
Louisville, KY	56	41	176
New Orleans, LA	69	54	87
Baltimore, MD	55	43	110
Detroit, MI	50	31	154
Minneapolis, MN	44	25	84
Jackson, MS	66	51	80
Kansas City, MO	57	34	105
St. Louis, MO	55	35	148
Omaha, NB	52	28	151
Concord, NH	46	39	39
New York, NY	55	42	127
Charlotte, NC	61	43	130
Cincinnati, OH	55	40	110
Cleveland, OH	50	35	139
Columbus, OH	52	37	96
Oklahoma City, OK	60	31	53
Portland, OR	53	37	79
Philadelphia, PA	54	42	125
Pittsburgh, PA	50	36	180
Providence, RI	50	42	92
Sioux Falls, SD	46	25	69
Memphis, TN	62	50	81
Nashville, TN	60	45	116
Dallas, TX	66	35	76
Houston, TX	69	46	79
Burlington, VT	43	33	44
Norfolk, VA	60	45	101
Seattle, WA	51	39	73
Milwaukee, WI	45	30	165

We can calculate the correlation coefficient, $\rho(T, R)$ as shown on page 168. Notice the arrangement in our work: from the totals in the first two columns, we arrive at the expected values of T and R; the next two columns lead us to the variances; finally, the column for the product helps with finding the covariance.

Calling to mind an observation often made in a less quantitative sense that tropical rainstorms occur frequently, we are not surprised that the data show some positive correlation between temperature and precipitation.

$n = 38$

	T	R	T^2	R^2	TR
	50	43	2500	1849	2150
	54	45	2916	2025	2430
	57	41	3249	1681	2337
	75	60	5625	3600	4500
	61	47	3721	2209	2867
	51	33	2601	1089	1683
	52	39	2704	1521	2028
	49	30	2401	900	1470
	57	28	3249	784	1596
	56	41	3136	1681	2296
	69	54	4761	2916	3726
	55	43	3025	1849	2365
	50	31	2500	961	1550
	44	25	1936	625	1100
	66	51	4356	2601	3366
	57	34	3249	1156	1938
	55	35	3025	1225	1925
	52	28	2704	784	1456
	46	39	2116	1521	1794
	55	42	3025	1764	2310
	61	43	3721	1849	2623
	55	40	3025	1600	2200
	50	35	2500	1225	1750
	52	37	2704	1369	1924
	60	31	3600	961	1860
	53	37	2809	1369	1961
	54	42	2916	1764	2268
	50	36	2500	1296	1800
	50	42	2500	1764	2100
	46	25	2116	625	1150
	62	50	3844	2500	3100
	60	45	3600	2025	2700
	66	35	4356	1225	2310
	69	46	4761	2116	3174
	43	33	1849	1089	1419
	60	45	3600	2025	2700
	51	39	2601	1521	1989
	45	30	2025	900	1350
Σ	2098	1480	117,826	59,964	83,265
$\frac{\Sigma}{n}$	55.21	38.95	3101	1578	2191

$$\mu_T = \frac{2098}{38} \doteq 55.21; \quad \text{var}(T) \doteq 3101 - 3048 = 53$$

$$\sigma_T \doteq 7.28$$

$$\mu_R = \frac{1480}{38} \doteq 38.95; \quad \text{var}(R) \doteq 1578 - 1517 = 61, \quad \sigma_R \doteq 7.81$$

$$\text{cov}(T, R) \doteq 2191 - 2150 = 41$$

$$\rho(T, R) = \frac{1}{\sigma_T \sigma_R} \text{cov}(T, R) \doteq \frac{41}{(7.28)(7.81)} \doteq 0.72$$

Example 6.5.2 From the same data of Example 6.5.1, we can look into the correlation between temperature and pollution. First, we construct the graph:

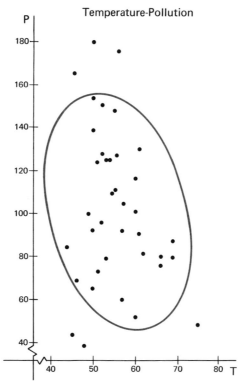

By comparison with the temperature-precipitation graph of Example 6.5.1, this temperature-pollution graph seems to show more scattering.

Using the table on the next page, we compute $\rho(T, P)$. In our computation we can take advantage of our previous work in Example 6.5.1, omitting the column for T^2, since we have already found var(T) and σ_T.

$$\mu_T \doteq 55.21; \qquad \sigma_T \doteq 7.28$$

$$\mu_P = \frac{3904}{38} \doteq 102.74; \quad \text{var}(P) \doteq 11847 - 10556 = 1291, \quad \sigma_P \doteq 35.93$$

$$\text{cov}(T, P) \doteq 5619 - 5672 = -53$$

$$\rho(T, P) = \frac{1}{\sigma_T \sigma_P} \text{cov}(T, P) \doteq -\frac{53}{(7.28)(35.93)} \doteq -0.20$$

$n = 38$

	T	P	P²	TP
	50	67	4489	3350
	54	125	15625	6750
	57	92	8464	5244
	75	49	2401	3675
	61	90	8100	5490
	51	124	15376	6324
	52	128	16384	6656
	49	100	10000	4900
	57	60	3600	3420
	56	176	30976	9856
	69	87	7569	6003
	55	110	12100	6050
	50	154	23716	7700
	44	84	7056	3696
	66	80	6400	5280
	57	105	11025	5985
	55	148	21904	8140
	52	151	22801	7852
	46	39	1521	1794
	55	127	16129	6985
	61	130	16900	7930
	55	110	12100	6050
	50	139	19321	6950
	52	96	9216	4992
	60	53	2809	3180
	53	79	6241	4187
	54	125	15625	6750
	50	180	32400	9000
	50	92	8464	4600
	46	69	4761	3174
	62	81	6561	5022
	60	116	13456	6960
	66	76	5776	5016
	69	79	6241	5451
	43	44	1936	1892
	60	101	10201	6060
	51	73	5329	3723
	45	165	27225	7425
Σ	2098	3904	450,198	213,512
$\frac{\Sigma}{n}$	55.21	102.74	11,847	5619

The correlation coefficient for Example 6.5.2 is closer to 0 than that in Example 6.5.1. This confirms our feeling that the dots are less clustered. Notice that the correlation coefficient is negative. This means that there is a slight tendency for higher temperature to be associated with lower pollution.

Example 6.5.3 Finally, we can examine the correlation between precipitation and pollution. Intuitively, we might feel that the greater amount of precipitation is associated with less pollution (like a good washing).

For the computations, again we take advantage of the previous work, listing this time, only R, P, and the product.

$n = 38$

R	P	RP
43	67	2881
45	125	5625
41	92	3772
60	49	2940
47	90	4230
33	124	4092
39	128	4992
30	100	3000
28	60	1680
41	176	7216
54	87	4698
43	110	4730
31	154	4774
25	84	2100
51	80	4080
34	105	3570
35	148	5180
28	151	4228
39	39	1521
42	127	5334
43	130	5590
40	110	4400
35	139	4865
37	96	3552
31	53	1643
37	79	2923
42	125	5250
36	180	6480
42	92	3864
25	69	1725
50	81	4050
45	116	5220
35	76	2660
46	79	3634
33	44	1452
45	101	4545
39	73	2847
30	165	4950
Σ 1480	3904	150,293
$\dfrac{\Sigma}{n}$ 38.95	102.74	3955

$\mu_R \doteq 38.95$
$\sigma_R \doteq 7.81$

$\mu_P \doteq 102.74$
$\sigma_P \doteq 35.93$

$\text{cov}(R, P)$
$\doteq 3955 - 4002$
$= -47$

$\rho(R, P)$

$= \dfrac{1}{\sigma_R \sigma_P} \text{cov}(R, P)$

$\doteq -\dfrac{47}{(7.81)(35.93)}$

$= -0.17$

The negative coefficient obtained here confirms our intuition that more precipitation is associated with less pollution (though not pronouncedly so). Comparing this coefficient with that of Example 6.5.2, we suspect that that graph would show more scattering of the dots because ρ is closer to 0. The graph below indicates that this could be the case, although the difference is so slight it is difficult to judge visually.

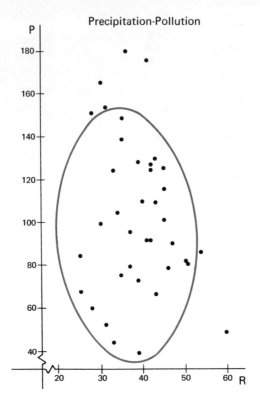

We used the data for temperature, precipitation, and pollution just to demonstrate the techniques used for obtaining coefficients of correlations. For a study of pollution, other factors might have been considered, such as proximity to various industries and to urban centers.

The observations that we have made may be generalized. The usual pattern in such "scattergrams" is elliptical. If the correlation coefficient is close to 1, the ellipse is narrow and long; if ρ is about 0.5, then the pattern is football-shaped; if close to 0, the dots are more randomly scattered with little discernible pattern.

We can see that for the kinds of calculations involved here, a table of squares, as is found at the back of this book, and a desk calculator are extremely handy. Of course, if a computer is accessible, the operations can be programmed directly.

★ COMPUTER INVESTIGATIONS

If you have access to computing equipment, try to carry out the following investigations.

1. Use a desk calculator to compute

$$\frac{\Sigma T}{n}, \quad \frac{\Sigma R}{n}, \quad \frac{\Sigma P}{n}, \quad \frac{\Sigma T^2}{n}, \quad \frac{\Sigma R^2}{n}, \quad \frac{\Sigma P^2}{n}, \quad \frac{\Sigma TR}{n}, \quad \frac{\Sigma TP}{n}, \quad \frac{\Sigma RP}{n}$$

for the first 10 cities in the table on page 167 and make a note of the time it takes.

2. Write a program for an electronic computer and find the values listed in Problem 1. Note the time required.

3. Compare the time required in Problem 2 with that in Problem 1. How would the times probably compare if the computations were made for more than 10 cities?

Chapter Summary

1. **a.** If $Z = \phi(X)$, then $E(Z) = \sum_{i=1}^{n} \phi(x_i) \cdot f(x_i)$.

   b. If X is a random variable with expected value $E(X)$, or μ_X, then
   $$E(aX + b) = aE(X) + b,$$
   or
   $$\mu_{aX+b} = a\mu_X + b,$$
   where a and b are arbitrary constants.

2. **a.** The *variance of a random variable* X is
   $$\sigma_X^2 = \text{var}(X) = E((X - \mu_X)^2) = E(X^2) - \mu_X^2 = \mu_{X^2} - \mu_X^2.$$

   b. The *standard deviation of a random variable* X is
   $$\sigma_X = \sqrt{\sigma_X^2}.$$

   c. *Chebyshev's Inequality:* If X is a random variable, then at most $\frac{1}{k^2}$ of the probability is distributed outside the interval $[\mu - k\sigma_X, \mu + k\sigma_X]$, $k \geq 1$. That is,
   $$P(|X - \mu_X| > k\sigma_X) \leq \frac{1}{k^2}.$$

3. **a.** If X and Y are any two random variables, then

$$E(X + Y) = E(X) + E(Y),$$

or

$$\mu_{X+Y} = \mu_X + \mu_Y.$$

b. If X and Y are independent random variables, then

$$\text{var}(X + Y) = \text{var}(X) + \text{var}(Y),$$

or

$$\sigma^2_{X+Y} = \sigma^2_X + \sigma^2_Y.$$

c. For n independent Bernoulli trials,

$$E(X_1 + X_2 + \cdots + X_n) = np$$

and

$$\text{var}(X_1 + X_2 + \cdots + X_n) = npq.$$

4. **a.** If X is a random variable, then

$$\text{var}(aX + b) = a^2 \text{var}(X) = a^2 \sigma^2_X$$

and

$$\sigma_{aX+b} = |a|\sigma_X.$$

b. If X is a random variable, then

$$Z = \frac{X - \mu_X}{\sigma_X}$$

is a random variable with $\mu_Z = 0$ and $\sigma_Z = 1$. Z is called a *standardized random variable*.

★ 5. **a.** If X and Y are two random variables, then

$$E((X - \mu_X)(Y - \mu_Y)), \quad \text{or} \quad E(XY) - \mu_X\mu_Y,$$

is called the *covariance* of the random variables X and Y and is denoted by $\text{cov}(X, Y)$.

b. If $\text{cov}(X, Y) = 0$, then X and Y are said to be *uncorrelated*.

c. For random variables X and Y,

$$\frac{1}{\sigma_X \sigma_Y} \text{cov}(X, Y)$$

is called the *correlation coefficient* and is denoted by the symbol $\rho(X, Y)$.

Chapter Review

1. A game is played by tossing a die. The player wins, in cents, 4 times the number thrown, plus 3¢. If the player is charged 20¢ per game:
   a. What is the expected value of the game in cents? (Hint: If X is a random variable representing the number tossed, $E(X) = \frac{7}{2}$.)
   b. In one hour, a certain player participated in 100 of the above games. If he pays 20¢ per game, how much does he expect to gain or lose in 100 games?

2. A spinner dial has 3 congruent sectors, marked 1, 2, 3, respectively. If X is a random variable corresponding to the number of the sector, determine whether or not the following is true:
$$[E(X)]^2 = E(X^2)$$

3. Let X be a random variable having probability distribution defined by the following table for $f(x)$.

x	−1	0	3	5

Let Z be a random variable related to X by $Z = 2X + 3$.
   a. Expand the given table to show the corresponding values of X, Z, X^2, Z^2.
   b. Find $E(X)$, $E(Z)$.
   c. Find $E(X^2)$.
   d. Find $E(Z^2)$.

4. For the probability distribution in Problem 3, find $\text{var}(X)$:
   a. by $E(X - \mu)^2$
   b. by using the results of Problems 3b and 3c

5. An urn contains 3 red marbles and 1 blue marble. Two marbles are drawn, one at a time without replacement. Let X be a random variable corresponding to the number of red marbles obtained on the first draw, Y the number on the second draw, and Z the total number of red marbles obtained in the two draws.
   a. Draw a tree diagram assigning probabilities to the outcomes.
   b. Construct tables showing the probability distributions for X, Y, Z.
   c. Verify that
$$\mu_Z = \mu_X + \mu_Y.$$
   d. Determine whether or not it is true that
$$\text{var}(Z) = \text{var}(X) + \text{var}(Y).$$

6. If X is a random variable with expected value 100 and $E(X^2) = 10225$, find σ_X.
7. X and Y are random variables related by $Y = aX + b$, having expected values, $\mu_X = 5$, $\mu_Y = -5$.
   a. If Y has value 1 when X has value 2: find the values of a and b.
   b. If $\sigma_X = 3$, find σ_Y.
8. If $\mu_X = 12$, $\sigma_X = 4$, and $Y = \dfrac{X - 8}{3}$, find:
   a. μ_Y b. σ_Y^2 c. σ_Y
9. If $\mu_X = 12$ and $\sigma_X = 4$, find a related random variable, $Z = X + b$ such that $\mu_Z = 0$.
10. A certain Bernoulli random variable, X, has $p = \frac{4}{7}$. Find the standardized random variable, Z, for this distribution.

Chapter 7

The Binomial Distributions

At several places in the preceding chapters, we have met the idea of a Bernoulli random variable (defined in Section 5.2). Such a variable has two values (usually 0 and 1) and a probability distribution given by:

x	0	1
$f(x)$	q	p

$+ q = 1$

We have seen that the variable has expected value p (Section 5.5) and variance pq (Section 6.2).

Moreover, we have discussed the meaning of repeated, independent Bernoulli trials (first mentioned in Section 3.5). If $X_1, X_2, \ldots, X_n$ are independent Bernoulli variables with the same distribution, then we have learned (Section 6.3) that the sum, X, of these variables is itself a random variable and that

$$X = X_1 + X_2 + \cdots + X_n \quad \text{has} \quad E(X) = np, \quad \text{var}(X) = npq.$$

We have not, however, investigated the nature of the probability distribution of such an X. This distribution is an especially important discrete distribution, and the present chapter is devoted to its development.

7.1 Definition of a Binomial Distribution

Repeated, independent Bernoulli trials arise in a wide variety of situations. The simplest experiment is an urn problem of repeated drawings with replacement. We recall that a Bernoulli trial must result in exactly one of two outcomes, often described by values 0 and 1. It is usual to refer to the result which yields 1 as a "success." The alternate outcome then becomes a "failure."

We are interested in making repeated trials and in examining the probability distribution of the *sum X* of the independent Bernoulli variables. A value x, of X, is the number of successes obtained in a series of trials. For a particular case, with only a few trials, we may determine the distribution of X by using a tree diagram.

Example 7.1.1 An urn contains 6 red and 2 blue marbles. We draw three times, with replacement. We use X_i as the random variable corresponding to the ith draw. $X_i = 1$ if a red marble is selected on the ith draw. To find the probability distribution of $X = X_1 + X_2 + X_3$, we proceed as follows, remembering that the trials are independent.

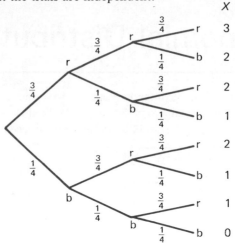

The outcomes of the three trials may be written as

$$\{rrr, rrb, rbr, rbb, brr, brb, bbr, bbb\}$$

or as

$$\{111, 110, 101, 100, 011, 010, 001, 000\}$$

(with the understanding, of course, that by "101," for example, we mean red on the first and third trials, blue on the second).

It is not difficult to assign probabilities. For example,

$$P(\{101\}) \text{ means } P(X_1 = 1 \text{ and } X_2 = 0 \text{ and } X_3 = 1).$$

Since the trials are independent,

$$P(X_1 = 1 \text{ and } X_2 = 0 \text{ and } X_3 = 1)$$
$$= P(X_1 = 1) \cdot P(X_2 = 0) \cdot P(X_3 = 1).$$

So

$$P(\{101\}) = \tfrac{3}{4} \cdot \tfrac{1}{4} \cdot \tfrac{3}{4} = \tfrac{9}{64}.$$

In tabular form, we have:

Outcome	111	110	101	100	011	010	001	000
Probability	$\tfrac{27}{64}$	$\tfrac{9}{64}$	$\tfrac{9}{64}$	$\tfrac{3}{64}$	$\tfrac{9}{64}$	$\tfrac{3}{64}$	$\tfrac{3}{64}$	$\tfrac{1}{64}$

Finally:

$$P(X = 3) = P(\{111\}) = \tfrac{27}{64}$$
$$P(X = 2) = P(\{110, 101, 011\}) = 3 \cdot \tfrac{9}{64} = \tfrac{27}{64}$$

$$P(X = 1) = P(\{100, 010, 001\}) = 3 \cdot \tfrac{3}{64} = \tfrac{9}{64}$$
$$P(X = 0) = P(\{000\}) = \tfrac{1}{64}$$

Check: $\tfrac{27}{64} + \tfrac{27}{64} + \tfrac{9}{64} + \tfrac{1}{64} = 1$

The probability distribution of X is thus:

x	3	2	1	0
$f(x)$	$\tfrac{27}{64}$	$\tfrac{27}{64}$	$\tfrac{9}{64}$	$\tfrac{1}{64}$

We may generalize. Suppose that we have n independent Bernoulli trials. There will be 2^n outcomes (see Problem 4, Exercises 2.2), each consisting of an ordered array of 0's and 1's. For $n = 8$, there would be 256 outcomes, one of which is $\{01101000\}$, and

$$P(\{01101000\}) = q \cdot p \cdot p \cdot q \cdot p \cdot q \cdot q \cdot q = p^3 q^5.$$

It is easy to see that every outcome which contains three 1's and five 0's has probability $p^3 q^5$. If $X = X_1 + X_2 + \cdots + X_8$, the outcome $\{01101000\}$ leads to $X = 3$.

If we visualize a tree diagram for $n = 8$, we see that there is a total of 256 paths. The particular path 01101000 yields $X = 3$. Our problem is to find $P(X = 3)$. Since each path leading to $X = 3$ has probability $p^3 q^5$, our problem is reduced to counting the number of these paths.

In general, for a fixed value of n (n independent trials), $P(X = x)$ will be found by multiplying the number of paths leading to $X = x$ by $p^x q^{n-x}$.

We shall use the symbol

$$\binom{n}{x}$$

to denote the number of paths (the number of outcomes) that lead to $X = x$.

So

$$P(X = x) = \binom{n}{x} p^x q^{n-x}, \qquad x \in \{0, 1, \ldots, n\}.$$

For example, $\binom{8}{3}$ is the number of outcomes in the set of $2^8 = 256$ outcomes that lead to $X = 3$. Thus, for $n = 8$,

$$P(X = 3) = \binom{8}{3} p^3 q^5.$$

Our task is to learn how to find values of $\binom{n}{x}$. For small values of n, we have the following results:

Figure 7.1

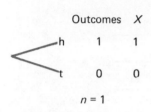

$n = 1$

Figure 7.2

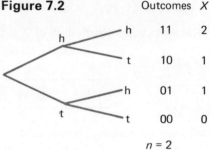

$n = 2$

$n = 1$: the outcomes are $\{0, 1\}$ (see Figure 7.1); the total number of outcomes in the set is 2^1:

1 of these outcomes leads to $X = 0$; so $\binom{1}{0} = 1$;

1 of these outcomes leads to $X = 1$; so $\binom{1}{1} = 1$.

$n = 2$: the outcomes are $\{00, 01, 10, 11\}$ (see Figure 7.2); the total number of outcomes in the set is 2^2:

1 of these leads to $X = 0$; $\binom{2}{0} = 1$;

2 of these lead to $X = 1$; $\binom{2}{1} = 2$;

1 of these leads to $X = 2$; $\binom{2}{2} = 1$.

$n = 3$: the outcomes are $\{000, 001, 010, 100, 011, 101, 110, 111\}$; the total number of outcomes in the set is 2^3:

1 of these leads to $X = 0$; $\binom{3}{0} = 1$;

3 of these lead to $X = 1$; $\binom{3}{1} = 3$;

3 of these lead to $X = 2$; $\binom{3}{2} = 3$;

1 of these leads to $X = 3$; $\binom{3}{3} = 1$.

Let us consider $n = 4$. It is convenient to relate the results for 4 trials to the known results for 3 trials:

To obtain 0 successes in 4 trials we must have

0 successes in 3 trials followed by another failure (or 3 zeros followed by a 0).

This can happen only along the path 0000. Hence,

$$\binom{4}{0} = 1.$$

To obtain 1 success in 4 trials, we need either

(a) 0 successes in 3 trials followed by a success

or

(b) 1 success in 3 trials followed by a failure.

That is,

$$\binom{4}{1} = \binom{3}{0} + \binom{3}{1} = 1 + 3 = 4.$$

To obtain 2 successes in 4 trials, we need either

(a) 1 success in 3 trials followed by a success

or

(b) 2 successes in 3 trials followed by a failure.

This leads to

$$\binom{4}{2} = \binom{3}{1} + \binom{3}{2} = 3 + 3 = 6.$$

Similarly,

$$\binom{4}{3} = \binom{3}{2} + \binom{3}{3} = 3 + 1 = 4.$$

Of course $\binom{4}{4}$ means that we have the single outcome, $\{1111\}$.

$$\binom{4}{4} = 1.$$

The argument is perfectly general:

To obtain x successes in n trials, we need either

(a) $x - 1$ successes in $n - 1$ trials followed by a success

or

(b) x successes in $n - 1$ trials followed by a failure.

So we have the formula:

$$\binom{n}{x} = \binom{n-1}{x-1} + \binom{n-1}{x}, \quad \text{for } 0 < x < n$$

This formula enables us to build a table of values of $\binom{n}{x}$ such as that shown in Figure 7.3 on the next page. A table developed in this way is said to be worked out *recursively*.

Figure 7.3 shows such a table extended to $n = 10$.

Figure 7.3

Values of $\binom{n}{x}$

n \ x	0	1	2	3	4	5	6	7	8	9	10
1	1	1									
2	1	2	1								
3	1	3	3	1							
4	1	4	6	4	1						
5	1	5	10	10	5	1					
6	1	6	[15]	20	15	6	1				
7	1	7	28	[35]	35	21	7	1			
8	1	8	28	56	70	56	28	8	1		
9	1	9	36	84	126	126	84	36	9	1	
10	1	10	45	120	210	252	210	120	45	10	1

Let us review how the entries in the table were obtained.

First, we note that by our previous argument, $\binom{n}{0} = 1$ and $\binom{n}{n} = 1$ for every value of n.

Second, we recall that we previously computed directly the first four lines of the table. We work from these entries, using the formula

$$\binom{n}{x} = \binom{n-1}{x-1} + \binom{n-1}{x}$$

to fill in new lines. Thus

$$\binom{5}{2} = \binom{4}{1} + \binom{4}{2} = 4 + 6 = 10.$$

Having completed line 5, we use the formula again and again to get lines 6, 7, 8, 9, 10, ...

As a final example, $\binom{7}{3} = 35$ is obtained by noting that

$$\binom{7}{3} = \binom{6}{2} + \binom{6}{3} = 15 + 20.$$

These entries are ringed in the table.

Using the table of Figure 7.3, we can now compute the probabilities of $X = x$ for any number of trials up to 10, provided of course, that the value of p is known. For example, if $n = 7$, $p = 0.4$, then

$$P(X = 3) = \binom{7}{3}(0.4)^3(0.6)^4 = 35(0.064)(0.1296) \doteq 0.2903.$$

To summarize:

If $X_1, X_2, \ldots, X_n$ are independent Bernoulli variables, each having $P(X_i = 1) = p$ and $P(X_i = 0) = q$, then the probability function of $X = X_1 + X_2 + \cdots + X_n$ is defined by

$$f(x) = P(X = x) = \binom{n}{x} p^x q^{n-x}, \qquad x \in \{0, 1, \ldots, n\},$$

and

$$E(X) = np \quad \text{and} \quad \text{var}(X) = npq.$$

This distribution is called a *binomial distribution*.

Example 7.1.2 A marksman shooting at a target knows from previous experience that his probability of hitting the bull's-eye is 0.20. He takes 6 shots. If we assume that these shots constitute independent trials, then the probability distribution of the random variable representing the number of bull's-eye is:

x	0	1	2	3	4	5	6
$f(x)$	0.2621	0.3932	0.2458	0.0819	0.0154	0.0015	0.0001

Check:

$$0.2621 + 0.3932 + 0.2458 + 0.0819 + 0.0154 + 0.0015 + 0.0001 = 1$$

A typical computation is:

$$f(2) = P(X = 2) = \binom{6}{2}(0.2)^2(0.8)^4$$

$$= 15(0.04)(0.4096) = 0.24576$$

That is, the probability of getting 2 bull's-eyes in 6 shots is about $\frac{1}{4}$.

Example 7.1.3 To find the probability that the marksman of Example 7.1.2 makes at most one bull's-eye, we have:

$$P(X \leq 1) = P(X = 0) + P(X = 1) = 0.2621 + 0.3932 = 0.6553$$

The probability that he makes two or more bull's-eyes is

$$P(X \geq 2) = 1 - P(X \leq 1) = 1 - 0.6553 = 0.3447$$

Check: $0.2458 + 0.0819 + 0.0154 + 0.0015 + 0.0001 = 0.3447$

It is clear that the distribution depends on the values of n and p. Each choice of n and p gives rise to a different distribution. (n and p are often called the **parameters** of the distribution.) To emphasize this, we use this notation for the binomial distribution:

$$b(n, p; x) = \binom{n}{x} p^x q^{n-x}.$$

In Section 7.4 we shall discuss such distributions in more detail. Before doing so, however, it is necessary to examine the numbers $\binom{n}{x}$ more carefully. How, for example, may we find the value of $\binom{14}{5}$, $\binom{26}{12}$, $\binom{52}{5}$, or $\binom{100}{98}$? Enlarging the table of Figure 7.3 to $n = 100$ is hardly practical if another method can be found.

After developing some preliminary ideas in the next section, we shall again turn our attention to $\binom{n}{x}$, which is called a **binomial coefficient**. Meanwhile, we may use Figure 7.3 if we are dealing with $n \leq 10$.

EXERCISES 7.1

- **1.** Find the probability distribution of X, when X represents the number of heads obtained if a fair coin is tossed
   a. 5 times **b.** 10 times
- **2.** Let X represent the number of sixes that are obtained if 3 fair dice are thrown.
   a. Find the probability distribution of X.
   b. We know that $E(X) = np = 3 \cdot \frac{1}{6} = \frac{1}{2}$. Verify this by computing $E(X)$ directly from part a.
   c. Verify that $\text{var}(X) = npq$.
- **3.** An urn contains 6 red and 2 blue marbles. Four marbles are drawn, one at a time, with replacement.
   a. Find the probability distribution of X if X represents the number of red marbles drawn.
   b. Determine $E(X)$.
   c. Determine σ_X.
- **4.** Repeat Problem 3a for Y, representing the number of blue marbles drawn. Compare.
- **5.** A baseball player has a batting average of 0.300. He comes to bat 4 times in a game. Assuming each "at bat" constitutes an independent Bernoulli trial, find the probability distribution of X, representing the number of hits he makes.
- **6.** A quiz consists of 6 multiple-choice questions, each offering 5 alternate answers. If a student guesses each answer at random, find
   a. the probability distribution of the number of questions answered correctly;
   b. the probability that the student "passes" if 4-or-more-correct is needed to pass.
- **7.** Verify that $\sum_{x=0}^{6} \binom{6}{x} = 2^6$.

8. In Figure 7.3, find:
   a. $\binom{7}{3}, \binom{7}{4}$ b. $\binom{6}{2}, \binom{6}{4}$ c. $\binom{9}{3}, \binom{9}{6}$
9. Find $P(X = 2)$ when $n = 5$ and $p = \frac{2}{5}$.
10. Find $P(X = 3)$ when $n = 7$ and $p = \frac{2}{7}$.
11. A baseball player has a batting average of 0.200. Using the assumption of Problem 5, find the probability that in 4 times at bat during the game, he would have:
    a. exactly one hit   b. at least one hit

7.2 Permutations and Combinations

In order to develop a formula for finding $\binom{n}{x}$, we need the ideas that we shall develop in this section.

Example 7.2.1 In Example 3.1.1 we considered a 12-member congressional committee. A chairman was to be chosen by lot, and so the chairman may be any one of the 12. Now suppose that a vice-chairman is to be chosen from the remainder. Since the chairman has already been selected, there are 11 possibilities for the vice-chairmanship. Using the multiplication principle (Section 2.2), we conclude that there are $12 \cdot 11 = 132$ ways in which the two offices may be filled. If a secretary is to be chosen, there are 10 possibilities remaining for this choice. Hence, there are $12 \cdot 11 \cdot 10 = 1320$ ways in which the three offices may be filled.

The argument of Example 7.2.1 can be made more general. If, of n objects, we choose 3 to be placed in a particular order, this may be done in

$$n(n - 1)(n - 2) \text{ ways.}$$

Similarly, an ordered selection of x of the n objects may be made in

$$n(n - 1)(n - 2) \cdots (n - x + 1) \text{ ways.}$$

In order to write such products conveniently, we define

$$n! = n \cdot (n - 1) \cdot (n - 2) \cdot \cdots \cdot 3 \cdot 2 \cdot 1,$$

where the symbol $n!$ is read "n factorial." For example:

$$3! = 3 \cdot 2 \cdot 1 = 6$$

$$4! = 4 \cdot 3 \cdot 2 \cdot 1 = 24$$

We also shall define

$$0! = 1.$$

(The reason for this last definition will become apparent shortly.) Notice that $(n + 1)! = (n + 1)(n!)$ for $n \geq 1$, and with the definition $0! = 1$ we have
$$1! = 1 \cdot 0! = 1,$$
which follows the same pattern.

Now
$$12 \cdot 11 \cdot 10 = \frac{12 \cdot 11 \cdot 10 \cdot 9 \cdot 8 \cdot 7 \cdot 6 \cdot 5 \cdot 4 \cdot 3 \cdot 2 \cdot 1}{9 \cdot 8 \cdot 7 \cdot 6 \cdot 5 \cdot 4 \cdot 3 \cdot 2 \cdot 1}$$
$$= \frac{12!}{9!} = \frac{12!}{(12 - 3)!}$$

In general, we see that
$$n(n - 1)(n - 2) \cdots (n - x + 1) = \frac{n!}{(n - x)!}.$$

An ordered arrangement of objects is often called a *permutation* of the objects. A symbol for the number of permutations is
$$_nP_x,$$
where *x* represents the number of objects to be arranged in order, and *n* reminds us that these objects are chosen from an original set of *n* objects.

(Here, *P* does *not* refer to a probability.) Thus, we can now write:
$$_nP_x = \frac{n!}{(n - x)!}.$$

In particular, we have
$$_nP_n = \frac{n!}{0!} = n! \quad \text{(since } 0! = 1\text{)}.$$

Example 7.2.2 In an 8-man race (if we ignore the possibility of ties) there are
$$_8P_8 = 8! = 8 \cdot 7 \cdot 6 \cdot 5 \cdot 4 \cdot 3 \cdot 2 \cdot 1 = 40{,}320$$
possible orders of finish.

If gold, silver, and bronze medals are to be given to the first three in the race, there are
$$_8P_3 = \frac{8!}{5!} = 8 \cdot 7 \cdot 6 = 336 \text{ ways}$$
in which the medals may be awarded.

Suppose now that we wish to make a selection without specifying the order.

A selection of *x* objects taken from a set of *n* objects without regard for the order of the selection is called a *combination*. A symbol for the number of combinations is
$$_nC_x.$$

Example 7.2.3 In Example 7.2.1 we obtained the number of ways an *ordered* selection may be made for the set of three officers chosen from a 12-member committee as $\frac{12!}{9!}$; that is,

$$_{12}P_3 = \frac{12!}{9!}.$$

We can also approach this task a bit differently. We can find the number of ways to select 3 names out of 12 without specifying the order. This is denoted by $_{12}C_3$. For *each* such selection of 3, there are $_3P_3 = 3!$ ways in which the names may be ordered. Therefore, the jobs of chairman, vice-chairman, and secretary may be filled in $3! \cdot {}_{12}C_3$ ways. But we know the total number of ways is $\frac{12!}{9!}$. It follows that

$$3! \cdot {}_{12}C_3 = {}_{12}P_3 = \frac{12!}{9!},$$

or

$$_{12}C_3 = \frac{_{12}P_3}{3!} = \frac{12!}{3! \cdot 9!} = \frac{12 \cdot 11 \cdot 10}{3 \cdot 2 \cdot 1} = 220.$$

The argument in Example 7.2.3 is general:

The number of ways to select x objects out of n objects without regard for the order of selection is

$$_nC_x = \frac{_nP_x}{x!} = \frac{n!}{x! \cdot (n-x)!}.$$

In summary, $_nP_x$ represents the number of permutations (ordered selections) of x objects out of n. $_nC_x$ represents the number of combinations (unordered selections) of x objects out of n. These ideas, together with the fundamental multiplication principle, enable us to do "sophisticated counting" of outcomes.

Example 7.2.4 A 3-man executive committee is to be selected at random from the 5 Republicans and 7 Democrats of Example 3.1.1. We have found that there are $_{12}C_3$, or 220, equally likely outcomes. To find the probability that 2 Republicans and 1 Democrat are selected, we need to count the outcomes favorable to that event. We may think of this in two steps. Two Republicans out of 5 must be selected. This may be done in $_5C_2$, or 10, ways. Then one Democrat out of 7 must be selected. This may be done in $_7C_1$, or 7, ways. By the multiplication principle, both jobs may be done in

$$_5C_2 \cdot {}_7C_1 = 10 \cdot 7 = 70 \text{ ways.}$$

Therefore,

P(2 Republicans and 1 Democrat are selected)

$$= \tfrac{70}{220} = \tfrac{7}{22}.$$

Example 7.2.5 There are $_{52}C_5$ possible 5-card hands from an ordinary deck of cards:

$$_{52}C_5 = \frac{52 \cdot 51 \cdot 50 \cdot 49 \cdot 48}{5 \cdot 4 \cdot 3 \cdot 2 \cdot 1} \; (= 2{,}598{,}960)$$

There are $_{13}C_5$ possible 5-card selections from the subset of 13 spades:

$$_{13}C_5 = \frac{13 \cdot 12 \cdot 11 \cdot 10 \cdot 9}{5 \cdot 4 \cdot 3 \cdot 2 \cdot 1} \; (= 1287)$$

Hence,

$$P(5 \text{ spades}) = \frac{_{13}C_5}{_{52}C_5} = \frac{13 \cdot 12 \cdot 11 \cdot 10 \cdot 9}{52 \cdot 51 \cdot 50 \cdot 49 \cdot 48} \doteq 0.0005.$$

A "flush" is 5 cards of any suit. To count the total number of flushes, we multiply $P(5 \text{ spades})$ by 4, since we may have a flush in any one of the four possible suits.

$$P(\text{flush}) \doteq 4(0.0005) = 0.002$$

A formula would be:

$$P(\text{flush}) = \frac{_4C_1 \cdot _{13}C_5}{_{52}C_5}$$

EXERCISES 7.2

- 1. Find $_7P_2$, $_7P_5$, $_7C_2$, $_7C_5$.
- 2. Find $_{13}P_2$, $_{13}P_{11}$, $_{13}C_2$, $_{13}C_{11}$.
- 3. In a certain card game a player is dealt 3 cards from a 52-card deck. How many different hands are possible if:
  a. the order in which the cards are received is important?
  b. the order is not important?
- 4. It is desired to make up some 5-letter code words using only the first 10 letters of the alphabet. If no repetitions of letters are allowed, how many such code words can be made up?
- 5. Suppose that a library has 12 books on a "special interest" shelf. How many pairs of those books may be selected?
- 6. Suppose that boxes are to be filled with 3 different kinds of candy. If 6 kinds of candy are available, in how many different ways can the boxes be filled? (Order is not important.)
- 7. The individual books of a four-volume set of Shakespeare are placed, at random, next to each other on a shelf. What is the probability that Volumes I, II, III, IV are in proper order?
- 8. A box contains 10 articles of which 3 are defective. Five articles are selected from the box at random. What is the probability that:
  a. none of the five is defective?
  b. exactly one is defective?
  c. exactly two are defective?
  d. exactly three are defective?

- **9.** Continue Example 7.2.4 to determine the probability that the executive committee consists of:
   a. 3 Republicans
   b. 1 Republican and 2 Democrats
   c. 3 Democrats
- **10.** An old slang term for a person who overstates his abilities, wealth, or resources is "four-flusher." Compare the probability of drawing four cards of one suit and a fifth card of a different suit with the probability of drawing five cards from one suit, that is, P(flush). (Hint: Choose a suit, choose four cards from this suit, choose one card from the other three suits.)

  11. A man is dealt a hand of 5 cards from a standard deck of 52 cards. He got a 7, 9, 10, J, Q (mixed suits). He is allowed to trade any of his cards for an equal number randomly selected from the rest of the deck.
   a. If he discards the 7, what is the probability of his getting a "straight" (five consecutive cards)?
   b. If he discards the queen, what is the probability of his getting a straight?
   c. Three of his cards are spades and two are hearts. If he discards the two hearts, what is the probability of his getting a flush?
   d. If he discards the two hearts, what is the probability of his getting another two hearts?
   e. What is the ratio of the probability in part d to the probability in part c?

   12. A rooming house has 8 single rooms available for rent. 8 men and 6 women applied for the rooms. If the applicants are selected by lot:
   a. What is the probability that the rooms will be rented to the men only?
   b. What is the probability that the rooms will be rented to 4 men and 4 women?

   13. The toll on the San Francisco-Oakland Bay Bridge was 25¢. In a coin box are kept seven coins: 1 quarter, 2 dimes, 3 nickels, and 1 penny. The driver reaches into the coin box and draws out 3 coins at random. What is the probability that the amount drawn is less than enough to pay the toll?

★ COMPUTER INVESTIGATIONS

If you have access to an electronic computer, try to carry out the following investigations.

1. Verify that the formula for the number of combinations of n objects taken x at a time may be written as:

$$_nC_x = \frac{n(n-1)(n-2)\cdots[n-(x-1)]}{x!}$$

Write a computer program for this formula.

2. Verify that the formula for the number of combinations of n objects taken x at a time may be written as:

$$_nC_x = \frac{n}{1} \cdot \frac{n-1}{2} \cdot \frac{n-2}{3} \cdot \ldots \cdot \frac{n-(x-1)}{x}$$

Write a computer program for this formula.

3. Test the formulas in Problems 1 and 2 with increasingly large values of n and x and compare the results.

7.3 The Binomial Coefficients

An array such as that of Figure 7.3 is usually called Pascal's Triangle. Although it was known before his time, Blaise Pascal (1623–1662), a French mathematician, discovered many remarkable properties of the numbers $\binom{n}{x}$ in it. Several of these properties are important for us. We have already noted (Section 7.1) three of these:

1. The sum of the entries for each value of n is 2^n.
2. The values of $\binom{n}{0}$ and $\binom{n}{n}$ are both 1 regardless of the size of n.
3. $\binom{n}{x} = \binom{n-1}{x-1} + \binom{n-1}{x}$, $\quad 0 < x < n$.

A fourth property is also apparent from the table. We note, for example, that

$$\binom{8}{0} = \binom{8}{8}, \quad \binom{8}{1} = \binom{8}{7}, \quad \binom{8}{2} = \binom{8}{6}, \quad \binom{8}{3} = \binom{8}{5}.$$

That is, for example, the number of paths leading to 3 ones and 5 zeros is the same as the number of paths leading to 5 ones and 3 zeros (recall Section 7.1). In general:

4. $\binom{n}{x} = \binom{n}{n-x}$ (For proof, see Problem 4, Exercises 7.3.)

Our immediate task is to determine a method of finding the value of $\binom{n}{x}$ for arbitrary n and x ($0 \le x \le n$). In Section 7.2 we developed a formula for $_nC_x$. How is $_nC_x$ connected to $\binom{n}{x}$?

Suppose that we wish to find $\binom{12}{3}$. We are concerned with those outcomes (each a string of 12 zeros and ones) which consist of 9 zeros and 3 ones. In how many ways may we place the 3 ones in the 12 available positions? The answer, of course, is $_{12}C_3$. But this is also the desired value of $\binom{12}{3}$—the number of outcomes containing precisely 3 ones.

Therefore,
$$\binom{12}{3} = {}_{12}C_3 = \frac{12!}{3! \cdot 9!} = 220.$$

Our formula for $\binom{n}{x}$ is thus:

$$\binom{n}{x} = {}_nC_x = \frac{n!}{x! \cdot (n-x)!}.$$

Remark

We have found a formula for $\binom{n}{x}$. In theory, we can find the value of, say, $\binom{100}{50}$ by performing the necessary arithmetic. The computations, however, would be extremely tedious except for a computer. However, certain approximation techniques have been developed, and in Chapter 8 we shall discuss a method of approximating

$$b(n, p; x) = \binom{n}{x} p^x q^{n-x}$$

for large values of n.

Perhaps you have been wondering why we have been using the familiar term "binomial" in this context. (The word "binomial" comes from Latin and means "two names." You have, no doubt, used it in algebra to refer to a sum of two terms, such as $x + y$, $r + b$, $4x^2 - 3$, $a + 2b$, $p + q$.)

Let us now examine $(r + b)^3$. We may express this differently by expanding it or "multiplying out." To do this, we make repeated use of the distributive property. Thus:

$$(r + b)^3 = (r + b)(r + b)(r + b)$$
$$= r(r + b)(r + b) + b(r + b)(r + b)$$
$$= r \cdot r(r + b) + r \cdot b(r + b) + b \cdot r(r + b) + b \cdot b(r + b)$$
$$= rrr + rrb + rbr + rbb + brr + brb + bbr + bbb$$

Compare this result with the tree diagram in Figure 7.4.

Figure 7.4

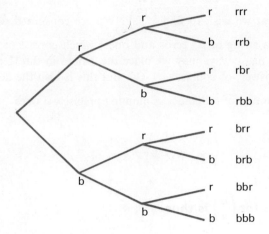

It is clear that our tree has 2^3, or 8, paths. Each path leads to a three-letter "word" consisting of r's and b's. The similarity to the tree diagram of Example 7.1.1 is obvious. We need not repeat the argument that the number of paths that pass through two r's and one b is precisely $\binom{3}{2} = 3$.

In general, the expansion of $(r + b)^n$ results in a sum of 2^n terms, $\binom{n}{x}$ of which are the product of x r's and $(n - x)$ b's. Thus:

$$(r + b)^n = \binom{n}{n} r^n + \binom{n}{n-1} r^{n-1} b + \cdots + \binom{n}{1} r b^{n-1} + \binom{n}{0} b^n$$

Since $\binom{n}{x}$ is the coefficient of the term $r^x b^{n-x}$ in the expansion of the binomial $(r + b)$ raised to the nth power, $\binom{n}{x}$ is called a *binomial coefficient*.

In particular, the binomial probability distribution may be obtained by considering

$$(p + q)^n = \binom{n}{n} p^n + \binom{n}{n-1} p^{n-1} q + \cdots$$

$$+ \binom{n}{x} p^x q^{n-x} + \cdots + \binom{n}{0} q^n.$$

EXERCISES 7.3

- 1. Extend the table of Figure 7.3 for $n = 11$ and $n = 12$. Thus verify that $\binom{12}{3} = 220$. (Compare Example 7.2.3.)

- 2. Compute $\binom{11}{6}$ directly and compare with your tabulated entry of Problem 1.

- 3. Find $\binom{100}{2}$, $\binom{100}{98}$, $\binom{52}{3}$, $\binom{52}{49}$.
- 4. Prove: $\binom{n}{x} = \binom{n}{n-x}$.
- 5. Find $\binom{25}{0}$, $\binom{50}{50}$.
- 6. Verify from the formula that $\binom{n}{0} = \binom{n}{n} = 1$.
- ★ 7. Suppose that $\binom{7}{x} = \binom{5}{x}$ for some number x.
  - a. Find this number, using the formula for $\binom{n}{x}$.
  - b. This problem leads to a quadratic equation, one solution of which is acceptable. Why is the other not permissible?
- 8. Show that, for a fixed n,

$$\sum_{x=0}^{n} \binom{n}{x} = 2^n.$$

(Hint: Choose $r = b = 1$.)
- 9. Expand $(w + r)^5$.
- 10. Expand $(3r - 2b)^4$.
- 11. Expand $(\frac{1}{3} + \frac{2}{3})^6$.
- 12. Show that $\sum_{x=0}^{n} \binom{n}{x} p^x q^{n-x} = 1$ if $p + q = 1$. (This verifies that axiom A_2 of Section 2.7 is satisfied by a binomial distribution.)

7.4 More about Binomial Distributions

In Section 7.1 we found that for fixed values of n and p, we have

$$b(n, p; x) = \binom{n}{x} p^x q^{n-x}, \qquad x \in \{0, 1, \ldots, n\},$$

as the probability function of the variable

$$X = X_1 + X_2 + \cdots + X_n,$$

where the X_i's are independent Bernoulli variables. The distribution of such an X we have called a binomial distribution.

Extensive tables have been constructed for binomial distributions. Table B at the back of the book shows values of $b(n, p; x)$ for $n = 1, 2, \ldots, 20$ and for $p = 0.05, 0.10, 0.20, 0.25, 0.30, 0.40, 0.50$. The entries are recorded to four decimal places.

Example 7.4.1 In an extrasensory perception test, a set of 100 cards are prepared. There are 20 cards bearing the number 1, 20 numbered 2, ..., 20 numbered 5. Eight cards are drawn (one at a time with replacement) by the experimenter. The subject is asked to name the numbers on the cards. If the guesses are actually made at random, then the probability of correctly naming any given card is $p = 0.2$. To find the probability of correctly naming 5 or more of the 8 cards by guessing, we use Table B.

$$P(X \geq 5) = P(X = 5) + P(X = 6) + P(X = 7) + P(X = 8)$$
$$= b(8, 0.2; 5) + b(8, 0.2; 6) + b(8, 0.2; 7) + b(8, 0.2; 8)$$
$$= 0.0092 + 0.0011 + 0.0001 + 0.0000$$
$$= 0.0104$$

In Problem 4, Exercises 7.3, we proved that

$$\binom{n}{x} = \binom{n}{n-x}.$$

This fact enables us to use Table B for values of $p > 0.50$. The idea is to interchange the roles of p and q. (See Problems 3 and 4, Exercises 7.1.) Thus:

$$b(n, p; x) = \binom{n}{x} p^x q^{n-x} = \binom{n}{n-x} q^{n-x} p^x$$
$$= b(n, q; n-x)$$

Example 7.4.2 Ninety percent of all articles produced by a certain manufacturing process are satisfactory, while 10% are faulty. The probability of obtaining exactly 4 good items out of 6 randomly selected articles is

$$b(6, 0.9; 4) = \binom{6}{4}(0.9)^4(0.1)^2$$
$$= \binom{6}{2}(0.1)^2(0.9)^4 = b(6, 0.1; 2)$$

From Table B,
$$b(6, 0.1; 2) = 0.0984,$$

which is the desired probability.

The probability of obtaining at least 4 good items out of 6 is:

$$P(X = 4) + P(X = 5) + P(X = 6)$$
$$= b(6, 0.9; 4) + b(6, 0.9; 5) + b(6, 0.9; 6)$$
$$= b(6, 0.1; 2) + b(6, 0.1; 1) + b(6, 0.1; 0)$$
$$\doteq 0.0984 + 0.3543 + 0.5314 = 0.9841$$

Since, as we have seen (page 183), a particular binomial distribution depends on the value of two parameters n and p, it is interesting to study binomial distributions by fixing the value of one of these parameters and considering various values of the other. We shall first fix the value of n and consider various values of p. Later, we shall fix the value of p and consider various values of n.

In any case, however, we should keep in mind that

$$\mu_X = E(X) = np,$$

$$\text{var}(X) = npq,$$

and

$$\sigma_X = \sqrt{npq}.$$

Let us choose a particular value of n, say $n = 8$. In order to compare distributions having different values of p, we examine the graphs of Figure 7.5 on page 196. The values of $b(8, p; x)$ have been rounded to two decimal places from the values given in Table B. We are able to draw several important conclusions.

1. μ_X moves to the right as p increases.
2. σ_X increases until it reaches its largest value for $p = \frac{1}{2}$; then it decreases.
3. The increased spread results in a "flattening" of the graphs as p increases to $\frac{1}{2}$.
4. If $p = \frac{1}{2}$, the graph is symmetric about μ_X.
5. For values of p "close to" $\frac{1}{2}$, the graphs are "nearly" symmetric about μ_X.
6. For a particular value c, the graphs for $p = c$ and for $p = 1 - c$ are mirror images of each other.

If we choose a particular value of p, we may investigate the graphs of the binomial distribution for differing values of n. (Figure 7.6 on page 197.)

As in Figure 7.5, we observe in Figure 7.6 the moving of μ_X to the right and the flattening of the graph as n increases. The tendency towards symmetry about μ_X is present, although less easy to see than in Figure 7.5.

Another method of graphing binomial distributions is to make use of the idea of a histogram (Chapter 4).

Example 7.4.3 Suppose that 8 fair coins are to be tossed. The probability distribution of X, where X represents the number of heads, is given by this table:

x	0	1	2	3	4	5	6	7	8
$b(8, \frac{1}{2}; x)$	0.004	0.031	0.109	0.219	0.274	0.219	0.109	0.031	0.004

Check: $0.004 + 0.031 + 0.109 + 0.219 + 0.274 + 0.219 + 0.109$
$+ 0.031 + 0.004 = 1.000$

(We have used Table B, rounding to 3 decimals.)

Figure 7.5 *Binomial Distributions (n = 8)*

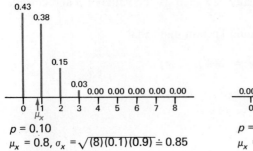
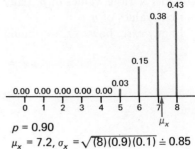
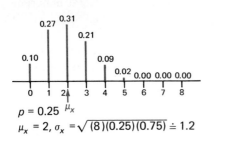
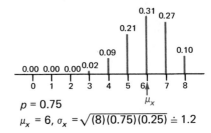
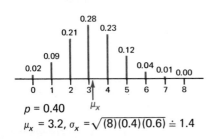
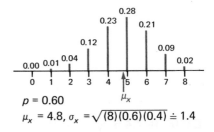
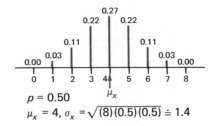

Figure 7.6 *Binomial Distributions (p = 0.25)*

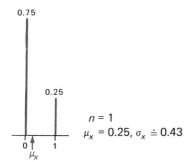

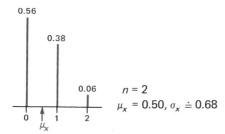

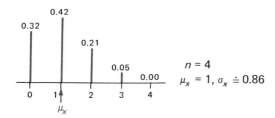

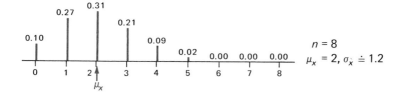

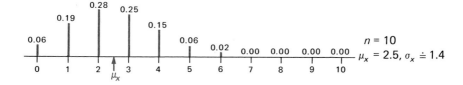

Suppose that we consider 1000 experiments, each consisting of 8 tosses. Taking the long-run interpretation of probability, we would anticipate 0 heads in about 4 cases, 1 head in about 31 cases, and so on. Hence, we may construct a *theoretical frequency* table as follows:

number of heads	0	1	2	3	4	5	6	7	8
theoretical frequency	4	31	109	219	274	219	109	31	4

Check: $4 + 31 + 109 + 219 + 274 + 219 + 109 + 31 + 4 = 1000$

We may graph the frequency table of this example in the form of a histogram (Figure 7.7).

Figure 7.7

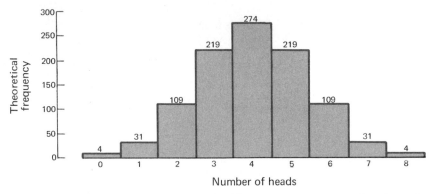

Since the theoretical frequencies in Figure 7.7 are proportional to the probabilities $b(8, \frac{1}{2}; x)$, we may simply relabel the vertical scale and obtain Figure 7.8.

Figure 7.8

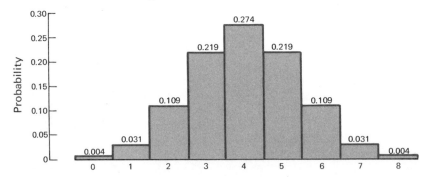

Remark

Notice that $\quad b(8, \frac{1}{2}; 3) = P(X = 3) = P(2.5 \leq X \leq 3.5) = 0.219.$

Our histogram enables us to associate probabilities with certain *intervals*. This idea will be useful in Section 8.2, where we consider approximating values of $b(n, p; x)$ by a continuous distribution (if n is large).

EXERCISES 7.4

1. Use Table B to graph $b(n, \frac{1}{2}; x)$ for
   a. $n = 1$ b. $n = 2$ c. $n = 4$ d. $n = 8$ e. $n = 16$
2. Calculate μ_X and σ_X for each distribution of Problem 1.
3. Use Table B to graph $b(n, \frac{1}{4}; x)$ for the values of n listed in Problem 1.
4. Calculate μ_X and σ_X for each distribution of Problem 3.
5. Draw histograms for the distributions of Problems 1c and 1e.
6. Suppose that the histogram of Figure 7.7 was obtained as the actual data from an experiment (Chapter 4). Find
   a. $\bar{x}$ b. s_x
7. In a particular experiment involving tossing 8 coins 1000 times, it is suspected that some of the coins are *not* fair but favor heads. If the suspicion is correct, would the actual distribution of heads agree with the theoretical binomial distribution of Example 7.4.4? Why not?
8. Assuming that all 8 coins are fair, would we anticipate that the distribution of heads for the 1000 tosses would coincide exactly with the distribution of Example 7.4.4? Why, or why not?
9. Table B does not contain values of $b(n, p; x)$ for $p = \frac{1}{3}$ nor for $p = \frac{1}{6}$. Find the following binomial distributions, leaving the probabilities in fractional rather than decimal form:
   a. $n = 6, p = \frac{1}{3}$ b. $n = 3, p = \frac{1}{6}$
10. Find μ_X and σ_X for:
    a. Problem 9a   b. Problem 9b
11. For Problem 9a, the expected value of X is 2. What is the most probable value of X?
12. For Problem 9b the expected value of X is 0.5. What is the most probable value of X?

7.5 What Do You Think?

Binomial distributions are exceedingly useful. They find application in a wide variety of practical experiments. On the other hand, there is a temptation to apply a binomial distribution to cases for which it is not truly appropriate.

To use a binomial distribution, we need to know that

(a) each trial gives rise to a Bernoulli random variable;
(b) the value of p does not change from trial to trial; and
(c) the trials are independent.

Only if all *three* conditions are fulfilled may we use a binomial distribution with confidence. In Example 7.1.2, the six shots of the marksman were assumed to be independent Bernoulli trials. In a real world situation is this assumption justified? What do you think?

EXERCISES 7.5a

- 1. Consider the marksman of Example 7.1.2.
  a. Give a plausible argument why the value of p might *increase* as successive shots are fired.
  b. Would it ever be reasonable that the successive values of p might *decrease*?
- 2. Consider the baseball player of Problem 5, Exercises 7.1. Explain why the assumption of independence is probably *not* valid.
- 3. For each of these experiments, determine whether all three conditions for a binomial distribution are fulfilled. If not, which ones are not satisfied?
  a. An urn contains 5 red, 3 blue marbles. Two marbles are drawn, without replacement. (Example 4.4.2.) Consider the distribution of the random variable which represents the total number of red marbles drawn.
  b. A marksman has $p = 0.6$ of hitting the bull's-eye on the first shot. If he makes the bull's-eye, then he has $p = 0.7$ of hitting the bull's-eye on the second shot; otherwise, he has $p = 0.45$. Consider the distribution of the random variable which represents the total number of bull's-eyes.
  c. A fair die is thrown twice and X is the random variable representing the number tossed on the second throw. Consider the distribution of the random variable representing the sum of the faces of the two throws.

There are practical problems that do not precisely satisfy the three conditions we have specified. In a drawing-without-replacement experiment, condition (c) is not met. Would it ever be reasonable to use a binomial distribution in a drawing without replacement experiment? What do you think?

EXERCISES 7.5b

- 1. A city has 100,000 voters. 55,000 are Democrats. Three voters are selected at random (one at a time, without replacement!)
  a. Find the probability that all three are Democrats. (Set up, but do not carry out the multiplications.)
  b. Is it reasonable to use $(0.55)^3$ in place of the exact probability of 1a?
  c. What is the approximate probability that exactly two Democrats are selected?
- 2. Consider the experiment: Choose two whole numbers, one number at a time without replacement from the set $\{1, 2, 3, \ldots, 9, 10\}$. Let $X_i = 1$ if the ith draw is an even number; $X_i = 0$ otherwise.

a. Find $P(X_2 = 1)$
  b. Find $P(X_2 = 1 \mid X_1 = 1)$
  c. Are the events $\{X_1 = 1\}$, $\{X_2 = 1\}$ independent?

● 3. Suppose in Problem 2 we do not limit ourselves to a finite set, but select two whole numbers, one at a time without replacement from the set $N = \{1, 2, 3, \ldots\}$. Do you think the events $\{X_1 = 1\}$, $\{X_2 = 1\}$ are independent?

Chapter Summary

1. a. For n independent Bernoulli trials, $\binom{n}{x}$ denotes the number of paths (in a tree diagram) that lead to $X = x$.

   b. $\binom{n}{x} = \binom{n-1}{x-1} + \binom{n-1}{x}$, $0 < x < n$; $\binom{n}{0} = \binom{n}{n} = 1$.

   c. If $X_1, X_2, \ldots, X_n$ are independent Bernoulli variables, each having $P(X_i = 1) = p$ and $P(X_i = 0) = q$, then the probability function of $X = X_1 + X_2 + \cdots + X_n$ is defined by

   $$f(x) = P(X = x) = \binom{n}{x} p^x q^{n-x}, \quad x \in \{0, 1, \ldots, n\},$$

   and

   $$E(x) = np, \quad \operatorname{var}(X) = npq, \quad \sigma_x = \sqrt{npq}$$

   This distribution is called a *binomial distribution*. Another notation is:

   $$b(n, p; x) = \binom{n}{x} p^x q^{n-x}$$

2. a. The number of *permutations* of n objects taken n at a time:

   $$_nP_x = \frac{n!}{(n-x)!}$$

   b. The number of *combinations* of n objects taken n at a time:

   $$_nC_x = \frac{_nP_x}{x!} = \frac{n!}{x! \cdot (n-x)!}$$

3. a. $\binom{n}{x} = \binom{n}{n-x}$ b. $\binom{n}{x} = {_nC_x}$

   c. $\binom{n}{x}$ is called a *binomial coefficient*.

Chapter Review

1. Seven cards are drawn, one at a time, with replacement from a standard deck of playing cards.
   a. Find $E(X)$ if X represents the number of picture cards (jack, queen, king) drawn.
   b. If Y represents the number of cards drawn that are not picture cards, how is $E(Y)$ related to $E(X)$?
   c. Find σ_X^2 and explain why $\sigma_Y^2 = \sigma_X^2$.
2. Along a main street, a signal is timed 40 seconds on GO and 20 seconds on STOP.
   a. If a man drives along this street once going to work each day, find the probability distribution of X if X represents the number of times he gets a GO signal at this intersection in three days.
   b. Find $E(X)$ for this distribution.
   c. Suppose that this man has to pass through three such signals with the same timing cycle on his way to work. If Y represents the number of times he gets a GO signal at these intersections on a given day, explain why $E(Y)$ is not likely to be the same as $E(X)$ of part b.
3. The suit of hearts is taken from a standard deck of playing cards and three cards are drawn from this suit, one at a time, without replacement.
   a. Find $E(X)$ if X represents the number of honor cards (ace, king, queen, jack, ten) in the three draws.
   b. Suppose that three cards are drawn at random simultaneously from this suit. What is the probability all three will be honor cards?
   c. Compare the probability in part b with the probability of $X = 3$ in part a.
4. A test has five multiple-choice questions, each offering five alternate answers.
   a. If a student guesses each answer at random, what is the probability that he answers correctly exactly 3 questions?
   b. How many times as great is the probability of obtaining exactly 2 questions (by random guessing) as it is to obtain exactly 3?
5. A player is dealt a five-card hand from a standard deck of playing cards. What is the probability of getting a full house (a triplet and a pair)? [Leave the answer in factorial form.]
6. For each of the following, give a reason why the probability distribution is not binomial. In each case, six slips, numbered 1, 2, 3, 4, 5, 6, respectively, are placed in a container and two draws are made.
   a. Drawings are made with replacement. Let $X = X_1 + X_2$ where X_i represents the number obtained on the ith draw.
   b. Drawings are made without replacement. Let $X = X_1 + X_2$, where $X_i = 1$ if an even number is drawn; $X_i = 0$ otherwise.
   c. Drawings are made with replacement. Let $X = X_1 + X_2$ where $X_1 = 1$ if an even number is drawn in the first draw; $X_2 = 1$ if the number drawn in the second draw is less than 3. $X_i = 0$ otherwise.

Chapter 8

Using Continuous Distributions

The binomial distributions are extremely important examples of discrete distributions. However, we discovered in Chapter 7 that finding some of the values may require complicated computations. In this chapter, we shall learn how we may use continuous distributions as approximations to binomial distributions, subject to certain conditions.

8.1 Standardizing a Normal Distribution

Up to now, we have used the expression "normal distribution" only in reference to the *standard* normal distribution (Section 5.4). There are, however, many "normal" distributions that are related to the standard normal distribution as we shall show below. Their importance stems from two sources.

First, many experiments give rise to distributions of data whose histograms appear to approximate the bell-shaped "normal" curve. This fact was discovered by early statisticians who investigated such data as the distribution of heights of soldiers, strength of grip, performance of individuals on a variety of mental and physical tests, growth rates, and biological processes. Thus, normal distributions may often be used as mathematical models for such distributions.

Secondly, a remarkable theorem links normal distributions to other distributions of random variables. We shall state (but not prove) one form of this theorem in Section 8.2.

We shall now show how other normal distributions may be derived from the standard normal distribution by using some properties of linear functions of a random variable that we studied in Chapter 6.

Suppose that Z is a random variable whose probability distribution is standard normal. It follows that $\mu_Z = 0$, $\sigma_Z = 1$ (Section 6.4). Let

$$X = aZ + b \quad (a > 0).$$

Then X is a random variable with (see Remark 3, page 159)

$$\mu_X = a\mu_Z + b = a \cdot 0 + b = b,$$

and

$$\sigma_X = |a|\sigma_Z = a \cdot 1 = a \quad (a > 0).$$

Thus, we may write

$$X = \sigma_X Z + \mu_X.$$

From this we observe that the expected value of the distribution of X as compared with Z has moved μ_X units from 0. Also, the coefficient, σ_X, acts as a "scaling" factor changing the spread of the distribution. However, the essential character of the distribution, its "normality," is not changed. We define:

If Z has a standard normal distribution ($\mu_Z = 0$, $\sigma_Z = 1$), and if $X = aZ + b$ with $a > 0$, then X has a *normal* distribution with $\mu_X = b$, $\sigma_X = a$.

If Z has a standard normal distribution, we sometimes say more briefly, "Z is standard normal" or "Z is normal ($\mu_Z = 0$, $\sigma_Z = 1$)."

Example 8.1.1 Let Z be standard normal. Suppose that $X = 2Z + 3$. Then X is normal with $\mu_X = 3$, $\sigma_X = 2$. The graphs of the corresponding distributions are given below. In the right-hand graph the vertical coordinates have been adjusted to keep the area under the curve equal to 1. (Compare Section 8.2.)

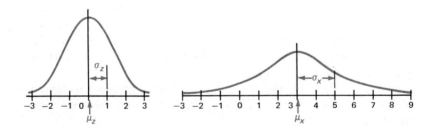

In practice, there are many occasions when we are "given" (from experimental evidence, perhaps) that a certain random variable X is normal with μ_X and σ_X known. We then solve $X = \sigma_X Z + \sigma_X$ to conclude that

$$Z = \frac{X - \mu_X}{\sigma_X},$$

and that Z is standard normal. After making this transformation, we may use Table N. The technique is shown in the following examples.

Example 8.1.2 Suppose that a random variable X has a normal distribution ($\mu_X = 15$, $\sigma_X = 5$). To determine probabilities such as $P(X < 20)$, $P(X > 7.5)$, or $P(9 < X < 18)$, we first recognize that if

$$Z = \frac{X - \mu_X}{\sigma_X},$$

then Z is normal ($\mu_Z = 0$, $\sigma_Z = 1$). If $X = 20$, then

$$Z = \frac{20 - 15}{5} = 1.$$

Therefore, from Table N,

$$P(X < 20) = P(Z < 1) \doteq 0.8413.$$

Similarly,

$$P(X \geq 7.5) = P(Z \geq -1.5) = P(Z \leq 1.5) \doteq 0.9332$$
$$P(9 \leq X \leq 18) = P(-1.2 \leq Z \leq 0.6)$$
$$= P(Z \leq 0.6) - P(Z \leq -1.2)$$
$$\doteq 0.7257 - 0.1151 = 0.6106.$$

Example 8.1.3 A survey shows that the gasoline consumption for new automobiles of a certain make appears to be normally distributed with expected value 23.3 mpg (miles per gallon) and standard deviation 2.5 mpg. If a new car of this make is purchased, the probability that it will get at least 20 mpg may be estimated as follows. We first let

$$Z = \frac{X - 23.3}{2.5}.$$

Thus, for $X = 20$, we have $Z = -1.32$, and so

$$P(X \geq 20) = P(Z \geq -1.32) = P(Z \leq 1.32) = 0.9066.$$

The probability that a new car of this make will perform within one standard deviation of the mean is

$$P(|Z| \leq 1) = P(Z \leq 1) - P(Z \leq -1) \doteq 0.8413 - 0.1587 = 0.6826.$$

EXERCISES 8.1

- 1. Suppose that X is normal ($\mu_X = 100$, $\sigma_X = 15$). Find:
    a. $P(X \leq 122.5)$ b. $P(X \geq 90)$
    c. $P(85 \leq X \leq 125)$ d. $P(|X - 100| \leq 12)$
 2. Suppose that Y is normal ($\mu_Y = 50$, $\sigma_Y = 10$). Find:
    a. $P(Y \leq 45)$ b. $P(Y \geq 55)$
    c. $P(45 \leq Y \leq 55)$ d. $P(|Y - 50| \leq 10)$
- 3. In Problem 1, give an estimate for $P(X \leq 150)$.
 4. In Problem 2, give an estimate for $P(Y \geq 85)$.

- 5. If X is normal ($\mu_X = 100$, $\sigma_X = 15$), determine the value of k such that:
  a. $P(X \leq k) = 0.5000$ b. $P(X \leq k) \doteq 0.8997$
  c. $P(|X - 100| \leq k) \doteq 0.3108$ d. $P(-k \leq X - 100 \leq k) \doteq 0.5000$

- 6. If Y is normal ($\mu_Y = 50$, $\sigma_Y = 10$), determine the value of k such that:
  a. $P(Y \leq k) = 0.5000$ b. $P(Y \leq k) \doteq 0.6480$
  c. $P(Y \geq k) \doteq 0.4207$ d. $P(-k \leq Y - 50 \leq k) \doteq 0.8000$

- 7. For the make of car in Example 8.1.3:
  a. What is the probability that a new car selected at random will get from 25.8 to 28.3 miles per gallon?
  b. What is the probability that a new car selected at random will exceed 28.3 mpg?
  c. What is the probability that two new cars selected at random each exceeds 25.8 mpg?

- 8. In Problem 7, which has the greater probability: (i) one car selected at random to perform at least 2 standard deviations better than the mean, or (ii) two cars selected at random, each to perform at least 1 standard deviation better than the mean?

- 9. For Example 8.1.3, what is the probability that a new car selected at random will get less than 18 mpg?

★ 10. The mean and standard deviation of an achievement test are, respectively, 550 and 125. If two individuals who have taken the test are selected independently at random, what is the probability that at least one of them got more than 675?

8.2 Normal Approximation to a Binomial Distribution

In Chapter 7 we encountered the difficulty of computing $b(n, p; x)$ for large values of n. We shall now develop a method of estimating values that are not included in Table B.

We have suggested, in Chapter 1 and elsewhere, that it is sometimes possible to approximate a continuous distribution by a discrete one. Occasionally, there are advantages in doing the opposite, that is, in approximating a discrete distribution by a continuous one. We now set about the task of showing how we may approximate a binomial distribution (which is discrete) by the standard normal distribution (which is continuous). Let us begin by considering the distribution of the binomial random variable X for $n = 16$, $p = \frac{1}{2}$ for which Table B does provide the values. We construct a histogram of the distribution of X as shown in Figure 8.1.

Figure 8.1

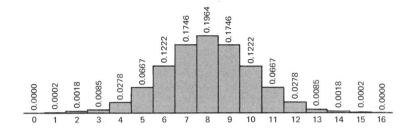

We have noted (Section 5.3) that for continuous distributions, intervals have probabilities. Although a binomial distribution is discrete, we shall find it natural to associate probabilities with certain intervals. For example,

$$P(9.5 \leq X \leq 10.5) = P(X = 10) = b(16, \tfrac{1}{2}; 10) = 0.1222.$$

Since each cell of the histogram is of width one, the *area* of each cell as well as its height is equal to the probability associated with the corresponding interval, and the total area is 1.

Since we plan to use the standard normal distribution as an approximating distribution, our next step is to standardize the binomial variable. We make the transformation (recall Section 7.1):

$$Y = \frac{X - \mu_X}{\sigma_X} = \frac{X - np}{\sqrt{npq}}$$

In this case,

$$Y = \frac{X - 8}{2}.$$

For example, $X = 8$ corresponds to $Y = 0$, and $X = 9$ corresponds to $Y = 0.5$. Thus, each interval is now 0.5:

$$P(7.5 \leq X \leq 8.5) = P(-0.25 \leq Y \leq 0.25)$$

$$P(8.5 \leq X \leq 9.5) = P(0.25 \leq Y \leq 0.75)$$

Figure 8.2

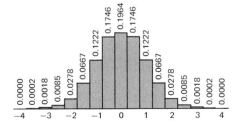

Since each interval is now of width 0.5, the areas of the rectangles in Figure 8.2 are no longer equal to the probabilities. However, by doubling the vertical scale as shown in Figure 8.3, we may restore this feature. Thus, in Figure 8.3, the *areas*, not the heights, of the rectangles are equal to the probabilities.

Figure 8.3

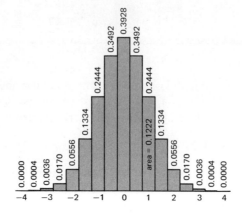

We now superimpose on the histogram of Figure 8.3 the standard normal curve, using the values given in the table below. The composite figure is shown in Figure 8.4. Notice that the values for the standard normal curve are "close to" the corresponding heights of the cells of Figure 8.3.

Values of Z	0	±0.5	±1.0	±1.5	±2.0
Height of curve	0.3989	0.3521	0.2420	0.1295	0.0540

Values of Z	±2.5	±3.0	±3.5	±4.0
Height of curve	0.0175	0.0044	0.0009	0.0001

Figure 8.4

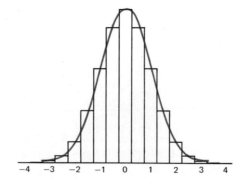

We see that the distribution of

$$Y = \frac{X - 8}{2}$$

is rather closely approximated by the distribution of Z, the standard normal random variable. For example, the exact probability of obtaining 10 successes in 16 trials is found from Table B:

$$b(16, \tfrac{1}{2}; 10) \doteq 0.1222$$

The corresponding normal approximation is found from Table N:

$$b(16, \tfrac{1}{2}; 10) = P(X = 10) = P(9.5 \leq X \leq 10.5)$$
$$= P(0.75 \leq Y \leq 1.25)$$
$$\doteq 0.8944 - 0.7734 = 0.1210$$

Figure 8.5 shows the comparison between the distribution of the standard normal variable Z and the standardized binomial variable Y for $n = 16$, $p = \tfrac{1}{2}$. The comparison between corresponding areas can be seen clearly from the last two columns.

Figure 8.5

Binomial X	Standardized Y	Interval of values (Y or Z)	Standardized binomial P(Y in interval)	Standard normal P(Z in interval)
0	−4.0	−4.25, −3.75	0.0000	0.0000
1	−3.5	−3.75, −3.25	0.0002	0.0002
2	−3.0	−3.25, −2.75	0.0018	0.0024
3	−2.5	−2.75, −2.25	0.0085	0.0092
4	−2.0	−2.25, −1.75	0.0278	0.0279
5	−1.5	−1.75, −1.25	0.0667	0.0655
6	−1.0	−1.25, −0.75	0.1222	0.1210
7	−0.5	−0.75, −0.25	0.1746	0.1747
8	0	−0.25, 0.25	0.1964	0.1974
9	0.5	0.25, 0.75	0.1746	0.1747
10	1.0	0.75, 1.25	0.1222	0.1210
11	1.5	1.25, 1.75	0.0667	0.0655
12	2.0	1.75, 2.25	0.0278	0.0279
13	2.5	2.25, 2.75	0.0085	0.0092
14	3.0	2.75, 3.25	0.0018	0.0024
15	3.5	3.25, 3.75	0.0002	0.0002
16	4.0	3.75, 4.25	0.0000	0.0000

It is rather clear that if we consider larger values of n, the process described above would yield even closer approximations.

Example 8.2.1 To approximate $b(50, \frac{1}{2}; 30)$, recall that

$$\mu_X = np = 25, \qquad \sigma_X = \sqrt{npq} = \sqrt{\tfrac{50}{4}}.$$

Thus, we have:

$$Y = \frac{X - 25}{\frac{5\sqrt{2}}{2}} = \frac{\sqrt{2}(X - 25)}{5} \doteq 0.283(X - 25)$$

$$\begin{aligned}
P(X = 30) &= P(29.5 \leq X \leq 30.5) \\
&\doteq P(1.27 \leq Y \leq 1.56) \\
&\doteq P(1.27 \leq Z \leq 1.56) \\
&= P(Z \leq 1.56) - P(Z \leq 1.27) \\
&= 0.9406 - 0.8980 \\
&= 0.0426
\end{aligned}$$

Therefore,

$$P(X = 30) \doteq 0.0426.$$

Example 8.2.2 Suppose that we are interested in finding the probability of obtaining from 40 to 60 heads in 100 tosses of a fair coin ($p = \frac{1}{2}$). Here

$$\mu_X = 50, \; \sigma_X = \sqrt{\tfrac{100}{4}} = 5,$$

and

$$Y = \frac{X - 50}{5}.$$

$$\begin{aligned}
P(39.5 \leq X \leq 60.5) &= P(-2.1 \leq Y \leq 2.1) \\
&= P(|Y| \leq 2.1) \\
&\doteq P(|Z| \leq 2.1) \\
&\doteq 0.9821 - 0.0179 \\
&= 0.9642
\end{aligned}$$

Our argument leads us to believe that for $p = \frac{1}{2}$, the standardized binomial distribution is closely approximated by the standard normal distribution for values of n which are sufficiently large. Even for n as small as 16, we have found a reasonably close agreement.

One reason that we are led to believe that this conclusion is correct is that the binomial distributions for $p = \frac{1}{2}$ are *symmetric* about $\mu_X = \frac{n}{2}$. It is surprising that our conclusion is valid even if $p \neq \frac{1}{2}$. We state without proof a theorem regarding this:

De MOIVRE'S* THEOREM If $X = X_1 + X_2 + \cdots + X_n$, where the X_i's are independent Bernoulli variables each having $\mu = p$, then if n is sufficiently large,

$$P\left(a \leq \frac{X - np}{\sqrt{npq}} \leq b\right) \doteq P(a \leq Z \leq b)$$
$$= P(Z \leq b) - P(Z \leq a),$$

where Z is a random variable having the standard normal distribution.

Notice that for this theorem, the three conditions discussed in Sections 7.1 and 7.4 must be satisfied:

(a) each trial gives rise to a Bernoulli random variable;
(b) the value of p does not change from trial to trial; and
(c) the trials are independent.

The accuracy of the approximation depends on both the size of n and the value of p. For values of p close to $\frac{1}{2}$, the approximation is good even for values of n as small as 10. If p is close to 0 or to 1, then quite large values of n are necessary to assure a satisfactory approximation by the standard normal distribution. A convenient rule of thumb is: The standard normal distribution may be used as an approximation to a standardized binomial distribution if the product np is at least 5.

Remark

De Moivre's Theorem, stated above, emphasizes the importance of the standard normal distribution. This theorem justifies the introduction and discussion of the "normal curve." Even more importantly, De Moivre's Theorem is itself merely a special case of an even more remarkable and powerful theorem which we shall discuss in Chapter 9.

EXERCISES 8.2

- 1. In Problem 4, Exercises 2.6, we gave the approximation $P(50$ heads, 100 tosses$) \doteq 0.08$. Verify this approximation.
- 2. In Problem 1, Exercises 2.6, 16 heads were obtained from 20 tosses of a fair coin. We argued that "something rather unusual has occurred" is a reasonable conclusion. Using the normal approximation, compare the probability of this event with the probability of getting 10 heads in 20 tosses.
- 3. Compare the event of getting 10 heads in 20 tosses with getting 50 heads in 100 tosses (Problems 1 and 2).

*Abraham De Moivre (1667–1754) was a French-born mathematician who worked in England and was a friend of Isaac Newton.

- **4.** Compare $P(4 \leq X \leq 6)$ for 10 tosses with $P(40 \leq X \leq 60)$ for 100 tosses (Example 8.2.2 and Problem 2b, Exercises 1.6).
- **5.** Compare the standard normal approximations for Problem 2c, Exercises 1.6: 45 to 55 heads in 100 tosses, against 450 to 550 heads in 1000 tosses.
- **6.** Repeat Problem 5 for Problem 2d, Exercises 1.6: 8 or more heads in 10 tosses, against 80 or more heads in 100 tosses.
- **7.** Using $n = 12$ and $p = \frac{1}{4}$, prepare a table corresponding to that in Figure 8.5.
- **8.** From the data in Problem 7, construct a diagram corresponding to that in Figure 8.4.

★ COMPUTER INVESTIGATIONS

If you have access to an electronic computer, try to carry out the following investigations.

1. Write a program that will represent tossing a coin N times and will report the number of heads and the ratio of the number of heads to N.
2. **a.** Use the program of Problem 1 to represent tossing a fair coin 100 times.
   b. Repeat several times and compare the results.
3. Repeat Problem 2 for $N = 1000$.
4. Modify the program of Problem 1 so that it will print "H" for each "toss" that results in "heads" and "T" for each "toss" that results in "tails."
5. **a.** Use the program of Problem 4 to represent tossing a fair coin 50 times. Observe the lengths of the runs of H's.
   b. Repeat several times and compare the results.
6. **a.** Use the program of Problem 1 to represent tossing a biased coin, $P(\text{heads}) = 0.4$, 100 times.
   b. Repeat several times and compare the results.

★ 8.3 Poisson Approximation to a Binomial Distribution

If $np \geq 5$, we have seen that it is possible to use the standard normal distribution to approximate a standardized binomial distribution. Suppose that $np < 5$; is there any way of approximating the values of $b(n, p; x)$ for large n? (If n is small, of course we may compute $b(n, p; x)$ directly or use Table B.)

The answer to our question is in the affirmative. There are other distributions which are helpful. These distributions, the Poisson distributions, have the probability functions given in Table P. If $np \leq 5$, this table provides approximations to the values of $b(n, p; x)$. We shall not attempt to explain the construction of Table P other than to say that the approximation to $b(n, p; x)$ is given by the probability function

$$h(x) = \frac{e^{-\lambda}\lambda^x}{x!},$$

where

$$\lambda = np.*$$

Instead, we shall be content to illustrate how it may be used in practice.

We first note that if np is small, the probability of obtaining a large number of "successes" is quite small. For example, if $n = 20$ and $p = 0.05$, then the binomial distribution is found in Table B to be:

x	b(20, 0.05; x)
0	0.3585
1	0.3774
2	0.1887
3	0.0596
4	0.0133
5	0.0022
6	0.0003

For $x > 6$,

$b(20, 0.05; x) = 0.0000$

to four decimal places

If we now compare this table with the appropriate ($\lambda = np = 1$) portion of Table P, we have:

x	Binomial probability	Poisson probability
0	0.3585	0.3679
1	0.3774	0.3679
2	0.1887	0.1839
3	0.0596	0.0613
4	0.0133	0.0153
5	0.0022	0.0031
6	0.0003	0.0005
7	0.0000	0.0001

* The symbol λ (lambda) is a letter of the Greek alphabet corresponding to the English "l."

Example 8.3.1 Suppose that 1% of the population of the United States suffer from some particular eye defect. A hundred individuals are selected at random. To compute the probability of finding exactly 2 individuals with the defect, we must calculate

$$b(100, 0.01; 2) = \binom{100}{2}(0.01)^2(0.99)^{98}.$$

Using the Poisson approximation ($\lambda = np = 1$), we see that

$$b(100, 0.01; 2) \doteq 0.1839.$$

Whether we use the standard normal or the Poisson distribution, we are concerned with *approximating* the binomial distribution. Values of $b(n, p; x)$ may be calculated by using logarithms or a computer.

★ **EXERCISES 8.3**

1. For each of the following, explain whether you would use the binomial, the standard normal, or the Poisson distribution in determining the probability.
   a. $b(100, 0.02; 50)$ b. $b(100, 0.2; 50)$
   c. $b(10, 0.5; 5)$ d. $b(100, 0.05; 3)$

2. Compare the results using the standard normal and the Poisson distributions for Problem 1d.

8.4 What Do You Think?

De Moivre's Theorem (Section 8.2) is useful in enabling us to use the standard normal distribution for approximating a binomial distribution. This theorem is a special case of the quite powerful theorem which we mentioned in Section 8.1. In this section, we shall present some situations which will lead us to the statement of this theorem in Chapter 9.

Suppose that we have six marbles, numbered 1, 2, 3, 4, 5, 6, in an urn. If one marble is drawn at random, we may naturally associate a random variable, X, with this experiment:

x	1	2	3	4	5	6
$f(x)$	$\frac{1}{6}$	$\frac{1}{6}$	$\frac{1}{6}$	$\frac{1}{6}$	$\frac{1}{6}$	$\frac{1}{6}$

From Example 5.5.2 and Problem 2, Exercises 6.2, we know that

$$\mu_X = E(X) = \tfrac{7}{2}, \quad \sigma_X^2 = \tfrac{35}{12}.$$

Let us change the experiment by drawing two marbles at random. There are, of course, $\binom{6}{2} = 15$ ways to make such a draw, each with probability $\frac{1}{15}$. We now introduce a new random variable by computing the average (the mean) of the two numbers drawn. Thus, if 2 and 6 are drawn, the mean is 4. We shall denote this new random variable as $\overline{X}$ (the bar reminds us that we are finding the mean of the numbers drawn).

It is now possible to find the probability distribution of $\overline{X}$. We may also compute $\mu_{\overline{X}}$ and $\sigma_{\overline{X}}^2$. It would appear to be a reasonable guess that these values are related to μ_X and σ_X^2. What do you think? Will it turn out that $\mu_{\overline{X}} = \mu_X$? Do you think that $\sigma_{\overline{X}}^2$ will be equal to, less than, or greater than σ_X^2?

EXERCISES 8.4

- 1. a. List the 15 possible draws of the two marbles and their means.
  b. Verify the following graph of the probability distribution of $\overline{X}$.

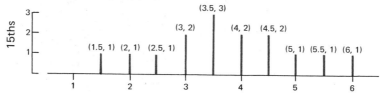

  c. Find $\mu_{\overline{X}}$.
  d. Verify that $\sigma_{\overline{X}}^2 = \frac{4}{10}\sigma_X^2$.

- 2. If we draw *three* marbles we may consider the new random variable $\overline{X}$ whose values are obtained by finding the mean of the three numbers.
  a. How many different equally likely 3-marble draws are possible?
  b. What is the smallest possible value of $\overline{X}$?
  c. The largest?
  d. How do the smallest and largest values of $\overline{X}$ for three draws compare with those of X and of $\overline{X}$ for two draws?
  e. Will $\sigma_{\overline{X}}^2$ for three draws be greater than or less than $\sigma_{\overline{X}}^2$ for two draws?

- 3. The graph of the distribution of $\overline{X}$ for 3 draws is given below.

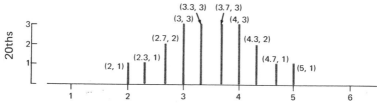

  a. What is the value of $\mu_{\overline{X}}$?
  b. Compare the shapes of the graphs of X, $\overline{X}$ for two draws, and $\overline{X}$ for three draws. What generalizations can you make?

4. We may define $\bar{X}$ for drawing 4 marbles at a time in a similar manner. The graph of its distribution is shown below.

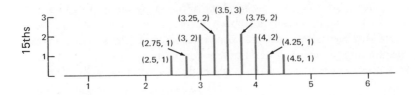

   a. What is the value of $\mu_{\bar{X}}$, drawing 4 marbles at a time?
   b. Is $\sigma_{\bar{X}}^2$ for drawing 4 at a time greater than or less than $\sigma_{\bar{X}}^2$ for drawing 3 at a time?
   c. The shape of the distribution of $\bar{X}$ for drawing 4 at a time is suggestive. Suppose we had originally placed marbles numbered 1, 2, 3, ..., 1000 in our urn. Are you willing to make some guesses about the distribution of $\bar{X}$ for drawing 30 marbles?

5. For the urn containing 6 marbles:
   a. How might we interpret $\bar{X}$ for drawing one marble?
   b. $\bar{X}$ for drawing 6 marbles?
   c. Find $\mu_{\bar{X}}$ and $\sigma_{\bar{X}}^2$, for drawing one marble.
   d. Find $\mu_{\bar{X}}$ and $\sigma_{\bar{X}}^2$, for drawing six at a time.

6. Suppose that a discrete random variable takes on the values

$$x_1, x_2, \ldots, x_N,$$

each with probability $\dfrac{1}{N}$. If n of the values are selected we may define the random variable $\bar{X}$ whose values are the means of the numbers selected. It may be shown that

$$\sigma_{\bar{X}}^2 = \sigma_X^2 \cdot \frac{N-n}{n(N-1)} \qquad (1 \leq n \leq N)$$

   a. Verify the result of Problem 1d. (For this problem, $N = 6$, $n = 2$.)
   b. Using this formula, compute $\sigma_{\bar{X}}^2$ for $N = 6$, and for $n = 3, 4, 5, 6$.
   c. Does the formula hold for $n = 1$?

Chapter Summary

1. a. If Z has a *standard normal distribution* ($\mu_Z = 0$, $\sigma_Z = 1$), and if
$$X = aZ + b$$
with $a > 0$, then X has a *normal distribution* with $\mu_X = b$, $\sigma_X = a$.

   b. If X has a normal distribution, then
$$Z = \frac{X - \mu_X}{\sigma_X}$$
is standard normal.

2. *De Moivre's Theorem:* If $X = X_1 + X_2 + \cdots + X_n$, where the X_i's are independent Bernoulli variables each having $\mu = p$, then if n is sufficiently large ($np \geq 5$),
$$P\left(a \leq \frac{X - np}{\sqrt{npq}} \leq b\right) = P(a \leq Z \leq b)$$
$$= P(Z \leq b) - P(Z \leq a),$$
where Z is a random variable having the standard normal distribution.

Chapter Review

1. Let X be a random variable having a normal distribution with $\mu = 100$ and $\sigma = 15$. Derive the standard normal random variable Z associated with X.

2. Suppose that the distribution of scores on a nationally administered test is normal with $\mu = 100$ and $\sigma = 15$.
   a. An individual is randomly selected from those taking this test. What is the probability that he scored more than 115 points?
   b. Two individuals from this group are randomly selected. What is the probability that both scored more than 115 points?
   c. In part b, what is the probability that either of the two randomly selected individuals scored more than 115?

3. Standardize the random variable X if X represents the number obtained on a throw of a die.

4. Using the standard normal table, find $P(3 \leq X \leq 5)$ if $\mu_X = \frac{7}{2}$ and $\sigma_X^2 = \frac{35}{12}$.

5. If a thumbtack is tossed, it can land with its point either in an "up" position (U) or in a "down" position (D).

Let $P(U) = \frac{2}{3}$. If this tack is tossed 900 times:
a. What is the probability that it lands "up" less than 575 times?
b. Find x such that $P(X \leq x) = \frac{1}{2}$.
c. Find x such that $P(|X - 600| \leq x) = 0.5$.
6. Suppose that another tack has $P(U) = 0.7$. Use the binomial distribution table to find the probability that the tack lands 14 times out of 20 tosses.

Chapter 9

Sampling

This chapter and the remaining two chapters are devoted to the study of certain aspects of *statistical inference*. The fundamental idea is important in a variety of situations. Suppose, for example, that a decision must be made in the face of uncertainty. If we can associate with the problem a random variable whose distribution is *known*, then often we may be able to reach a decision by applying probability theory. (For example, our decision might be based on the size of the expected value of the associated random variable.)

In practice, however, we are frequently in the position of *not* knowing the relevant probability distribution. In such cases, a reasonable procedure is to conduct some experiments or to make some observations. We can then use this partial information to make intelligent guesses, or inferences, about the probability distribution. In particular, we may be able to estimate μ. Our partial information comes from our *sample* observations. The process of selecting the observations to be made is called *sampling*.

To illustrate these two situations—of a known and an unknown distribution—consider the following example.

The legislature of a certain large state has passed by a 60–40 vote a strict antismog law, which has reached the governor's desk. He must decide in two weeks whether to sign or to veto the law. If the law is enacted, smog levels will be reduced, but the citizens of the state will be inconvenienced and the cost to the taxpayers will be high. Before making his decision, the governor wishes to know the opinions of the voters of the state.

Since the measure received a 60–40 vote, if the governor restricts his attention to the legislature, then he knows all the relevant data. He can think of a "yes" vote as being recorded as 1 and a "no" vote as 0. Thus, the vote of the legislature might be considered a binomial distribution (or *population*) with $p = 0.6$ and $n = 100$.

Even though the legislature is assumed to represent the voters, the governor wants a more accurate indication of the will of the voters. It is impossible to ask every one of the 10,000,000 voters his opinion. Imagine a binomial distribution with p unknown and $n = 10{,}000{,}000$. It is more practical to consider getting the opinions of 1000 voters, and, on the basis of the results of this *sample*, estimate *p*. An important question, of course, is whether or not this sample is representative of the voters at large, and this is, in turn, a question relating to the selection of the 1000 voters.

In this chapter, we shall concentrate on procedures for selecting a sample and on certain important conclusions about samples. Chapter 10 takes up the problem of estimation in more detail. Finally in Chapter 11, we shall investigate how estimates are used in deciding between two courses of action.

9.1 Population Random Variable

For any given population, the entire collection of data we wish to study is called the *population*. The values of the mean and variance of these data are called the *population mean* and the *population variance*.

Example 9.1.1 The data for the test scores of 50 students of Section 4.1 (see also Figure 4.3) were as follows:

Score	0	1	2	3	4	5	6	7	8	9	10
Frequency	1	0	2	6	2	11	6	10	8	3	1
Relative frequency	0.02	0	0.04	0.12	0.04	0.22	0.12	0.20	0.16	0.06	0.02

Suppose that we are interested only in these 50 scores. We have found, in Chapter 4, that $\bar{x} = 5.84$ and $s_x^2 = 4.49$. Thus, the 50 scores form our population, and the population mean and variance are known.

We may interpret the data of Example 9.1.1 in terms of a *population random variable* if we think of the experiment "choose one student's score at random." The score chosen, x, is then one value of the random variable X whose possible values are 0, 1, 2, ..., 10. The probability function for X is found by referring to the relative frequencies of the possible scores:

x	0	1	2	3	4	5	6	7	8	9	10
$f(x)$	0.02	0	0.04	0.12	0.04	0.22	0.12	0.20	0.16	0.06	0.02

For this random variable, we are able to see that $\mu = 5.84$ and $\sigma^2 = 4.49$. Thus μ has the same value as $\bar{x}$, and σ^2 has the same value as s_x^2. μ and σ^2 were computed by considering a random variable, and $\bar{x}$ and s_x^2 were obtained by the methods of Chapter 4.

Having made this identification between

$$\bar{x} \quad \text{and} \quad \mu,$$

and between

$$s_x^2 \quad \text{and} \quad \sigma^2$$

for a *population*, we shall henceforth use μ and σ^2 as symbols for the population mean and the population variance, respectively. We shall reserve the symbols $\bar{x}$ and s_x^2 as measures which relate to the partial information obtained from a *sample*.

If a population is small, as in Example 9.1.1, it is usually practical to obtain all the pertinent data. In such cases, there is no need to study a sample. Our major concern, then, is with large populations. We may consider three types of populations, illustrated by the following examples.

Example 9.1.2 The population considered in the introduction to this chapter is large ($n = 10,000,000$), but *finite*.

Example 9.1.3 Consider repeated trials of the experiment "choose a real number in the interval [3, 15]" (see Example 5.3.1). The population is *infinite*. It is important to notice that for any particular value x of X,

$$P(X = x) = 0.$$

Therefore, if one particular number is removed, this does not affect the probabilities associated with the events (intervals) of the experiment.

Example 9.1.4 Coin tossing presents a situation which gives rise to an imaginary, or theoretical, population. Suppose that we are handed a particular coin and are asked whether or not $P(\text{heads}) = \frac{1}{2}$ for that coin. What population concerns us? We might toss 10 times, getting 5 heads. We could then assert that if one of these 10 tosses is chosen at random, then

$$P(\text{heads}) = \tfrac{1}{2}.$$

The question, however, actually refers to the probability of getting heads on any one of the (theoretically) infinite number of tosses that might be made with the coin. Making a single toss and getting "heads" in no way reduces the supply of heads available for future tosses. We may treat this case as though the population were infinite.

Sometimes in ordinary conversation we speak of a "population" as a collection of objects or persons. For example, we might refer to the population of all students at a certain college. This would not be a statistical population, however. For a statistical population we must have in mind some collection of numbers and an associated random variable. Note that in Example 9.1.1 the statistical population is the set of 50 test scores, not the set of 50 students.

9.2 Random Sampling

In earlier chapters, we have had considerable experience with *sampling* even though we have not been using that term. The development of the binomial distribution (Chapter 7) was based on the central idea of repeated, independent trials of experiments involving a Bernoulli random variable. Such trials (observations) illustrate a special case of random sampling. In Chapter 7 the underlying population (sometimes theoretical) had a Bernoulli distribution.

In general, we may consider any population random variable. If we make a single (random) observation, the possible values associated with this observation are the values of a random variable—call it X_1. The probability distribution of X_1 is identical with the distribution of the population random variable. We refer to X_1 as a **random sample of size one.** A second observation gives rise to another random variable, X_2. Under certain circumstances X_2 will also have a distribution identical with that of the population. This will be the case, for example, if we make our observations with replacement from a finite population. Furthermore, it is possible that X_1 and X_2 will be independent random variables. Again, observing with replacement provides an illustration. Similarly, we may consider n random variables $X_1, X_2, \ldots, X_n$ and make the following definition.

If

  (a) $X_1, X_2, \ldots, X_n$ **are random variables, each having the same probability distribution as the population,**

and if

  (b) $X_1, X_2, \ldots, X_n$ **are independent,**

then

$$\{X_1, X_2, \ldots, X_n\}$$

is a *random sample of size n.*

We need to distinguish between the random variables $X_1, X_2, \ldots, X_n$ and a *particular* sample with values $x_1, x_2, \ldots, x_n$. A specific collection of numbers *cannot* be called either random or not random. The important matter is: were these numbers obtained by a procedure which guarantees that conditions (a) and (b) above are satisfied?

We shall refer to

$$\{x_1, x_2, \ldots, x_n\}$$

as a *particular sample*. The numbers making up a particular sample may be treated by the techniques of Chapter 4. For example, we may compute $\bar{x}$ and s_x^2.

How do we satisfy the conditions for a random sample? There are two obvious possibilities.

1. The population is finite. We make selections at random with replacement.
2. The population is infinite (Example 9.1.3) or "imaginary" (Example 9.1.4).

In practice, a common situation is one in which we are concerned with a very large, but still finite, population. Fortunately in these cases, the computation for sampling without replacement is not too different from that for sampling with replacement. See the following example.

Example 9.2.1 Suppose that, of the 10,000,000 voters of Example 9.1.2, there are actually 7,000,000 who favor the antismog legislation. If one voter is to be selected at random, the sample random variable X_1 has a Bernoulli distribution with $p = 0.7$ (assuming that "$X_1 = 1$" represents a favorable opinion). If we now choose a second voter (without replacing the first one!), this voter is being selected from a population of 9,999,999. $P(X_2 = 1)$ is still 0.7, but X_1 and X_2 are not independent, since

$$P(X_2 = 1 \mid X_1 = 1) = \frac{6,999,999}{9,999,999} \neq 0.7.$$

It is clear, however, that the value of the fraction is very close to 0.7. We would be justified in working with X_1 and X_2 as though they were independent.

Generalizing Example 9.2.1, we see that if the population size is quite large compared with the sample size, then we may treat sampling without replacement as though it were sampling with replacement. In the examples and exercises that follow, we shall assume that our samples may be treated as though they were random samples.

Example 9.2.2 In an investigation of the study habits of 10,000 students at State College, a random sample of 200 students is chosen. These students report on the number of hours they studied during the week preceding final examination. The mean of the particular sample is 24.7. This figure gives some indication of the mean studying time of all students at the college.

There are many extremely complex questions which arise in connection with actually selecting a random sample. Let us assume we have a large finite population (as in Example 9.2.2). One method of choosing a random sample is to assign each student a number. These numbers may be placed on slips of paper (or marbles, or table-tennis balls) and placed in a container (an urn, perhaps). The slips of paper are thoroughly mixed and then drawn one at a time, mixing again after each draw. Although this sounds reasonable, the procedure is not so simple in practice.

Example 9.2.3 In 1969, a draft "lottery" was conducted by associating each eligible individual with his birthdate. The dates were then placed in a drum, mixed, and drawn. The results of the draw showed that an exceptionally large number of December dates were drawn very early in the process. Some statisticians have concluded that perhaps the December dates were placed in the drum last and that the mixing was not thorough. In 1970, both the dates and the order numbers were drawn in an effort to make the results more nearly random.

To help insure randomness in making drawings, some techniques have been developed which eliminate the physical problem of mixing. Tables of "random numbers" have been produced. Sometimes a computer is used to make "random" selections of numbers. We shall not discuss these techniques here. It is sufficient for us to realize that while in our theoretical work with samples we assume that the samples are random, in practice we must often deal with samples which are only approximately random.

Sometimes a population can be separated into subpopulations. We may then use the process of **stratified** random sampling. Example 9.2.4 illustrates this procedure.

Example 9.2.4 In Example 9.2.2, choosing 200 of 10,000 students was considered. Suppose that the 10,000 were distributed as follows:

	Male	Female
Lower Division	1500	1200
Upper Division	3400	2000
Graduate	1100	800

Instead of selecting 2% of the 10,000 at random, we might choose 2% (again at random) from each of the subpopulations. Thus we would choose 68 upper division males, and so forth. There are three advantages in this procedure. It is somewhat easier physically to make random selections from the smaller subpopulations. Furthermore, we are assured that each category is "proportionately represented" in our sample. Finally, we would be able to compare the differences in study hours between the subpopulations.

One point must be emphasized. The goal of the sampling processes is to minimize the probability of obtaining a **biased** (nonrepresentative) sample. Stratifying the sample is one step in this direction. In Example 9.2.4, our procedure insures that we will not choose all 200 students from the graduate category (they might study more hours than would undergraduates). However, no procedure can guarantee that the sample will be representative of the population. The very nature of uncertainty and of randomness tells us that unusual events can occur! If an urn contains 1 red and 99 blue marbles and if we make two random draws with replacement, we *may* get red both times. The probability of this is only $(0.01)^2 = 0.0001$, but it *may* happen.

Another serious problem in sampling is to decide on the size of the sample. As we have said, it is generally true that the larger the value of n, the more information we obtain about the population. The cost, in time and money, of drawing samples and in making tests and observations must be considered. We shall not attempt a full treatment of this difficult question, but Chapter 10 will offer some insights regarding appropriate sample sizes.

EXERCISES 9.2

● 1. A particular die is thrown 60 times. We are interested in whether or not the *a priori* probability of throwing a five is $\frac{1}{6}$ on subsequent throws of this die. Our 60 throws yield 15 fives.
   a. What is the sample size?
   b. What is the value of $\bar{x}$?
   c. What population is under consideration?
   d. If "$P(\text{five}) = \frac{1}{6}$" is true, what is the expected number of fives in 60 throws of the die?

● 2. In order to test the mathematical abilities of third graders, it was decided to select a representative sample of 60,000 third graders in a particular state. Suppose that it has been determined that 0.6 of the children in that state live in urban centers, 0.25 in the suburbs, and 0.15 in rural districts. Which of the following sampling procedures would you choose? Give the reasons for your choice.
   a. Take a random sample of 20,000 children in each of the 3 categories.
   b. Take random samples of 36,000 urban children, 15,000 suburban and 9,000 rural children.
   c. Take a random sampling of the schools, including enough classes so that there will be 60,000 children in the sample.

● 3. Mechanical methods for choosing samples are frequently used. Explain how each of the following devices might be used:
   a. A fair coin to make a random selection of one of two urns.
   b. A red and a green die to select one of a population of 36 or less.
   c. A coin and a die to select one of 12 objects.

● 4. Suggest a procedure for choosing a sample of size four from your class.

5. The number of samples of size 2 that can be drawn from a population of 10 with replacement is 10×10, or 100. The number of samples that can be drawn without replacement is
$$\binom{10}{2} = \frac{10 \cdot 9}{2 \cdot 1} = 45.$$
Find the number of samples (1) with replacement and (2) without replacement for:
   a. sample, 2; population, 20
   b. sample, 2; population, 100
   c. sample, 3; population, 10
   d. sample, 3; population, 100

6. Suppose that an urn contains 3 blue marbles and 6 red marbles. The possible samples of size 2, drawn with replacement, are represented in the following diagram:

(3 blue, 6 red marbles)

Will any one sample reflect the proportion of blue and red marbles in the urn?

7. Suppose that samples of size 3 are drawn from the urn in Problem 6. Will any one sample reflect the proportion of blue and red marbles in the urn?

9.3 The Sample Mean, $\overline{X}$

Suppose that we have a population with μ and σ known. Consider choosing a sample of size n with $X_1, X_2, \ldots, X_n$ as random variables (Section 9.2). Then we define

$$\overline{X} = \frac{X_1 + X_2 + \cdots + X_n}{n},$$

which is also a random variable. If $x_1, x_2, \ldots, x_n$ are particular values of $X_1, X_2, \ldots, X_n$, we write

$$\overline{x} = \frac{x_1 + x_2 + \cdots + x_n}{n}.$$

Here $\overline{x}$ is the *mean of the particular sample*, and $\overline{x}$ is a value of $\overline{X}$.

$$\overline{X} = \frac{X_1 + X_2 + \cdots + X_n}{n} \text{ is called the } \textit{sample mean.}$$

It is clear that the value of $\overline{x}$ for a particular sample gives some indication as to the value of μ, the population mean. In Chapter 10 we shall discuss the question of estimating μ in some detail.

Our immediate goal is to investigate the probability distribution of the random variable

$$\overline{X} = \frac{X_1 + X_2 + \cdots + X_n}{n},$$

where the X_i's are independent random variables, each having the same distribution with μ and σ known.

As a first step, we shall show how to find the expected value and the variance of $\overline{X}$. To do this, we prove the following theorem:

THEOREM If $\overline{X}$ is the random variable defined by

$$\overline{X} = \frac{X_1 + X_2 + \cdots + X_n}{n},$$

then

(a) $\mu_{\overline{X}} = E(\overline{X}) = \mu$

and

(b) $\sigma_{\overline{X}}^2 = \text{var}(\overline{X}) = \frac{\sigma^2}{n},$

where μ and σ^2 are the mean and variance, respectively, of the population; $\mu_{\overline{X}}$ is the mean of the means of all samples of size n, and $\sigma_{\overline{X}}^2$ is the variance of those means.

Proof:

(a) $\mu_{\overline{X}} = E(\overline{X}) = E\left(\dfrac{X_1 + X_2 + \cdots + X_n}{n}\right)$

$= \dfrac{1}{n}[E(X_1) + E(X_2) + \cdots + E(X_n)]$ (Section 6.3)

$= \dfrac{1}{n}\underbrace{(\mu + \mu + \cdots + \mu)}_{n \text{ terms}}$ (since $E(X_i) = \mu$)

$= \dfrac{1}{n}(n\mu) = \mu$

(b) $\sigma_{\overline{X}}^2 = \text{var}(\overline{X}) = \text{var}\left(\dfrac{X_1 + X_2 + \cdots + X_n}{n}\right)$

$= \dfrac{1}{n^2}[\text{var}(X_1 + X_2 + \cdots + X_n)]$ (Section 6.4)

$= \dfrac{1}{n^2}[\text{var}(X_1) + \text{var}(X_2) + \cdots + \text{var}(X_n)]$

(since the X_i's are independent)

$= \dfrac{1}{n^2}\underbrace{(\sigma^2 + \sigma^2 + \cdots + \sigma^2)}_{n \text{ terms}}$

$= \dfrac{1}{n^2}(n\sigma^2) = \dfrac{\sigma^2}{n}$

From part (b) of this theorem, it follows that

$$\sigma_{\overline{X}} = \frac{\sigma}{\sqrt{n}}.$$

Remark 1

The assumption of independence was not used in the proof of part (a) of the preceding theorem. "$\mu_{\overline{X}} = \mu$" is true even if the X_i's are *not* independent, provided that they have identical probability distributions. Look again at the solutions of Problems 1, 3, 4, and 5 of Exercises 8.4, and notice that $\mu_{\overline{X}} = \mu$ in each case.

Remark 2

If the sampling is made from a finite population *without* replacement, then the X_i's are not independent. If the population size is N and the sample size is n, it may be shown that

$$\sigma_{\overline{X}}^2 = \frac{N-n}{n(N-1)} \cdot \sigma^2.$$

(We considered such a case in Section 8.4 with $N = 6$.)

We may write $\dfrac{N-n}{n(N-1)}$ in a different form:

$$\frac{N-n}{n(N-1)} = \frac{1}{n}\left(1 - \frac{n-1}{N-1}\right)$$

For a fixed sample size n, if N is very large in relation to n, the fraction $\dfrac{n-1}{N-1}$ is very small, and so

$$\frac{N-n}{n(N-1)} \doteq \frac{1}{n} \quad \text{if } N \text{ is very large.}$$

Therefore

$$\sigma_{\overline{X}}^2 \doteq \frac{\sigma^2}{n} \quad \text{(see Problem 5, Exercises 9.3).}$$

This remark helps to justify the assertion that if a finite population is very large, then sampling without replacement resembles sampling with replacement.

Example 9.3.1 Let the given population be standard normal. (That is, we consider an infinite population, with an associated random variable which has the standard normal distribution. In particular, $\mu = 0$ and $\sigma = 1$.) If we consider all possible random samples of size 100, the distribution of the means of these samples has (from the theorem above)

$$\mu_{\overline{X}} = \mu = 0 \quad \text{and} \quad \sigma_{\overline{X}}^2 = \frac{\sigma^2}{100} = 0.01.$$

Example 9.3.2 Let $X_1, X_2, \ldots, X_n$ be identically distributed Bernoulli variables with $P(X_i = 1) = p$. Then

$$\overline{X} = \frac{X_1 + X_2 + \cdots + X_n}{n}$$

has

$$\mu_{\overline{X}} = \mu = p \quad \text{and} \quad \sigma_{\overline{X}}^2 = \frac{\sigma^2}{n} = \frac{pq}{n}.$$

Example 9.3.3 It is of interest to verify the result of Example 9.3.2 directly for small values of n. Suppose that $n = 2$.

Possible samples (x_1, x_2)	(1, 1)	(1, 0)	(0, 1)	(0, 0)
$\bar{x} = \dfrac{x_1 + x_2}{2}$	1	$\frac{1}{2}$	$\frac{1}{2}$	0
$P(\overline{X} = \bar{x})$	p^2	pq	qp	q^2

$$\mu_{\overline{X}} = E(\overline{X}) = 1 \cdot p^2 + \tfrac{1}{2}(2pq) = p^2 + pq = p(p + q) = p$$

$$\sigma_{\overline{X}}^2 = 1^2(p^2) + (\tfrac{1}{2})^2(2pq) - (p)^2 = p^2 + \frac{pq}{2} - p^2 = \frac{pq}{2}$$

Example 9.3.4 In Section 8.4 we considered a population of six marbles numbered 1, 2, 3, 4, 5, 6. The mean and the variance were found to be

$$\mu = \tfrac{7}{2} \quad \text{and} \quad \sigma^2 = \tfrac{35}{12}.$$

Now suppose that samples of size two are drawn with replacement. The possible number of samples is

$$6 \times 6 = 36.$$

The distribution of sample means is:

$\bar{x}$	1	$\frac{3}{2}$	2	$\frac{5}{2}$	3	$\frac{7}{2}$	4	$\frac{9}{2}$	5	$\frac{11}{2}$	6
$f(\bar{x})$	$\frac{1}{36}$	$\frac{2}{36}$	$\frac{3}{36}$	$\frac{4}{36}$	$\frac{5}{36}$	$\frac{6}{36}$	$\frac{5}{36}$	$\frac{4}{36}$	$\frac{3}{36}$	$\frac{2}{36}$	$\frac{1}{36}$

$$\mu_{\overline{X}} = \tfrac{1}{36}[1 + 2(\tfrac{3}{2}) + 3(2) + 4(\tfrac{5}{2}) + 5(3) + 6(\tfrac{7}{2})$$
$$+ 5(4) + 4(\tfrac{9}{2}) + 3(5) + 2(\tfrac{11}{2}) + 1(6)] = \tfrac{126}{36} = \tfrac{7}{2} = \mu$$

$$\sigma_{\overline{X}}^2 = \tfrac{1}{36}[1 + 2(\tfrac{9}{4}) + 3(4) + 4(\tfrac{25}{4}) + 5(9) + 6(\tfrac{49}{4}) + 5(16)$$
$$+ 4(\tfrac{81}{4}) + 3(25) + 2(\tfrac{121}{4}) + 1(36)] - \tfrac{49}{4}$$

$$= \frac{1}{36}\left(355 + \frac{277}{2}\right) - \frac{49}{4} = \frac{35}{24} = \frac{\sigma^2}{2}$$

The solution of Problem 1, Exercises 8.4, shows that for the $\binom{6}{2}$, or 15, samples of size two drawn without replacement

$$\mu_{\overline{X}} = \mu \quad \text{and} \quad \sigma_{\overline{X}}^2 = \frac{1}{2}\left(\frac{6-2}{6-1}\right)\sigma^2 = \frac{4}{10}\sigma^2.$$

EXERCISES 9.3

- 1. Assume that I.Q. scores are normally distributed with $\mu = 100$ and $\sigma = 15$. Find $\mu_{\overline{X}}$ and $\sigma_{\overline{X}}$ for random samples of size:
  a. $n = 1$ b. $n = 100$ c. $n = 10{,}000$

- 2. An urn contains 2 red marbles and 1 blue marble. With red taken as a success, 72 draws are made with replacement. Determine $\mu_{\overline{X}}$ and $\sigma_{\overline{X}}$.

- 3. A fair die is thrown 320 times. A success is recorded if a 3 appears.
  a. Find $\mu_{\overline{X}}$ and $\sigma_{\overline{X}}$.
  b. What is the expected number of successes?
  c. Find the standard deviation of the expected number of successes.

★ 4. (See Example 9.3.3.) Verify that for random sampling from a Bernoulli population, $\mu_{\overline{X}} = p$ and $\sigma_{\overline{X}}^2 = \dfrac{pq}{n}$ for $n = 3$.

- 5. If $N = 10{,}000$ and $n = 100$, compute $\dfrac{N - n}{n(N - 1)}$ to four decimal places, and compare the result with $\dfrac{1}{n}$.

6. Repeat Problem 5 with $N = 220{,}000{,}000$ and $n = 2000$.

7. An urn contains three slips of paper bearing numbers 2, 4, 6.
  a. Find μ and σ^2 for this population.
  b. If samples of size two are drawn with replacement, find the distribution of sample means and verify that

$$\mu_{\overline{X}} = \mu \quad \text{and} \quad \sigma_{\overline{X}}^2 = \frac{\sigma^2}{n}.$$

  c. If samples of size two are drawn without replacement, find the distribution of sample means and verify that

$$\mu_{\overline{X}} = \mu \quad \text{and} \quad \sigma_{\overline{X}}^2 = \frac{1}{n}\left(\frac{N - n}{N - 1}\right)\sigma^2.$$

8. Repeat Problem 7 for an urn containing four slips of paper bearing numbers 3, 6, 9, 12, and the samples are of size three.

9.4 The Law of Large Numbers

We are now able to prove a theorem of considerable importance. The theorem is a special form of what is frequently called the *Law of Large Numbers*. The result is the basis for the long-run interpretation of probability and the so-called "law of averages" (see Section 2.8). Before stating the conclusion, we recall Chebyshev's Inequality (Section 6.2):

If Y is a random variable with μ_Y and σ_Y^2 given, then

$$P(|Y - \mu_Y| > k\sigma_Y) \leq \frac{1}{k^2}$$

where k is any positive constant.

If we substitute $c = k\sigma_Y$, we obtain an alternate form of Chebyshev's Inequality:

$$P(|Y - \mu_Y| > c) \leq \frac{\sigma_Y^2}{c^2}$$

Now we replace Y by $\overline{X}$, μ_Y by $\mu_{\overline{X}} = \mu$, and σ_Y^2 by $\sigma_{\overline{X}}^2 = \frac{\sigma^2}{n}$:

$$P(|\overline{X} - \mu| > c) \leq \frac{\sigma^2}{nc^2}.$$

If we choose n (the sample size) sufficiently large, then $\frac{\sigma^2}{nc^2}$ can be made as close to 0 as we wish. That is, for large values of n

$P(|\overline{X} - \mu| > c)$ is close to 0, or

$P(|\overline{X} - \mu| \leq c)$ is close to 1.

Since c is any constant, we may take c as small as we wish. We have proved:

THE LAW OF LARGE NUMBERS If the size of a randomly chosen sample is "sufficiently large," then the probability that the mean of the sample differs by more than a small amount from the population mean is nearly 0.

In other words:

For a large random sample we can be "very confident" (not certain!) that the mean of the sample is "close to" the population mean.

Applying this result to the Bernoulli case, with $P(\text{success}) = p$, we have this:

If T is the total number of successes obtained in n independent Bernoulli trials, then the ratio $\frac{T}{n}$ is close to p if n is large.

Example 9.4.1 If a particular coin is fair, $P(\text{heads}) = \frac{1}{2}$. In making 1000 tosses, we anticipate that the total number of heads will be close to 500.

Example 9.4.2 We have a particular coin and wish to decide whether "$P(\text{heads}) = \frac{1}{2}$" is true or false. Suppose that we have made 1000 tosses and have obtained 492 heads. The law of large numbers enables us to conclude that 0.492 is "close to" the true value of p. We cannot be *certain* that $p = \frac{1}{2}$, but we may conclude that the true value of p does not differ greatly from 0.492.

We began Section 9.3 with the stated goal of investigating the probability distribution of $\overline{X}$. We have found that

$$\mu_{\overline{X}} = \mu \quad \text{and} \quad \sigma_{\overline{X}}^2 = \frac{\sigma^2}{n}.$$

These results, in turn, led us to the law of large numbers. The difficulty with using this theorem is the occurrence of the phrases "sufficiently large," "very confident," and "close to." These are descriptive phrases, but not very helpful unless we can answer the questions, "How large?" "How confident?" and "How close?" We might be able to answer these questions if we knew the precise distribution of $\overline{X}$.

EXERCISES 9.4

★ 1. An urn contains 3 red and 2 blue marbles. How many draws with replacement should be made (at least) so that the probability of not drawing a red marble between 0.5 to 0.7 of the time is not more than 0.05? Use $P(|\overline{X} - 0.6| > 0.1) \leq 0.05$, where $0.05 = \dfrac{\sigma^2}{nc^2}$.

★ 2. For the urn of Problem 1, let Y represent the number of blue marbles drawn. If $n = 500$, determine an interval of the form $|\overline{Y} - 0.4| < c$ such that the probability of obtaining a value of $\overline{Y}$ in the interval is greater than 0.99.

★ 3. Assume that I.Q. is normally distributed with $\mu = 100$ and $\sigma = 15$. Use Chebyshev's Inequality to find the sample size needed in order that the probability that the sample mean is between 95 and 105 is at least 0.99.

● 4. The English statistician R. A. Fisher remarked that the law of averages works by "swamping." In Section 2.8 we considered a fair coin which fell heads on the first 10 tosses. Interpret Fisher's remark for this situation.

9.5 The Distribution of $\overline{X}$

In studying the probability distribution of $\overline{X}$ in Section 9.3, we have found that

$$\mu_{\overline{X}} = \mu \quad \text{and} \quad \sigma_{\overline{X}}^2 = \frac{\sigma^2}{n},$$

where μ and σ^2 are the expected value and variance of the population. It might seem reasonable to assume that the pattern, or shape, of the distribution of $\overline{X}$ also would depend on the distribution of the population random variable. While this is true in general, our experience (Chapter 8) with the binomial distribution leads us to guess that if n is large, the various distributions of $\overline{X}$ might have the same general shape.

The binomial distribution itself arises from considering repeated, independent trials from a Bernoulli population. We may interpret this in terms of random sampling, since our definition of random sampling requires precisely these conditions. For large values of n we stated without proof De Moivre's Theorem (Section 8.2) that the standardized binomial variable

$$Z = \frac{X - np}{\sqrt{npq}}$$

has a distribution which is approximately standard normal. If we divide numerator and denominator by n and replace $\frac{X}{n}$ by $\overline{X}$, we have

$$Z = \frac{\overline{X} - p}{\sqrt{\frac{pq}{n}}}.$$

For a Bernoulli population, $\mu_{\overline{X}} = \mu = p$, $\sigma_{\overline{X}} = \frac{\sigma}{\sqrt{n}} = \sqrt{\frac{pq}{n}}$, and this conclusion may be restated:

For a Bernoulli population, the standardized random variable

$$Z = \frac{\overline{X} - \mu}{\frac{\sigma}{\sqrt{n}}}$$

has a distribution which is approximately standard normal if n is sufficiently large.

This is a special case of the following powerful theorem of statistics:

CENTRAL LIMIT THEOREM If X_1, X_2, ..., X_n are identically distributed random variables each having expected value μ and variance σ^2, then the probability distribution of

$$Z = \frac{\overline{X} - \mu}{\frac{\sigma}{\sqrt{n}}}$$

is approximately standard normal if n is sufficiently large.

Remark 1

The proof of the Central Limit Theorem would require more mathematical knowledge than we have assumed for this course. We have tried to make the conclusion plausible for the Bernoulli case. In Section 8.4 we saw the tendency toward normality even for a special case in which the sampling technique did not satisfy the condition of independence.

Remark 2

Again we encounter the phrase "if n is sufficiently large." It is rather accepted practice to use the standard normal approximation if $n \geq 30$.

Remark 3

If the population itself is normal, then it can be shown that the distribution of

$$\frac{\overline{X} - \mu}{\frac{\sigma}{\sqrt{n}}}$$

is standard normal (rather than only approximately standard normal).

Example 9.5.1 A random sample of size 100 is to be drawn from a population having $\mu = 8$ and $\sigma^2 = 25$. Before actually drawing the sample, we may answer such questions as: What is the (*a priori*) probability that the obtained value of $\overline{X}$ will:
a. be less than 9?
b. be more than 7.5?
c. lie between 7.8 and 8.2?

We have $\mu_{\overline{X}} = 8$, $\sigma_{\overline{X}} = \frac{5}{\sqrt{100}} = 0.5$, and $Z = \frac{\overline{X} - \mu}{\sigma_{\overline{X}}} = \frac{\overline{X} - 8}{0.5}$.

a. $P(\overline{X} < 9) = P(\overline{X} - 8 < 1) = P\left(\frac{\overline{X} - 8}{0.5} < 2\right)$
$= P(Z < 2) \doteq 0.9773$ (Table N)

b. $P(\overline{X} > 7.5) = P(\overline{X} - 8 > -0.5) = P\left(\frac{\overline{X} - 8}{0.5} > -1\right) = P(Z > -1)$
$= P(Z < 1) \doteq 0.8413$

c. $P(7.8 \leq \overline{X} \leq 8.2) = P\left(-0.4 \leq \frac{\overline{X} - 8}{0.5} \leq 0.4\right)$
$= P(|Z| \leq 0.4) \doteq 0.6554 - 0.3446 = 0.3108$

In connection with Example 9.5.1, we must remind ourselves that once the sample is drawn and $\bar{x}$ is computed, then the value of $\bar{x}$ is a *number*. We would not ask probability questions about the value of a definite number. "$P(\overline{X} < 9) \doteq 0.9773$" is a statement about the random variable $\overline{X}$. When $\bar{x}$ is known, then $P(\bar{x} < 9)$ is either 0 or 1.

EXERCISES 9.5

- 1. Assume that I.Q. scores of school children are normally distributed with $\mu = 100$ and $\sigma = 15$. A random sample of size 400 is selected. What is the (*a priori*) probability that the sample mean will be:
  a. less than 101?
  b. between 99 and 101?
  c. more than 102?

- 2. Using the information of Problem 1, suppose that two independent samples, each of size 100 are selected. What is the probability that:
  a. both sample means are less than 101 (recall Section 3.2)?
  b. at least one sample mean exceeds 101 (recall Section 2.6)?

3. An experiment consists of tossing a fair coin 100 times. In what percent of such experiments would we expect:
   a. more than 48 heads?
   b. between 45 and 55 heads?
4. An experiment consists of tossing a fair coin 1000 times. In what percent of such experiments would we expect:
   a. more than 480 heads?
   b. between 450 and 550 heads?
5. Recall that for a fair die, $\mu = 3.5$ and $\sigma^2 = \frac{35}{12}$ (Example 5.5.2 and Problem 6, Exercises 6.1). In a random sample of 100 throws, what is the probability that the total number of points made is between 345 and 355? (Hint: The average is between 3.45 and 3.55.)
6. For the fair die of Problem 5, in a random sample of 100 throws, what is the probability that the total number of points made is less than 200?
7. A manufacturer claims that the tires it produces have an average life of 40 months. Assuming a standard deviation of 3 months, what is the probability that a random sample of 36 of these tires will have an average life less than
   a. 3 years?   b. 39 months?
8. A claim is made that a product has an average life of 4.5 years with a standard deviation of 0.5 year. What is the probability that a random sample of n of these products will have an average life of at least 4 years when:
   a. $n = 36$?   b. $n = 100$?   c. $n = 1000$?

9.6 What Do You Think?

Our theme has been "making decisions in the face of uncertainty." Perhaps the most common situation is one where a decision *could* be made if the distribution of a population were known.

Example 9.6.1 A manufacturing company is interested in the fraction, p, of defective items produced. If p is small, then there is no need to install expensive new equipment. If p is larger than some acceptable value, a decision to install new machinery must be made. In practice, the value of p can be estimated by sampling. A particular random sample of size n will yield a certain fraction of defects *for that sample*.

The major problems we must tackle are:

a. Knowing a particular value, $\bar{x}$, of $\bar{X}$, what do we know about μ?
b. Is the size of the population variance a factor?
c. If the population variance is unknown, how may it be estimated?

What do you think?

EXERCISES 9.6a

For Problems 1–7 consider an urn which contains two marbles, either both red, both blue, or one of each color. That is, $p = P(\text{red})$ has one of the three values 1, 0, or $\frac{1}{2}$. We wish to determine which is the correct value by making repeated drawings with replacement.

For Problems 1–4:
a. Guess the correct value of p.
b. Describe your degree of belief in your guess (certain, very confident, not very confident).

- 1. 10 draws, yielding 3 red and 7 blue
- 2. 3 draws, yielding 3 reds
- 3. 300 draws, yielding 300 reds
- 4. 1 draw yielding blue
- 5. From the sample of Problem 4 we conclude that $p \neq 1$. Our choice, therefore, lies between "$p = 0$" and "$p = \frac{1}{2}$." What is the *a priori* probability of drawing blue if:
  a. $p = 0$?   b. $p = \frac{1}{2}$?
- 6. In Problem 2 we have drawn 3 reds. What is the *a priori* probability of 3 reds if:
  a. $p = 1$?   b. $p = \frac{1}{2}$?
- 7. Do the results of Problems 5 and 6 confirm your guesses for Problems 4 and 2?

(The preceding problems are simple illustrations of the "maximum likelihood" principle. Briefly stated, of two choices for p, we prefer the one that yields the highest *a priori* probability of obtaining the actual results.)

★ 8. An urn contains either 3 blue and 1 red or 2 blue and 2 red marbles. Three marbles are drawn (with replacement). The actual draws are: red, blue, red. Use the "maximum likelihood" principle to guess which mixture of marbles is in the urn.

- 9. Comparing Problems 2 and 3, we are reminded that the results from a large sample give us more confidence in our guess than the results from a small sample. Suppose that there is reason to surmise that a certain population, with σ known, has μ near to 50. Which of the following situations best supports this surmise?
  a. $\bar{x} = 51$ for $n = 100$
  b. $\bar{x} = 50$ for $n = 5$
  c. $\bar{x} = 49$ for $n = 102$

- 10. A population has known σ but unknown μ. A sample of size 100 yields $\bar{x} = 17$. We decide to use 17 as an approximation to μ. For which value of σ would we have the most confidence in our approximation?
  a. $\sigma = 5$ b. $\sigma = 50$

One of the problems that plague statisticians is the fact that unusual events do happen (see Section 2.8 and Section 9.2). Even though a sampling procedure is well planned and satisfies the criteria for randomness, a particular sample may grossly misrepresent the population.

Example 9.6.2 An urn contains 10 marbles, one numbered 0 and each of the other nine numbered 5. Hence $\mu = 4.5$. Three draws (with replacement) are made and the mean of these three draws is $\bar{x} = 0$. The probability that this will occur even though the marbles are thoroughly mixed after each drawing is $(\frac{1}{10})^3 = 0.001$. If $\bar{x}$ were used as an estimate of μ, the resulting approximation would be very poor.

A possible solution to this problem is to increase the sample size (see Section 9.4). The Law of Large Numbers comes to our rescue. However, even for large n it *may* happen that an "unusual" selection has been made purely by chance.

Another approach is to ignore a few cases of wildly divergent ("maverick") data.

Example 9.6.3 A biologist wishes to investigate the effect of a certain hormone on the health of mice. A hundred mice are treated, and after the treatment 98 mice are obviously more healthy, but 2 die within a few hours. Should the biologist ignore the two extreme cases and report that the hormone is beneficial? Perhaps. It is also possible, however, that the two special cases are really the ones of most interest.

Maverick data are a real problem in many practical situations. In the field of human medicine, for example, the exceptional cases cannot be ignored.

EXERCISES 9.6b

- 1. The probability for a particular sequence of two repeated digits in a table of random numbers is $(\frac{1}{10})^2$.
   a. On a particular page containing 850 pairs of digits were fifteen 99's, and on the next two pages a total of fourteen 99's. Would you say that something is wrong with this table? Explain.
   b. In a table of 1,000,000 triplets of digits, how many triplets 999 are expected? Would you be surprised to find 100 such triplets in such a table?
- 2. A man tossed 21 sevens in a row in a dice game. A lady spectator decided to bet against him tossing another seven on the next throw and she lost. Was she betrayed by the law of large numbers?

- 3. A surgeon knows that the rejection rate for a certain imperfect matching in a heart transplant is 0.01. His patient leaves the decision to operate entirely up to the doctor. If you were the doctor, would you ignore the 1% failure rate regardless of the circumstances (money is no object), would you never ignore the 1%, or sometimes ignore it?
- 4. A student's high school transcript shows 18 A's, 17 B's, and 1 F. Is it reasonable to think of this student as a B+ student, even though the F lowers his grade point average below B+?
- 5. The 20-year record for a man's (high) blood pressure was: 135, 138, 137, 140, 180, 141, 138, 135, 138, 138, 140, 137, 138, 135, 142, 136, 136, 138, 137, 137. The doctor of this patient presents these data to a new physician, asking for his evaluation of the "maverick" 180. If you were the new physician, how would you react?

Chapter Summary

1. The symbols μ and σ^2 will be used for the population mean and the population variance, respectively, while $\bar{x}$ and s_x^2 will be used for the corresponding measures obtained from a sample.

2. If
   (a) $X_1, X_2, \ldots, X_n$ are random variables, each having the same probability distribution as the population,
   and if
   (b) $X_1, X_2, \ldots, X_n$ are independent,
   then
   $$\{X_1, X_2, \ldots, X_n\}$$
   is a *random sample of size n*.

3. a. $\bar{X} = \dfrac{X_1 + X_2 + \cdots + X_n}{n}$ is called the *sample mean*.
   b. If μ is the population mean, then
   $$\mu_{\bar{X}} = \mu.$$
   c. If σ^2 is the population variance, then
   $$\sigma_{\bar{X}}^2 = \frac{\sigma^2}{n}.$$

4. *The Law of Large Numbers:* If the size of a randomly chosen sample is "sufficiently large," then the probability that the mean of the sample differs by more than a small amount from the population mean is nearly 0.

5. *Central Limit Theorem:* If $X_1, X_2, \ldots, X_n$ are identically distributed random variables each having expected value μ and variance σ^2, then the probability distribution of

$$Z = \frac{\overline{X} - \mu}{\frac{\sigma}{\sqrt{n}}}$$

is approximately standard normal if n is sufficiently large.

Chapter Review

1. An urn contains four slips of paper bearing the numbers 1, 3, 5, 7.
   a. Find μ and σ^2 for this population.
   b. Consider all possible samples of size two if we draw one slip at a time with replacement. Find the distribution of $\overline{X}$ and verify that

   $$\mu_{\overline{X}} = \mu$$

   and

   $$\sigma_{\overline{X}}^2 = \frac{\sigma^2}{n}.$$

   c. Consider all possible samples of size two if we draw two at a time. There are $\binom{4}{2} = 6$ such samples. Find the distribution of $\overline{X}$ and verify that

   $$\mu_{\overline{X}} = \mu$$

   and

   $$\sigma_{\overline{X}}^2 = \frac{1}{n}\left(\frac{N - n}{N - 1}\right)\sigma^2.$$

2. If we choose, at random, a single value of Z from a standard normal distribution, what is the sample size? the population size?

3. In Section 2.8, some questions were raised about the "law of averages." What is the proper interpretation of this phrase?

4. Assume that a random variable has the standard normal distribution. What is the *a priori* probability that a random sample of size 64 will yield a value of $\overline{X}$ that is
   a. between 0 and 0.2?
   b. between -0.2 and 0.2?
   c. more than 0.2?

5. A college librarian wishes to determine the average weekly number of hours of use of the reading room by students at the college during a given term.
   a. What population is he considering?
   b. He decides to conduct a "survey" by asking the first 100 students who enter the reading room on Monday morning how many hours they used the room the preceeding week. Will the data that he collects be representative of his population?
6. Three fourths of the seedlings of a certain plant will bear blossoms.
   a. If 200 seedlings are planted, what is the expected number that will bear blossoms?
   b. What is the variance of the number that will bear blossoms?
   c. Use Chebyshev's Inequality to find the maximum probability that the number bearing blossoms will be less than 140 or more than 160.

Chapter 10

Estimation

An important part of our study is concerned with making statistical inferences about a population on the basis of a sample drawn from that population. In line with this idea, the bulk of this chapter deals with estimating the value of a population mean by examining the mean of a sample.

The material of Chapter 9 led us to assert that if a particular random sample has mean $\bar{x}$, then we are "confident" that $\bar{x}$ is "close to" the value of μ. Our remaining task is to investigate the questions "How confident?" and "How close?"

10.1 Sample of Size One

Our major concern shall be the estimation of the value of μ based on a value $\bar{x}$ of $\bar{X}$ obtained from a "large" sample. We already know, from Chapter 9, a great deal about the distribution of $\bar{X}$ if n is large. Specifically, we know that for a large sample, $\bar{X}$ has a distribution which is approximately normal and

$$\mu_{\bar{X}} = \mu, \qquad \sigma_{\bar{X}} = \frac{\sigma}{\sqrt{n}},$$

where μ and σ are the population mean and standard deviation, respectively.

Before considering further the large sample situation, it is instructive to examine the extreme case of the smallest possible sample size, namely, $n = 1$.

Example 10.1.1 Suppose that we are interested in trying to determine, as nearly as we can, the value of μ for a population which is known to be normal with $\sigma = 2$. We are allowed to choose, at random, a single member, x, of the population. That is, we consider a sample of size 1. Thus, our particular x is some number; let us suppose that it is 70. What do we now know about μ? The answer to this question is important. We don't *know* anything at all about the value of μ, but we are now able to do some intelligent guessing.

Let us see what intelligent guesses we might make about μ, using the data of Example 10.1.1. We know a great deal about a normal distribution. For example, less than 1% of the distribution lies more than 2.58 standard deviations from μ. Since $\sigma = 2$, the probability that a randomly chosen x would be more than 2(2.58), or 5.16, away from μ is less than 0.01. Such a thing might happen "by chance," but we are 99% confident that our obtained value of 70 lies within 5.16 of the population mean.

Again referring to Example 10.1.1, suppose that we are asked, "What do you think is the exact value of μ?" We might try to avoid the question by replying that we believe μ is close to 70. If our inquirer insists on a single number as our guess as to the value of μ, it is reasonable for us to answer that we don't know the exact value of μ but, being forced to name a single number as our guess, we have no better guess than 70.

EXERCISES 10.1

In each case, assume that we have observed a randomly chosen member of the population. That is, we have a single value, x, obtained from a random sample of size 1. (For normal populations, use Table N.)

- 1. The length of adult fish of a certain species is normally distributed with $\sigma = 4$ centimeters. If one fish is selected at random from this species, what is the probability that its length will be within 7.84 centimeters of the mean length of all adult fish of this species?

- 2. Suppose that $\mu = 25$ centimeters for the population of Problem 1. What is the probability that a randomly selected fish has a length which falls outside the interval 31.56 to 18.44 centimeters?

- 3. The 2, 3, 4, ..., 10 of hearts are taken from an ordinary deck and placed in an urn. We shall draw one card and record its value.
  a. Find μ for this population.
  b. We may check that $\sigma \doteq 3.7$. What is the probability that the value of the card drawn at random will be more than 2σ away from the mean?

- 4. Suppose that a single Bernoulli trial yields a success. What definite conclusion may we draw about $P(X = 1)$?

- 5. If we have no information about a population distribution except that $\sigma = 2$, the probability that a randomly chosen member of this population is within 6 units of μ is at least 0.889. Why? (Hint: $6 = k\sigma$; use Chebyshev's Inequality.)

- 6. A single value randomly chosen from a normal distribution is 30. A second, independent, randomly chosen value is 20. Find the probability that:
  a. Both values are larger than μ.
  b. Both values are smaller than μ.
  c. One value is larger than μ and one value is smaller than μ.

- 7. If you had to choose a single number as a guess for μ, would you choose 30, 20, or 25 for Problem 6?
- 8. Suppose that the standard deviation of a normal distribution is 10 and the mean is 5. What is the probability that a single value drawn at random will:
   a. fall between -5 and 15?
   b. be greater than 25?
- 9. A sample of size 1 is drawn from the cards used in Problem 3. Suppose that it is 9.
   a. What is the mean of this sample?
   b. What is the variance for this sample?
- 10. In Problem 9:
   a. What is the mean of all samples of size 1?
   b. What is the variance of all samples of size 1?

10.2 Point Estimation

In the discussion of Example 10.1.1 we saw that we could do some intelligent guessing about the population mean even if we made only a single observation. Of course, a single observation provides only one piece of information. If we have a sample of size $n(n > 1)$, we have more information and can sharpen our guess.

For any particular random sample we can calculate its mean, $\bar{x}$. Now $\bar{x}$ is a particular value of $\bar{X}$. Since $E(\bar{X}) = \mu$, if we wish to choose a single number as an estimate for μ, we do the obvious thing and choose $\bar{x}$. A single number estimate is called a *point estimate*.

A frequently occurring situation is one in which the population is known to be Bernoulli. The various possible Bernoulli populations are determined by the values of p. We call p the population **parameter** of a Bernoulli distribution. The experimentally determined value of the ratio

$$\bar{X} = \frac{\text{number of successes}}{\text{number of trials}}$$

is a **sample statistic**. We use a particular value, $\bar{x}$, of $\bar{X}$ as a point estimate of p. The actual value of p, of course, is some real number in the interval $[0, 1]$. If we use $\bar{x}$ as an estimate of p, we are estimating the location of p by singling out one particular point in the interval $[0, 1]$.

Example 10.2.1 To determine the probability, p, of heads for a particular bent coin, we make 100 tosses and obtain 42 heads. As a reasonable guess for the value of p, we take $\bar{x} = \frac{42}{100} = 0.42$. The true value of p may be 0.4213, 0.40578, or any one of an infinite number of values. It may even be quite different from 0.42, say 0.8724 or 0.19462. However, we are inclined to believe that 0.42 is close to the true value of p.

A Bernoulli population depends only on the single parameter p. If a distribution (population) is normal, we are concerned with two **parameters**, μ and σ (or σ^2). For most populations, even though not necessarily normal, μ and σ are important parameters.

Example 10.2.2 Suppose that the data of Example 9.1.1 arose from a random sample of some population. For these test scores, $\bar{x} = 5.84$ and $s_x^2 = 4.49$. We would use 5.84 as our point estimate for μ.

The point estimate for μ in Example 10.2.2 might be interpreted as follows: if the test were to be given to a large number of students with the same training, background, and capacities as the students in the sample, the best guess for the population mean would be 5.84.

Unfortunately, the matter of deciding on a point estimate for σ^2 is not quite so simple. To decide on the best estimate of σ^2, we would have to conduct an investigation into the distribution of the random variable whose values are the values of s_x^2 from possible samples of size n. This is a rather complicated matter. Our intuition, perhaps, suggests that we use the obtained value of s_x^2 as the best point estimate of σ^2. It turns out to be the case that a better point estimate for σ^2 is

$$\frac{n}{n-1} s_x^2.$$

Of course, if n is large, the factor $\frac{n}{n-1}$ is close to 1, and, for simplicity, we shall assume that n is sufficiently large that we *may* use s_x^2 as an estimate for σ^2.

In practice, happily, it is often the case that we have reasonably accurate information about σ^2 even though the value of μ is unknown.

Example 10.2.3 A certain machine is designed to drill 1″-diameter holes in pieces of steel. Not every hole will be exactly 1″ in diameter. There will be a certain variability among the diameters. Considering the first day's run as a population, we might have $\mu = 1.00$ and $\sigma^2 =$ some small value. Several weeks later the cutting edge of the drill might have dulled so that a day's production will yield a different (smaller) value of μ. The new σ^2, however, probably will *not* differ much from the original value.

10.3 Confidence Intervals

The use of a point estimate for μ is rather unsatisfactory. First of all, for most situations, the probability that a particular sample will yield $\bar{x}$ exactly equal to μ is very small (or even zero). Secondly, while we know that an obtained $\bar{x}$ will probably be close to μ, the questions of "How close?" and "With what probability?" still remain to be answered.

In order to handle these difficulties, we turn to the idea of examining an *interval* of values and raising probability questions about that interval. We first consider cases for which the population variance is known. Our principal tool is the basic result of Chapter 9; that is, for large random samples, the variable

$$Z = \frac{\overline{X} - \mu}{\sigma_{\overline{X}}}$$

is approximately standard normal.

Example 10.3.1 The average gasoline consumption (μ) for cars of a certain make is 16 miles per gallon (mpg) with standard deviation of 5 mpg. Consider random samples of size 100 $\left(\sigma_{\overline{X}} = \dfrac{5}{\sqrt{100}} = 0.5\right)$ chosen from the population of all cars of this make. We can find an interval about μ that includes 95% of the means of all such samples. Since

$$P(|Z| \leq 1.96) = 0.95$$

(from Section 5.4), we have

$$P\left(\left|\frac{\overline{X} - 16}{0.5}\right| \leq 1.96\right) = 0.95.$$

This may be rewritten as

$$P\left(-1.96 \leq \frac{\overline{X} - 16}{0.5} \leq 1.96\right) = 0.95$$

or

$$P(-0.98 \leq \overline{X} - 16 \leq 0.98) = 0.95$$

or further

$$P(15.02 \leq \overline{X} \leq 16.98) = 0.95.$$

We conclude that 95% of all samples of size 100 will yield a value of $\overline{X}$ that falls within 0.98 of μ.

For Example 10.3.1 we have found an interval about $\mu = 16$ which includes 95% of all means of samples of size 100. This interval is [15.02, 16.98]. (We sometimes express this as 16 ± 0.98.)

In general:

For *n* sufficiently large:

$$P\left(|\overline{X} - \mu| \leq 1.96 \frac{\sigma}{\sqrt{n}}\right) = 0.95,$$

and the interval

$$\left[\mu - 1.96 \frac{\sigma}{\sqrt{n}},\ \mu + 1.96 \frac{\sigma}{\sqrt{n}}\right]$$

contains 95% of all sample means.

If we consider selecting one sample, the *a priori* probability that the mean of that sample will fall in this interval is 0.95. That is, we are 95% sure that $\bar{x}$ lies in the interval shown in Figure 10.1.

Figure 10.1

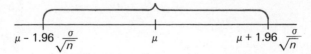

This is all very straightforward. If we know μ (as well as σ), we may make probability statements about the possible values of $\bar{X}$. It was this idea that was developed in Chapter 9. Our problem, however, is to estimate the value of μ, knowing only one particular value, $\bar{x}$, of $\bar{X}$. Let us rewrite our result from the bottom of page 145:

$$P\left(|\bar{X} - \mu| \leq 1.96 \frac{\sigma}{\sqrt{n}}\right) = P\left(|\mu - \bar{X}| \leq 1.96 \frac{\sigma}{\sqrt{n}}\right)$$

$$= P\left(\bar{X} - 1.96 \frac{\sigma}{\sqrt{n}} \leq \mu \leq \bar{X} + 1.96 \frac{\sigma}{\sqrt{n}}\right)$$

$$= 0.95.$$

We must be careful in interpreting this last equation. Since μ is a constant, this is *not* a probability statement about μ. It is, instead, a probability statement about the *random interval*

$$\left[\bar{X} - 1.96 \frac{\sigma}{\sqrt{n}}, \bar{X} + 1.96 \frac{\sigma}{\sqrt{n}}\right]$$

having random endpoints.

For a particular $\bar{x}$, we are 95% confident that the interval

$$\left[\bar{x} - 1.96 \frac{\sigma}{\sqrt{n}}, \bar{x} + 1.96 \frac{\sigma}{\sqrt{n}}\right]$$

contains μ. Such an interval is called a 95% *confidence interval*. The endpoints are called *confidence limits*.

We have estimated the location of μ in terms of an interval. Thus, we may say that we are 95% confident that the interval shown in Figure 10.2 contains μ. (We are assuming that n is sufficiently large and that the normal approximation is applicable.)

Figure 10.2

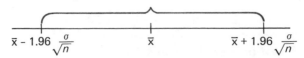

It is sometimes convenient to write a 95% confidence interval in the form

$$\bar{x} \pm 1.96 \frac{\sigma}{\sqrt{n}}.$$

Example 10.3.2 A certain population is known to have $\sigma = 6$. It is decided to estimate μ by finding a 95% confidence interval based on a random sample of size 64. For the sample, $\bar{x} = 30$. A 95% confidence interval is

$$\bar{x} \pm 1.96 \frac{\sigma}{\sqrt{n}} = 30 \pm 1.96 \frac{6}{\sqrt{64}} = 30 \pm 1.47,$$

or

$$[28.53, 31.47].$$

We are 95% confident that μ lies in this interval. We reason that only 5% of samples of size 64 would give rise to an interval of this width which would fail to include μ.

At long last, we have obtained some answers to our questions "How confident?" are we that $\bar{x}$ is "How close?" to μ. There are still some problems remaining. For example, what if the population standard deviation is not known? Also, how do we decide on an appropriate sample size? We shall turn to these and to related questions in Section 10.4.

Before a sample is drawn,

$$P\left(|\bar{X} - \mu| \le 1.96 \frac{\sigma}{\sqrt{n}}\right) = 0.95$$

is, as we have remarked, an *a priori* probability. *After* we have computed $\bar{x}$, we refer to

$$\bar{x} \pm 1.96 \frac{\sigma}{\sqrt{n}}$$

as the 95% *confidence interval*. Thus "95% confidence" may be thought of as our degree of belief that something happened, which we would expect (in advance) to happen 95% of the time.

EXERCISES 10.3

- **1.** Consider the confidence interval $\bar{x} \pm 1.64 \frac{\sigma}{\sqrt{n}}$. Of all such intervals, what fraction contain μ? (Hint: Find $P(|Z| \le 1.64)$.

- **2.** Consider the intervals $\bar{x} \pm a \frac{\sigma}{\sqrt{n}}$. Determine a such that:
  a. 98% of all such intervals contain μ.
  b. 99% of all such intervals contain μ.
  c. 68.26% of all such intervals contain μ.

- 3. We may consider the general idea of a confidence interval as follows: By establishing an interval $\bar{x} \pm a \dfrac{\sigma}{\sqrt{n}}$, we have constructed a "trap" of length $2a \dfrac{\sigma}{\sqrt{n}}$, and we have a certain confidence that we have captured μ in our interval. If $a = 1.96$, we have 95% confidence. Assuming μ is constant:
  a. If we desire to increase our confidence in having captured μ, do we lengthen or shorten our "trap"?
  b. If we desire to capture μ within a shorter interval, will our confidence in the success of our trap be increased or decreased?
- 4. Problem 3 points out that if we wish a narrow (accurate) interval estimate for μ, we must sacrifice "confidence." If we wish high confidence (probability) that our interval estimate is correct, we must lengthen the interval. If we desire to increase our accuracy without sacrificing confidence, what might be done?
- 5. a. What is the confidence that the interval $\bar{x} \pm 1.96 \dfrac{\sigma}{\sqrt{n}}$ will fail to include μ?
  b. What is the confidence that the interval $\bar{x} \pm 1.96 \dfrac{\sigma}{\sqrt{n}}$ lies entirely to the *left* of μ?
  c. Since $\bar{x} - 1.96 \dfrac{\sigma}{\sqrt{n}}$ is surely to the left of $\bar{x} + 1.96 \dfrac{\sigma}{\sqrt{n}}$, we could consider the set of real numbers to the left of $\bar{x} + 1.96 \dfrac{\sigma}{\sqrt{n}}$ as a "one-sided interval." From Table N, $P(Z \leq 1.96) \doteq 0.9750$, and so about 97.5% of all such intervals contain μ. What can be said about the set of real numbers to the right of $\bar{x} - 2.58 \dfrac{\sigma}{\sqrt{n}}$?
- 6. A certain population is known to have $\sigma = 10$. Samples of 100 are drawn. Find the 95% confidence intervals for μ if the obtained value of $\bar{x}$ is:
  a. 20   b. 22   c. 24
- 7. A certain population is known to have $\sigma = 10$. Find the 95% confidence intervals for the following situations:
  a. $\bar{x} = 21, n = 100$ b. $\bar{x} = 21, n = 400$ c. $\bar{x} = 21, n = 625$

10.4 Using Confidence Intervals

For large random samples, we establish confidence intervals for μ based on our knowledge of σ and on the obtained value of $\bar{x}$ (Section 10.3). The width of a confidence interval

$$\bar{x} \pm a \dfrac{\sigma}{\sqrt{n}} \quad \text{is} \quad 2a \dfrac{\sigma}{\sqrt{n}}.$$

Assuming that σ is known, we see that the width depends on a and on n. The factor a depends in turn on the desired degree of confidence, values of

a being found from Table N. Suppose that a 95% confidence is desired. Then a is fixed ($a = 1.96$) and the width is now determined by n, the sample size. By increasing the size of n, we are able to locate μ within a smaller interval. (Of course, we are not *sure* that μ is in this interval; we are only 95% confident.)

Example 10.4.1 In Example 10.3.2, ($\sigma = 6$, $a = 1.96$, $\bar{x} = 30$, $n = 64$), a 95% confidence interval was found to be

$$30 \pm 1.47, \text{ of width } 2.94.$$

By choosing a sample of size 400, we find that the corresponding 95% confidence interval is of width

$$2(1.96) \frac{6}{\sqrt{400}} \doteq 1.18.$$

Thus, if $\bar{x} = 30$ for the sample of size 400, we would be 95% confident that the value of μ was in the interval 30 ± 0.59, or between 29.41 and 30.59.

Thus, as a general rule: The greater the sample size, the narrower the estimated confidence interval. We have already pointed out, however, that sampling may be expensive and time-consuming. Therefore, it is often the practice to choose the sample size on the basis of the desired precision of estimate.

Example 10.4.2 Again taking $\sigma = 6$ and $a = 1.96$ as in Example 10.4.1, suppose that it is desired that our 95% confidence interval should be of width 0.5. That is, we wish to be 95% confident that μ lies within 0.25 of the obtained value $\bar{x}$. Thus, we wish to have

$$1.96 \frac{6}{\sqrt{n}} = 0.25.$$

Hence

$$\sqrt{n} = \frac{(1.96)(6)}{0.25} = \frac{11.76}{0.25} \doteq 47.$$

Since $47^2 = 2209$, we need a sample size of more than 2200.

Often there are practical limitations on the sample size. If we desire a particular interval width for a given sample size, then the degree of confidence may be calculated as in the following example.

Example 10.4.3 For $\sigma = 6$, $\bar{x} = 30$, and $n = 64$, if we wish to establish a confidence interval [29.5, 30.5], or 30 ± 0.5, we proceed as follows:

$$0.5 = a \frac{6}{\sqrt{64}}; \quad a \doteq 0.67.$$

Since only approximately 50% of the area under the normal curve lies between $Z = -0.67$ and $Z = 0.67$, we are only 50% confident that μ lies in the interval 30 ± 0.5.

To obtain a confidence interval for the value of p for a Bernoulli population, we face a peculiar problem. Up to this point we have assumed that the population standard deviation, σ, was known. However, for a Bernoulli population,

$$\sigma = \sqrt{pq} = \sqrt{p(1-p)}.$$

So if p is unknown, we cannot know the value of σ.

Example 10.4.4 For the bent coin of Example 10.2.1, 42 heads were obtained in 100 tosses. We accept $\bar{x} = 0.42$ as a point estimate for p. A 90% confidence interval for p is of the form

$$0.42 \pm 1.64 \frac{\sigma}{10}.$$

Since we cannot compute $\sigma = \sqrt{pq}$, a reasonable procedure is to use

$$\sqrt{\bar{x}(1-\bar{x})}$$

as a point estimate for σ. Our interval becomes

$$0.42 \pm 1.64 \frac{\sqrt{(0.42)(0.58)}}{10}, \quad \text{or}$$

$$0.42 \pm 0.083, \quad \text{or} \quad [0.34, 0.50].$$

The procedure of Example 10.4.4 is justified from two points of view:

1. The true value of p is not very likely to be much different from $\bar{x}$. Therefore $\sqrt{\bar{x}(1-\bar{x})}$ is close to $\sqrt{pq}$.

2. Even a rather large error in estimating p does not affect $\sqrt{pq}$ drastically. For instance, compare:

$$p = 0.30, \quad \sqrt{pq} \doteq 0.458$$
$$p = 0.42, \quad \sqrt{pq} \doteq 0.494$$
$$p = 0.50, \quad \sqrt{pq} \doteq 0.500$$

If the population under consideration is not Bernoulli, it is possible that previous experiments or studies of similar populations will enable us to estimate the value of σ. If we have only one sample with $\bar{x}$ and s_x computed from the sample, we may use s_x as a point estimate of σ provided the sample size is large (see Section 10.2). As noted in Remark 2 of Section 9.5, $n \geq 30$ is often accepted as adequate for this purpose. We should note, however, that confidence intervals $\bar{x} \pm a \frac{s_x}{\sqrt{n}}$ are not as reliable as $\bar{x} \pm a \frac{\sigma}{\sqrt{n}}$.

Example 10.4.5 A study of the handicaps of 80 golfers chosen at random from among a group of 46,000 golfers shows $\bar{x} = 15.1$ and $s_x = 3.4$. We could use

$$15.1 \pm 1.96 \frac{3.4}{\sqrt{80}}, \quad \text{or}$$

$$15.1 \pm 0.75, \quad \text{or} \quad [14.35, 15.85]$$

as a 95% confidence interval for the mean handicap of the 46,000 golfers.

Let us summarize our results about confidence intervals. We assume that we are dealing with a population with μ unknown and a random sample of size n from this population having mean $\bar{x}$ and standard deviation s_x. Furthermore, we assume that n is sufficiently large so that the distribution of $\bar{X}$ is approximately normal. Then:

1. Since $E(\bar{X}) = \mu$, the sample value $\bar{x}$ may be taken as a point estimate of μ. We are "confident" that $\bar{x}$ is "close to" the true value of μ.

2. The answers to the questions "How confident?" and "How close?" are related to each other and to the size of the sample. If σ is known:
   a. For a desired degree of confidence and given n, we find the confidence interval $\bar{x} \pm a \dfrac{\sigma}{\sqrt{n}}$, where a is found from Table N. The width of this interval is $2a \dfrac{\sigma}{\sqrt{n}}$. (Example 10.4.1)
   b. For a desired degree of confidence and a given width of interval, the value of a may be found and the appropriate value of n may be computed. (Example 10.4.2.)
   c. For a desired width of interval and given n, the corresponding value of a may be found, and the degree of confidence determined from Table N. (Example 10.4.3.)

3. If σ is unknown, we use s_x (or slightly more accurately, $\sqrt{\dfrac{n}{n-1}} \, s_x$) as a point estimate for σ. The corresponding confidence interval is of the form
$$\bar{x} \pm a \frac{s_x}{\sqrt{n}}.$$
If $\sqrt{\dfrac{n}{n-1}} \, s_x$ is used as a point estimate, the corresponding confidence interval is of the form
$$\bar{x} \pm a \frac{s_x}{\sqrt{n-1}}.$$

4. It is useful to bear in mind the following facts from Section 5.4:
$$P(|Z| \leq 1.64) \doteq 0.90$$
$$P(|Z| \leq 1.96) \doteq 0.95$$
$$P(|Z| \leq 2.58) \doteq 0.99$$

We shall next direct our attention to learning how we may use confidence intervals in decision making. We shall introduce some methods of application in Section 10.5 and shall expand on these in a slightly more formal way in Chapter 11.

EXERCISES 10.4

1. Given $\sigma = 20$, $n = 100$, and $\bar{x} = 50$, find the following confidence intervals for μ:
   a. 75%   b. 90%   c. 95%   d. 99%

2. Graph the intervals found in Problem 1.

3. Given $\sigma = 10$, find the sample size that will give 95% confidence intervals of the following widths:
   a. 20   b. 10   c. 5   d. 2.5

4. Given $\sigma = 10$, find the sample size that will give a confidence interval of width 5 for the following confidence levels:
   a. 68.26%   b. 90%   c. 95%   d. 99%

5. One hundred randomly chosen college freshmen obtain an average score of 32 on a memory test. It is known that $\sigma^2 = 64$ for this test. Find a 90% confidence interval for the average score of all college freshmen.

6. A survey of 1000 randomly selected homes in a large community reveals that a television set is turned on an average of 6.2 hours per day with standard deviation of 2.1 hours. Find a 95% confidence interval for the average number of television hours for the entire community.

7. Four hundred thumbtacks of a given size and shape are tossed and 240 land point up. Find a 95% confidence interval for p, the probability that a tack of this size and shape lands point up.

8. A random sample of motorcycle owners shows an average age of 24.7 years and standard deviation 4.2 years. Find a 99% confidence interval for the average age of all motorcycle owners.

9. A sample of size 100 yields a 90% confidence interval for μ of width 1.4.
   a. What was the value of σ?
   b. What sample size would be necessary to produce a 90% confidence interval of width 0.7?
   c. Do you need to know the value of σ in order to answer part b?

10. It is known that the standard deviation of the life of a certain type of light bulb is 40 hours. How large a random sample must be chosen in order to be 95% confident that the sample mean would be within 10 hours of mean life for the population of all light bulbs of this type?

11. A random sample of 900 domestic cats shows a life expectancy of 9.6 years. Assuming that $\sigma = 3.3$ years, what is the degree of confidence that the average life span of all domestic cats is between 9.5 and 9.7 years?

12. Assume that the weights of adult males in the United States are normally distributed with $\sigma = 35$ pounds. How large a random sample should be chosen in order to be 95% confident that the sample mean does not differ from the population mean by more than 5 pounds?

13. For Problem 12, how large should a random sample be in order to be 90% confident that the sample mean does not exceed the population mean by more than 5 pounds?

14. A new method of treating a disease proves successful in 72 out of 100 cases. If we assume that the 100 patients are a random sample of all those having the disease, establish a 99% confidence interval for the probability that the treatment will be effective in a given case.

15. Suppose that we know that $p = 0.6$ for a Bernoulli population. How large a sample size is necessary to be 95% confident that the obtained value of $\bar{x}$ lies in the interval [0.5, 0.7]?

★ COMPUTER INVESTIGATIONS

If you have access to an electronic computer, try to carry out the following investigations:

1. Find values of $\sqrt{pq}$, where $q = 1 - p$, for the following values of p:

    0, 0.05, 0.10, 0.15, ..., 0.85, 0.90, 0.95, 1

2. Plot the graph of $y = \sqrt{x(1-x)}$.
3. Plot the graph of $y^2 = x - x^2$.
4. For which values of p does $\sqrt{pq}$ differ from 0.5:
   a. by no more than 0.05   b. by no more than 0.005?

10.5 Significance

We have learned (Section 9.5), if μ and σ are known, how to assign probabilities concerning the location of sample means. Thus, the probability that the mean, $\bar{X}$, of a large random sample will fall in the interval $\mu \pm 1.96 \frac{\sigma}{\sqrt{n}}$ is 0.95.

By a change of viewpoint, we have seen (Sections 10.3 and 10.4) that if $\bar{x}$ is known, and μ unknown, we may construct confidence intervals for μ. Thus, if $\bar{x}$ is the mean of a large random sample, we are 95% confident that the population mean lies in the interval $\bar{x} \pm 1.96 \frac{\sigma}{\sqrt{n}}$.

One use of confidence intervals as a basis for statistical inference occurs when μ, σ, and $\bar{x}$ are all known and this question is raised: "Is the sample truly random? Does it 'represent' the population?" Of course, we can never answer such a question with certainty, but we can assign probabilities.

Example 10.5.1 A sample of size 36 is reputed to be randomly chosen from a population having $\mu = 40$ and $\sigma = 5$. If the sample were random, the *a priori* probability would be 0.99 that the obtained value of $\overline{X}$ lies in the interval

$$40 \pm 2.58 \frac{5}{\sqrt{36}}, \quad \text{or} \quad 40 \pm 2.15, \quad \text{or} \quad [37.85, 42.15].$$

Suppose now that the mean actually obtained from the 36 cases falls outside this interval. Then either

(a) something has happened that would have happened less than 1 time in 100 by chance, or
(b) the sample was not randomly chosen from the population.

We would be 99% confident that (b) is correct.

We may state our conclusions from Example 10.5.1 in the following form: The sample differs **significantly** from the population as regards average performance. Since the obtained value of $\overline{X}$ lies outside the 99% confidence interval, it is usual to say that the sample differs from the population at the **1% level of signifiance.**

Example 10.5.2 It is known that the average score on a reading-readiness test for all 5-year-olds in a large city school is 15 with a standard deviation of 5. The average score on the test for a certain group of 100 children was 16.4. We conclude that this group is significantly different from the population of all 5-year-olds at the 1% level. Our reasoning is based on confidence intervals. 99% of all random samples of size 100 would yield a value of $\overline{X}$ such that

$$|\overline{X} - 15| \leq 2.58 \frac{5}{\sqrt{100}} = 1.29.$$

For the group being considered,

$$|\bar{x} - \mu| = 16.4 - 15 = 1.4,$$

which is greater than 1.29. A difference of 1.4 between $\bar{x}$ and μ would occur less than 1% of the time by chance.

The argument of Example 10.5.2 is often used in making statistical inferences about experimental results. The 100 children might have been selected (originally) at random from among 4-year-olds and then exposed to special preschool training. As 5-year-olds, this especially trained group is significantly different from the population of 5-year-olds at the 1% level. It is tempting to conclude that the special training accounts for the difference.

Remark

The conclusion that the training accounts for the observed difference is the natural one to draw. It should be remembered, however, that several alternative explanations are available. We offer some here.

First of all, the difference $|\bar{x} - \mu| = 1.4$ *might* have been accidental, although there is less than 1 chance in 100 of this being the case.

Secondly, the original selection (from among 4-year-olds) might not have been random.

Thirdly, other factors than the special training might have affected the test results. (For example, the mere fact that these children had been selected for training might have increased their parents' concern with providing reading experience at home.)

For such reasons, experiments must be carefully designed in order to be able to identify a single factor as the cause of an observed significant difference.

We may consider significance levels other than 1%. For some decision-making purposes the 5% level is satisfactory. In particular situations the importance (and the cost) of the decision must be considered. In Chapter 11 we shall note that an early step in using statistical tests is to determine the acceptable significance level.

Example 10.5.3 The average blood pressure for a certain type of heart patient is a certain constant with a standard deviation of 10. A new drug is administered to a randomly selected group of 50 patients. The average blood pressure for this group differs from the population mean by 3 points.

$$P(|\bar{X} - \mu| \geq 3) = P\left(|Z| \geq \frac{\frac{3}{10}}{\sqrt{50}}\right)$$

$$\doteq P(|Z| \geq 2.12)$$

Since $1.96 < 2.12 < 2.58$, the observed difference is significant at the 5% level but not at the 1% level.

Our attention has been focused on the difference $|\bar{X} - \mu|$. There are correspondences between the confidence interval

$$\mu \pm a \frac{\sigma}{\sqrt{n}}$$

and the level of significance. For a particular sample,

$$a = \frac{|\bar{x} - \mu|}{\frac{\sigma}{\sqrt{n}}}.$$

The value of a is sometimes called the **critical ratio**. The values of a that are customarily used are repeated in Figure 10.3.

Figure 10.3

Confidence (Two-sided intervals)	Significance Level	Value of a (Critical ratio)
90%	10%	1.64
95%	5%	1.96
98%	2%	2.33
99%	1%	2.58

It is also useful to consider "one-sided" intervals (see Problem 5, Exercises 10.3) as shown in the following example.

Example 10.5.4 For the 5-year-olds of Example 10.5.2 we might ask whether the group of 100 performed significantly *better* than the population rather than significantly *differently*. Since 99% of the area under the standard normal curve lies to the left of $Z = 2.33$, 99% of all random sample means lie to the left of

$$\mu + 2.33 \frac{\sigma}{\sqrt{n}}.$$

In our case,

$$15 + 2.33(\tfrac{5}{10}) \doteq 16.17.$$

Since $\bar{x} = 16.4$, which is greater than 16.17, we see that the average performance of the group is significantly *better* than μ, at the 1% level.

Corresponding to Figure 10.3, we have a table for one-sided intervals as shown in Figure 10.4. A one-sided interval has either a lower limit,

$$\mu - a \frac{\sigma}{\sqrt{n}},$$

or an upper limit, $\mu + a \frac{\sigma}{\sqrt{n}}$. (See Problems 2 and 3, Exercises 10.5.) The values in Figure 10.4 may be read from Table N.

Figure 10.4

Confidence (One-sided interval)	Significance Level	Value of a (Critical ratio)
90%	10%	1.28
95%	5%	1.64
98%	2%	2.05
99%	1%	2.33

The following diagrams show a two-sided (Figure 10.5a) and a one-sided (Figure 10.5b) 90% confidence intervals for μ.

Figure 10.5

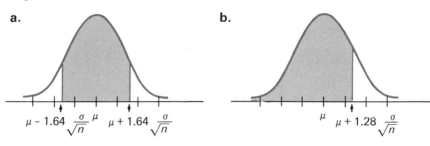

Remark 1

Confidence, since it arises from a probability statement, is "cumulative." If we are, for example, 95% confident of something, we are automatically 94%, ..., 90%, ..., 60%, ... confident.

Remark 2

It is important to note that our use of "significant" is a technical, statistical one. Saying that an observed difference in performance of two groups of children, for example, is "significant" in the technical sense means that the difference is measurable statistically. It does *not necessarily* mean that this difference is important enough to matter in the behavior of these children in the real world.

EXERCISES 10.5

- **1.** A nationwide mathematics test for high school seniors has $\mu = 52.3$ and $\sigma = 32$. Big City High School of Science offers special opportunities in mathematical training. For this school, 400 seniors attain an average score of 57.5.
  a. Is 57.5 significantly *different* from μ at the 1% level?
  b. Is 57.5 significantly *greater* than μ at the 1% level?

- **2. a.** If $|\mu - \bar{x}| \geq 1.96 \dfrac{\sigma}{\sqrt{n}}$, may we conclude that the sample mean is larger than μ at the 5% level of significance?
  b. If $\mu - \bar{x} \geq 1.96 \dfrac{\sigma}{\sqrt{n}}$, then the sample mean is smaller than μ at what level of significance?
  c. If $\bar{x} - \mu \geq 1.96 \dfrac{\sigma}{\sqrt{n}}$, what conclusion may be drawn?

- 3. a. If $\bar{x}$ is significantly *different* from μ at the 5% level, is it also either significantly *greater* or *less* at the 5% level?
  b. If $\bar{x}$ is significantly *greater* than μ at the 5% level, must it also be significantly *different* from μ at the 5% level?
  c. If $\bar{x}$ is significantly *greater* than μ at the 5% level, is it also significantly *different* from μ at the 10% level?

4. Suppose that a population has $\mu = 25$ and $\sigma = 10$. Find the lowest level at which these differences are significant.
  a. $\bar{x} - \mu = 3.6, n = 100$
  b. $\mu - \bar{x} = 0.5, n = 400$

- 5. A lathe produces items, of which 2.2% are defective. An adjustment is made in the lathe and of the next 1000 items produced, 12 are defective. Has the adjustment made a significant difference at the 2% level?

- 6. Use a one-sided interval to determine whether or not the adjusted lathe (Problem 5) is significantly *better* at the 2% level.

- 7. A tire manufacturer wants to produce tires with average life of 20,000 miles and would like at least 68% of his tires to come within 1,200 miles of this specification. (Recall Figure 5.8.) As a random sample, 64 tires were tested under simulated road testing equipment. The mean of the sample was 19,600 miles. Is the difference from the mean significant at either the 1% or the 5% level?

8. Assume that the I.Q. values are normally distributed with $\mu = 100$ and $\sigma = 15$. In a random sample of 400 individuals, how many points from 100 would the mean of the sample have to be in order for:
  a. its difference to be at least significant at the 2% level?
  b. it to be significantly *better* at the 2% level?

9. Compare Figure 10.3 and 10.4. Note that the value of a at the 5% level of a one-sided interval is the same as the value of a at the 10% level of a two-sided interval. Propose a plausible generalization.

10.6 What Do You Think?

In Chapter 9 we discussed the distribution of the random variable (the sample mean):

$$\bar{X} = \frac{X_1 + X_2 + \cdots + X_n}{n},$$

where the X_i's are independent random variables, each having the same probability distribution. On the basis of this discussion, we have developed in this chapter the idea of a confidence interval to estimate the value of μ.

It might seem natural to proceed to investigate the distribution of sample variances in order to determine confidence intervals for the value of σ^2. Unfortunately, this investigation would take us too far afield.

There is, however, one important statistic whose distribution is not too difficult to discover, using our earlier results. We shall have use for this statistic in Chapter 11.

Suppose that we have two populations with associated random variables X and Y. We consider large random samples of size n_X from the first population and of size n_Y from the second. We know about the distribution of $\overline{X}$ and $\overline{Y}$. The new random variable we want to introduce is

$$\overline{D} = \overline{X} - \overline{Y}.$$

It is reasonable to ask whether we can express $\mu_{\overline{D}}$, $\sigma^2_{\overline{D}}$, in terms of μ_X, μ_Y, σ^2_X, and σ^2_Y. Furthermore, it is also reasonable to ask whether the distribution of

$$\frac{\overline{D} - \mu_{\overline{D}}}{\sigma_{\overline{D}}}$$

has some "nice" shape. What do you think?

To illustrate how the distribution of $\overline{D}$ may arise in practice, we consider the following example, which proposes a problem of a type we shall solve in Chapter 11.

Example 10.6.1 Two methods of treatment of industrial waste at a certain manufacturing plant are proposed to reduce pollution. In a 64-day test of one method, measurements showed an average pollution index of 82.3 per day with $s = 5.6$. In a 36-day test of the second method, the corresponding values were 79.8 and $s = 7.2$. The statistical problem is to determine whether the observed difference in the averages (2.5) is "significant" or can be explained in terms of chance alone. If the difference is (statistically) significant, then the better method will be used. If the difference is not significant, the plant will adopt the cheaper of the two methods.

EXERCISES 10.6

● 1. Express $\mu_{\overline{D}}$ in terms of:
   a. $\mu_{\overline{X}}$ and $\mu_{\overline{Y}}$
   b. μ_X and μ_Y

● 2. Let us assume that the population random variables X and Y are independent. It can be shown that $\overline{X}$ and $\overline{Y}$ are also independent. Making this assumption, express $\sigma^2_{\overline{D}}$ in terms of:
   a. $\sigma^2_{\overline{X}}$ and $\sigma^2_{\overline{Y}}$
   b. σ^2_X and σ^2_Y

- 3. What is the standardized random variable corresponding to $\bar{D}$ (assuming that X and Y are independent)?
- 4. In view of our earlier experiences, what is your guess as to the nature of the distribution of the standardized variable of Problem 3? (Assume that n_X and n_Y are large.)
- 5. If σ_X and σ_Y are not known, how would it be possible to construct a random variable having a distribution approximately that of

$$\frac{\bar{D} - \mu_{\bar{D}}}{\sigma_{\bar{D}}} ?$$

Chapter Summary

1. A sample of size 1 gives limited information about a population that is normal with σ known.

2. a. The mean of a single random sample is called a *point estimate* of the mean of the population.
   b. The value $\dfrac{n}{n-1} s_x^2$, where s_x^2 is the variance of a single random sample, is taken as a *point estimate* of the population variance, σ^2. If n is large, s_x^2 may be used as a point estimate for σ^2.

3. If the population variance, σ^2, is known, then for a particular $\bar{x}$, we are 95% confident that the interval

$$\left[\bar{x} - 1.96 \frac{\sigma}{\sqrt{n}},\ \bar{x} + 1.96 \frac{\sigma}{\sqrt{n}} \right]$$

contains μ. Such an interval is called a *95% confidence interval*. The endpoints are called *confidence limits*.

4. a. The width of a confidence interval

$$\left[\bar{x} - a \frac{\sigma}{\sqrt{n}},\ \bar{x} + a \frac{\sigma}{\sqrt{n}} \right]$$

is $2a \dfrac{\sigma}{\sqrt{n}}$.

   b. The greater the sample size, the narrower the estimated confidence interval for the same degree of confidence.

   c. For a Bernoulli population,

$$\sigma \doteq \sqrt{\bar{x}(1-\bar{x})}.$$

5. a. If a sample mean, $\bar{x}$, lies outside the 99% confidence interval, then $\bar{x}$ is said to *differ* from μ at the 1% *level of significance*.
   b. The value
   $$a = \frac{|\bar{x} - \mu|}{\frac{\sigma}{\sqrt{n}}}$$
   is called the *critical ratio*.
   c. If a sample mean, $\bar{x}$, corresponds to a point that lies to the right of 99% of the area under the standard normal curve, then $\bar{x}$ is said to be *greater* than μ at the 1% level of significance.

6. If X and Y are *independent* random variables, and $\bar{D} = \bar{X} - \bar{Y}$, then the standardized random variable corresponding to $\bar{D}$ is
$$Z = \frac{\bar{D} - \mu_{\bar{D}}}{\sigma_{\bar{D}}} = \frac{(\bar{X} - \bar{Y}) - (\mu_X - \mu_Y)}{\sqrt{\frac{\sigma_X^2}{n_X} + \frac{\sigma_Y^2}{n_Y}}}.$$

Chapter Review

1. An urn contains a mixture of red and blue marbles. The fraction of red marbles, k, is known to be either $\frac{2}{3}$ or $\frac{1}{4}$. That is, the *a priori* probability that $k = \frac{2}{3}$ is $\frac{1}{2}$. A single marble, drawn at random, is red.
   a. Which value of k would you guess is correct?
   b. What is the probability that your guess is wrong?

2. We have seen that "confidence" is cumulative in that 95% confidence implies k% confidence for $k \leq 95$. In what sense is "significance" cumulative?

3. It has been said that confidence is a measure of a "bygone probability." Explain.

4. Consider a k% confidence interval for μ of the form $\bar{x} \pm a \frac{\sigma}{\sqrt{n}}$, where $\bar{x}, a, \sigma, \mu,$ and n are all given.
   a. If we replace a with b, where $b < a$, the resulting interval is an r% confidence interval. Is r less than or greater than k?
   b. If we use the same value of a, but have a random sample of size $m > n$, is $\bar{x} \pm a \frac{\sigma}{\sqrt{m}}$ also a k% confidence interval?
   c. Is the interval of part b shorter or longer than $\bar{x} \pm a \frac{\sigma}{\sqrt{n}}$?

5. A normal population is known to have $\sigma = 4.2$. A random sample of size 144 yields $\bar{x} = 23.7$. Find a 90% (two-sided) confidence interval for μ.

6. Using the data of Problem 5, we are 90% confident that μ is less than what number?

7. A sample of size 100 yields $\bar{x} = 62.4$ and $s_x^2 = 164.1$. Find a two-sided 95% confidence interval for μ.

Chapter 11

Decision Making

We have been interested in the process of decision making in the face of uncertainty. In the course of our study, we have developed a great deal of mathematical machinery, based on the theory of probability. We have obtained the following results.

I. If μ and σ of a population are known, we found (Section 9.5) how to assign probabilities concerning the location of sample means.

For example, the probability that the mean, $\overline{X}$, of a large random sample will fall in the interval

$$\mu \pm 1.96 \frac{\sigma}{\sqrt{n}}$$

is 0.95.

II. If σ of a population is known and if $\overline{x}$ is the mean of a particular large random sample, we found (Sections 10.3 and 10.4) how to estimate μ.

For example, we are 95% confident that μ lies in the interval

$$\overline{x} \pm 1.96 \frac{\sigma}{\sqrt{n}}.$$

III. If μ and σ of a population are known, as well as a sample mean, $\overline{x}$, we found (Section 10.5) how to determine whether or not $\overline{x}$ differs significantly from μ.

For example:

a. If

$$|\overline{x} - \mu| = 1.96 \frac{\sigma}{\sqrt{n}},$$

then $\overline{x}$ *differs* from μ at the 5% level of significance.

b. If

$$\overline{x} > \mu + 1.64 \frac{\sigma}{\sqrt{n}},$$

then $\overline{x}$ *is greater than* μ at the 5% level of significance.

We shall now complete our investigations by considering some tests of **statistical hypotheses**. The very nature of uncertainty prevents us from being sure that a given decision will be correct. (Unusual events do occur!) That is why we are concerned with such things as "confidence" and "significance."

We shall restrict ourselves to decisions based on statistical tests of a somewhat limited nature. Our concentration will be on tests connected with the location of a population mean, μ.

11.1 Statistical Hypotheses

The decision process, when based on a statistical test, begins with an assumption that is to be tested. The assumption must be statistical in nature. In general, a statistical hypothesis is a statement about a population parameter—for our purposes a statement about the value of μ.

Suppose that we are offered an opportunity to play a game involving the tossing of a coin. We have two courses of action: to play or to refuse to play. Suppose that we decide to play if and only if the coin is fair. Before making our decision we are allowed to experiment with the coin by observing the results of several tosses. Our decision then will be based on whether or not the experimental results indicate that "$P(\text{heads}) = 0.50$" is reasonable for this coin. We might formulate other hypotheses, but an obvious one for this situation is the hypothesis:

$$H: p = 0.50$$

This is a statement about a population parameter and is, therefore, a statistical hypothesis. Our decision depends on whether or not the experimental tosses lead us to **accept** or to **reject** the hypothesis.

We recall the discussion of Section 1.3 as well as the material of Chapter 10. No amount of experimenting will enable us to *prove* that "$p = 0.50$" is true. The best we can do is to be "confident" that p is "close to" 0.50. If a statistical hypothesis is *accepted*, we behave (make our decisions) as though it were, in fact, true.

In Section 1.3 we remarked that if p differs greatly from 0.50, experimenting will more easily reveal that H: $p = 0.50$ is false. For this reason, much statistical testing is worded in terms of **rejection**. We either "reject" or "fail to reject" a particular hypothesis. For the coin game, we decide to play if testing fails to reject the hypothesis H: $p = 0.50$.

For the coin, the population is Bernoulli and p is the expected value of the Bernoulli random variable. More generally, we may wish to test hypotheses of the form

$$H: \mu = \mu_0,$$

where μ is the expected value of the population random variable and μ_0 is some constant.

Example 11.1.1 A certain electronic device is designed to emit an average of 60 beats per second. We wish to buy the device if testing fails to reject the hypothesis H: $\mu = 60$.

Example 11.1.2 Before administering the reading-readiness test to the 100 five-year-olds of Example 10.5.2, the question is: Will the test results for this group lead us to reject the hypothesis H: $\mu = 15$? This is simply a rewording of the question: Is the sample significantly different (in average performance) from the population for which $\mu = 15$ is known?

Suppose, for our coin tossing game, that we will win if "heads" occurs. We might make a decision to play if and only if the coin were in our favor. We would then be interested in testing the hypothesis

$$H: p > 0.50.$$

For a non-Bernoulli population, we might be interested in

$$H: \mu > \mu_0 \quad \text{(or in H: } \mu < \mu_0\text{)}.$$

Example 11.1.3 For the group of five-year-olds in Example 11.1.2, we might be interested in determining whether the test results for this group will be significantly at least as good as the results for the population. The corresponding question is: Will the results lead us to reject the hypothesis H: $\mu < 15$?

Another type of hypothesis arises in comparing the means of two populations. Using X and Y to represent the random variables, we may often find it interesting to investigate whether

$$H: \mu_X = \mu_Y$$

is to be accepted or rejected. Since this hypothesis can be written in the form

$$H: \mu_X - \mu_Y = 0,$$

it is called the **null hypothesis.** In words, the hypothesis is that there is *no* difference between the population means. *Rejection of the null hypothesis* implies that we have decided that the populations *do differ* as regards the location of their means. Similarly, H: $p = p_0$ and H: $\mu = \mu_0$ may be called null hypotheses (see Problem 3, Exercises 11.1).

Example 11.1.4 In Example 10.6.1 we considered two methods of treating industrial wastes. To decide whether the methods provide significantly different results, we test the null hypothesis

$$H: \mu_X - \mu_Y = 0.$$

Of course, the results of our sampling will be used to decide whether we accept or reject H.

EXERCISES 11.1

- 1. Not all statements are statistical hypotheses. Which of the following are statistical?
    a. Jones is a better basketball player than Anderson.
    b. Jones is a better free-throw shooter than Anderson.
    c. This die is loaded in favor of "6."
    d. Girls are prettier than boys.
    e. There is no difference between men and women in intelligence test scores.
    f. Smith will defeat Robinson in the (two-man) race for Congress.
- 2. A coin is tossed 100 times and 50 heads are obtained. Explain why it is better to say "We will not reject H: $p = 0.5$" rather than "H: $p = 0.5$ is true."
- 3. Hypotheses of the form H: $\mu = \mu_0$ may also be considered as null hypotheses. Explain.
- 4. Formulate appropriate statistical hypotheses for the situations described in the following statements:
    a. The top card of a thoroughly shuffled deck is a spade.
    b. A new drug shortens the recovery time from a certain illness.
    c. Special weight-lifting practice does not help the performance of high-jumpers.

11.2 Two-Sided Statistical Tests

Suppose that we wish to determine whether

$$H: p = 0.50$$

is to be accepted or rejected for a particular coin, we will make n experimental tosses of the coin. This set of tosses is a random sample of the corresponding Bernoulli population (binomial distribution). If the hypothesis is true, then we know the distribution of

$$\overline{X} = \frac{\text{number of successes}}{n}.$$

Before making the tosses, we can establish a confidence interval for $\overline{X}$. Suppose that we choose a 95% confidence interval:

$$0.50 \pm 1.96 \frac{\sigma}{\sqrt{n}}$$

If the obtained value of $\overline{X}$ falls *outside* this interval, we *reject* H: $p = 0.50$ at the 5% significance level. The region outside a confidence interval is called the **rejection region**.

Example 11.2.1 A possible test for H: $p = 0.50$ would be: make 100 tosses; reject H if the obtained value of $\overline{X}$ lies outside the interval

$$0.50 \pm 1.96 \frac{\sqrt{(0.5)(0.5)}}{\sqrt{100}} = 0.50 \pm 0.098.$$

Our test may be reworded: reject H if we obtain less than 41 or more than 59 heads.

Example 11.2.2 If, after designing the test of Example 11.2.1, our 100 tosses yield 42 heads, we have $\overline{x} = 0.42$. Since 0.42 lies in the interval [0.402, 0.598], we do not reject H: $p = 0.50$.

Example 11.2.3 If our test results yield 61 heads, we reject H: $p = 0.50$, since $\overline{x} = 0.61$ lies outside the 95% confidence interval; that is, it lies in the rejection region.

The reasoning of the preceding example is familiar. If the obtained value of $\overline{X}$ lies outside the 95% confidence interval, then either

(1) H: $p = 0.50$ is correct but something unusual has occurred—an event whose *a priori* probability of occurring is less than 0.05,

or

(2) H: $p = 0.50$ is false.

In rejecting H (that is, preferring (2) to (1)), we are 95% confident that we have decided correctly. Less than 5% of all rejections based on this procedure will be wrong.

A statistical test is based on the results obtained from a random sample. Before examining the sample, we usually agree, *in advance*, on the level of significance and on the sample size.

The level of significance is usually denoted by α,* and is most frequently chosen as 0.50, 0.02, or 0.01. As we know (Section 10.4), the width of the confidence interval depends on the sample size, n. (In general, we wish n to be as large as possible, but many practical considerations often place a restriction on n.)

Example 11.2.4 Mendelian theory predicts that $\frac{3}{4}$ of a certain hybrid variety of pea will have smooth, as opposed to wrinkled, skins. A biologist wishes to determine whether or not special growing conditions will have any significant effect on this. He decides to test the null hypothesis,

$$\text{H: } p = 0.75,$$

at the $\alpha = 0.01$ (1%) level by growing 36 plants under the special conditions.

---

*The symbol α (alpha) is a letter of the Greek alphabet corresponding to the English "a."

The biologist of Example 11.2.4 might consider that he has established a **test statistic** (compare Section 10.3):

$$Z = \frac{\overline{X} - \mu}{\frac{\sigma}{\sqrt{n}}} = \frac{\overline{X} - 0.75}{\sqrt{\frac{(0.75)(0.25)}{36}}} \doteq \frac{\overline{X} - 0.75}{0.072}$$

If the observed value $\bar{x}$ of $\overline{X}$ yields a value z of Z such that

$$|z| > 2.58,$$

he will reject the null hypothesis at the 1% level. This interpretation corresponds to the idea of critical ratio introduced in Section 10.4 (Figure 10.3).

Example 11.2.5 It is decided to test a manufacturer's claim that the mean life of a certain kind of flashlight battery is 200 hours, with standard deviation of 15 hours. Assuming that "$\sigma = 15$" is correct, we might test the hypothesis

$$H: \mu = 200$$

by choosing $\alpha = 0.02$ and by measuring the mean life of a random sample of 64 batteries. Our test statistic is

$$Z = \frac{\overline{X} - 200}{\frac{15}{\sqrt{64}}} = \frac{\overline{X} - 200}{1.875}.$$

Since $\alpha = 0.02$, $a = 2.33$ (Figure 10.3), and we reject H if the sample mean falls outside the interval

$$200 \pm (2.33)(1.875) \doteq 200 \pm 4.4, \quad \text{or} \quad [195.6, 204.4].$$

Remark

If H: $\mu = \mu_0$ is to be tested and if σ is known, we proceed as in Example 11.2.5. If σ is not known, we use instead the value of s_x obtained from the sample. The test statistic is then

$$\frac{\overline{X} - \mu_0}{\frac{s_x}{\sqrt{n}}}.$$

This procedure is acceptable if n is large.

Statistical tests such as those used in this section are called **two-sided** or **two-tailed tests**. Figure 11.1 illustrates the two-tailed rejection region for testing H with $\alpha = 0.05$ under the assumption that X is normally distributed.

If the obtained value of the test statistic $\frac{\overline{X} - \mu_0}{\frac{\sigma}{\sqrt{n}}}$ falls in either of the shaded regions, we reject H: $\mu = \mu_0$.

Figure 11.1

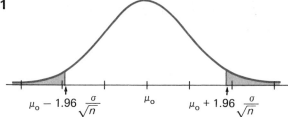

EXERCISES 11.2

1. It is decided to test H: $\mu = 0$ with $\alpha = 0.02$ and a sample of size 144. Determine the rejection region assuming that $\sigma = 6$.

2. It is decided to test H: $\mu = 0$ with $\alpha = 0.05$ and a sample of size 625. Determine the rejection region assuming that $\sigma = 10$.

3. To test H: $p = 0.5$ for a particular coin, 100 tosses are to be made, and H is to be rejected if the number of heads is less than 43 or more than 57. Determine α for this test. (Hint: Use the interval [0.425, 0.575].)

4. To test H: $p = 0.6$ for a particular coin, 100 tosses are to be made, and H is to be rejected if the number of heads is less than 55 or more than 65. Determine α for this test.

5. It is known that a certain population is approximately normal with $\sigma = 12$. If a random sample of size 400 yields $\bar{x} = 50$, would we accept or reject H: $\mu = 47$ at the 1% level?

6. It is known that a certain population is approximately normal with $\sigma = 10$. If a random sample of size 625 yields $\bar{x} = 33$, would we accept or reject H: $\mu = 30$ at the 1% level?

7. Children from two different school districts are given the same arithmetic test. The results are:

   District I: $n_x = 500$, $\bar{x} = 62.1$, $s_x = 15$
   District II: $n_y = 400$, $\bar{y} = 59.4$, $s_y = 16$

   a. What is the null hypothesis?
   b. Would we accept or reject the null hypothesis at the 1% level? (Hint: Recall Problem 5, Exercises 10.6.)

8. We wish to test H: $p = \frac{1}{6}$ for a given die where $p = P(\text{five occurs})$. In an experiment, 72 out of 600 throws led to "five." Do we accept or reject H for
   a. $\alpha = 0.02$?
   b. $\alpha = 0.01$?

11.3 One-Sided Statistical Tests

Our illustrations in the preceding section have involved two-sided, or two-tailed, tests of hypotheses of the form

$$H: \mu = \mu_0.$$

To test a hypothesis of the form

$$H: \mu > \mu_0,$$

it is appropriate to use a **one-sided**, or **one-tailed**, **test** and to consider a corresponding one-tailed rejection region. If the obtained value of

$$\frac{\overline{X} - \mu_0}{\frac{\sigma}{\sqrt{n}}}$$

is smaller than some specified negative value, we reject H: $\mu > \mu_0$. (An obtained value of $\overline{X}$ which is larger than μ_0 would persuade us to accept $\mu > \mu_0$.) Figure 11.2 illustrates the one-tailed rejection region for $\alpha = 0.05$. (Recall Figure 10.4.)

Figure 11.2

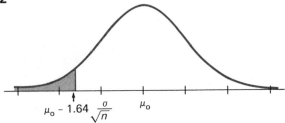

Remark

To test H: $\mu < \mu_0$ for $\alpha = 0.05$ we would use a rejection region to the right of $\mu_0 + 1.64 \frac{\sigma}{\sqrt{n}}$.

Example 11.3.1 A taxi company has been using tires which have an average life of 18,000 miles. It is decided to buy a slightly more expensive brand if a sample confirms for $\alpha = 0.05$ that the average life of these tires exceed 20,000 miles. Four hundred tires are tested with the results $\overline{x} = 19,700$ and $s_x = 1,000$. Since

$$\frac{19,700 - 20,000}{\frac{1000}{20}} = -6, \text{ which is less than } -1.64,$$

we *reject* H: $\mu > 20,000$ at the 5% level. (Observe, however, that we would accept

$$H: \mu > 19,000 \quad \text{as well as} \quad H: \mu > 18,000.)$$

In Section 10.6 we examined the random variable

$$\bar{D} = \bar{X} - \bar{Y}$$

where X and Y are independent random variables from two populations. Our informal argument (Problems 3 and 4, Exercises 10.6) led us to conclude that, for large values of n_X and n_Y, the standardized variable

$$Z = \frac{\bar{D} - \mu_{\bar{D}}}{\sigma_{\bar{D}}} = \frac{(\bar{X} - \bar{Y}) - (\mu_X - \mu_Y)}{\sqrt{\dfrac{\sigma_X^2}{n_X} + \dfrac{\sigma_Y^2}{n_Y}}}$$

has a distribution which is approximately standard normal ($\mu = 0$ and $\sigma = 1$).

We may make use of these ideas if we wish to test the null hypothesis

$$H: \mu_X - \mu_Y = 0.$$

Example 11.3.2 We repeat the information of Example 10.6.1: Two methods for treating industrial wastes yield sample values

$$n_X = 64, \quad \bar{x} = 82.3, \quad s_x = 5.6; \qquad n_Y = 36, \quad \bar{y} = 79.8, \quad s_y = 7.2.$$

Since σ_X and σ_Y are unknown, we use s_x and s_y as point estimates. Our hypothesis is the null hypothesis

$$H: \mu_X - \mu_Y = 0,$$

and so a two-tailed rejection region is appropriate. Let us choose $\alpha = 0.05$. The experimentally obtained value of Z is

$$z = \frac{82.3 - 79.8}{\sqrt{\dfrac{(5.6)^2}{64} + \dfrac{(7.2)^2}{36}}} \doteq \frac{2.5}{1.39} \doteq 1.80.$$

Since $|z| \doteq 1.80$, which is less than 1.96, we do not reject the null hypothesis. We conclude that the population means do not differ at the 5% level of significance.

However, we may approach Example 11.3.2 in a slightly different way. We might consider

$$Z = \frac{\bar{D} - \mu_{\bar{D}}}{\sigma_{\bar{D}}}$$

and establish a 95% confidence interval for $\mu_{\bar{D}}$. Our obtained value, $\bar{d}$, of $\bar{D}$ is 2.5. $\sigma_{\bar{D}} \doteq 1.39$. Thus, the corresponding confidence interval for $\mu_{\bar{D}}$ is

$$2.5 \pm 1.96(1.39) \doteq 2.5 \pm 2.72.$$

Since the interval $[-0.22, 5.22]$ contains 0, we do not reject

$$H: \mu_{\bar{D}} = \mu_X - \mu_Y = 0.$$

On the other hand, we may be interested in a hypothesis of the form H: $\mu_X > \mu_Y$. This hypothesis is the same as H: $\mu_{\bar{D}} > 0$, and so a one-sided test is appropriate as shown in the next example.

Example 11.3.3 In Example 11.3.2, our two-sided test did not reject the null hypothesis. More realistically, before any experimenting is done, it would be appropriate to argue as follows. If the cheaper method (Y) shows the better experimental results, we will use this method. If the more expensive method (X) turns out to be significantly better at the 5% level, we will adopt it. Since it turns out that $\bar{x} > \bar{y}$, we are interested in testing H: $\mu_X > \mu_Y$ (or $\mu_{\bar{D}} > 0$). We are 95% confident that the true value of $\mu_{\bar{D}}$ lies to the right of

$$\bar{d} - 1.64\sigma_{\bar{D}} = 2.5 - 1.64(1.39)$$
$$= 2.5 - 2.28 = 0.22$$

Since $0.22 > 0$, we accept (we cannot reject) H: $\mu_X > \mu_Y$. Therefore, we use the more expensive, but significantly better, method of pollution control.

Remark 1

It is not always obvious whether to use a one-tailed or a two-tailed test. If there is a theoretical, or *a priori*, reason to believe in the null hypothesis, a two-tailed test is usually appropriate. In 600 throws of a fair die, we would expect about 100 fives. If we actually obtain 72 fives we would be "surprised" and be tempted to test H: $p < \frac{1}{6}$ using a one-tailed test. Actually, we would have been just as surprised to observe 128 fives. To cover both cases, it is better to use a (more conservative) two-tailed test of H: $p = \frac{1}{6}$. (See Problem 8, Exercises 11.2 and Problems 2 and 3, Exercises 11.3.)

Remark 2

There is a further reason to prefer two-tailed tests in many statistical problems. In general, the hypothesis should be stated *before* experimentation is begun. Problem 7, Exercises 11.2, illustrates this idea. Simply examining the obtained values of $\bar{x}$ and $\bar{y}$, we might be tempted to test H: $\mu_X > \mu_Y$. *Before* the experimental results are found, we have no reason to state a hypothesis that one population has a larger mean than the other. The null hypothesis is the appropriate one.

EXERCISES 11.3

- **1.** A random sample of 100 lengths of wire rope of a certain size has a mean breaking stress of 9947 pounds and standard deviation 300 pounds. Would we reject, at the 5% level:
  a. H: $\mu = 10,000$?
  b. H: $\mu > 10,000$?
  c. H: $\mu < 10,000$?

- 2. Suppose that 72 out of 600 throws of a given die led to "five." For the one-tailed test H: $p < \frac{1}{6}$, find the rejection region for:
  a. $\alpha = 0.02$
  b. $\alpha = 0.01$
- 3. Remark 1 in this section refers to two-tailed tests as being more conservative. Explain. (See Problem 8, Exercises 11.2 and Problem 2 above.)
- 4. The standard deviation of a certain population is $\sigma = 2.42$. We wish to test H: $\mu > 7$. If a sample of size 64 yields $\bar{x} = 7.65$, do we reject H at:
  a. the 5% level?
  b. the 2% level?
  c. the 1% level?

11.4 Risk; Type I Errors

If we make a decision based on a statistical test of a hypothesis, we run the risk of deciding incorrectly. For a given hypothesis H, our test may lead us to make one of two types of errors:

a. Type I error: reject H when it is, in fact, true.
b. Type II error: accept H when it is, in fact, false.

For a given test it is appropriate to ask: What are the probabilities of making each type of error? Our experience with confidence intervals and significance levels enables us to discuss Type I errors without too much difficulty. In Section 11.5 we shall treat briefly Type II errors.

Ideally, tests should be designed so that the probabilities of making each type of error are quite small. In practice, it is not always possible to achieve this goal. We often face the necessity for determining which of the two possible errors gives rise to the more serious consequence. If we are able to make this choice, it is often possible to state a hypothesis in such a way that the more serious consequence results from a Type I error.

Example 11.4.1 A new surgical technique for organ transplants is proposed. It is claimed that the life expectancy of patients will be significantly increased. Let μ_X and μ_Y be the mean life expectancies resulting from the new and the old techniques, respectively. We would wish to be very careful not to reject H: $\mu_X > \mu_Y$ if it were, in fact, true.

Example 11.4.2 A new method for training certain skilled electronic workers involves considerable time and expense. If μ_X and μ_Y are the mean productivity of workers trained by the new and old methods, respectively, we would be careful not to accept hypothesis H: $\mu_X > \mu_Y$ if it were, in fact, false.

To illustrate the computation of the probability of making a Type I error, suppose that we test

$$H: \mu = \mu_0$$

by the rule:

"Reject H if the obtained value of $\overline{X}$ lies outside the interval

$$\mu_0 \pm 1.96 \frac{\sigma}{\sqrt{n}}."$$

If the hypothesis is *true* and assuming that the normal approximation is appropriate, we recognize that the probability that this test will lead to a (false) rejection of H is less than 0.05. Thus, $\alpha = 0.05$ is the probability of making a Type I error. Notice that this is the same as "level of significance." Let us use the notation

$$\alpha = P_H(\text{reject H})$$
$$= P \text{ (the value of the test statistic falls in the } 100\alpha\% \text{ rejection region, } given \text{ that H is true).}$$

As we have seen in Chapter 10 and in Section 11.2, it is often desirable to decide on the size of α before conducting the test. On the other hand, we may compute α if the test is specified (see Problem 3, Exercises 11.2).

Example 11.4.3 Let X be the random variable representing the number of successes in three independent trials. It is desired to test H: $p = 0.8$ for a Bernoulli population. We may compute α for each of the following possible one-tailed tests (recall Section 7.4):

Test 1. Reject H if $X = 0, 1$, or 2.

Test 2. Reject H if $X = 0, 1$.

Test 3. Reject H if $X = 0$.

For Test 1, $\alpha = P_{0.8}(X < 3) = 1 - P_{0.8}(X = 3) = 1 - (0.8)^3 = 0.488$

For Test 2, $\alpha = P_{0.8}(X < 2) = P_{0.8}(X = 0) + P_{0.8}(X = 1)$

$$= (0.2)^3 + \binom{3}{1}(0.8)^1(0.2)^2 = 0.104$$

For Test 3, $\alpha = P_{0.8}(X < 1) = P_{0.8}(X = 0) = (0.2)^3 = 0.008$

Remark

The results of Example 11.4.3 suggest a simple method of making $\alpha = 0$. We could decide to accept H regardless of the experimental results. This procedure not only makes $\alpha = 0$ but also has the advantage of not requiring the necessity of experimenting at all! One is reminded of the man who asserts: "My mind is made up (to accept H), don't bother me with facts."

Example 11.4.4 Suppose that we wish to test H: $p = 0.5$ for a Bernoulli population, again using three independent trials as in Example 11.4.3. A possible two-tailed test is: "Reject H if $X = 0$ or if $X = 3$." For this test,

$$\alpha = P_{0.5}(X = 0 \text{ or } X = 3) = (0.5)^3 + (0.5)^3 = 0.25.$$

EXERCISES 11.4

Determine α for each test described below. We are dealing with independent trials from a Bernoulli population.

- 1. H: $p = 0.5$. Test: Reject H unless we obtain 4, 5, or 6 successes in 10 trials. (Use Table B.)
2. H: $p = 0.5$. Test: Reject H if we obtain 0, 1, 9, or 10 successes in 10 trials.
- 3. H: $p = 0.5$. Test: Reject H unless we obtain from 40 to 60 successes in 100 trials. (Use normal approximation (Section 8.2).)
4. H: $p = 0.5$. Test: Reject H if we obtain 10 or less successes or 90 or more successes in 100 trials.
- 5. H: $p = 0.1$. Test: Reject H if we obtain 3 successes in 3 trials.
- 6. H: $p = 0.1$. Test: Reject H if we obtain 2 or more successes in 3 trials.
7. H: $p = 0.1$. Test: Reject H if we obtain 1 or more successes in 3 trials.
8. H: $\mu = 70$, $\sigma = 4$, $n = 64$. Test: Reject H if $|\bar{x} - 70| > 0.6$.
- 9. H: $\mu < 70$, $\sigma = 4$, $n = 64$. Test: Reject H if $\bar{x} - 70 > 0.6$.

11.5 Alternate Hypotheses; Type II Errors

For every statistical hypothesis, there can be formulated alternate hypotheses. As an example, the usual alternate to the null hypothesis

$$H: \mu = \mu_0$$

is the hypothesis

$$A: \mu \neq \mu_0.$$

The various tests of H we have been discussing have permitted us to compute the probability of making a Type I error:

$$\alpha = P_H(\text{reject H})$$

We may now define the probability of making a Type II error:

$$\beta = P_A(\text{accept H}).*$$

*The symbol β (beta) is a letter of the Greek alphabet corresponding to the English "b."

Acceptance of H implies rejection of A; hence,
$$\beta = P_A(\text{reject A}).$$

Remark

If we make a Type I error in falsely rejecting a true H, we have automatically falsely accepted A. These are not two distinct errors. To avoid confusion, remember that α and β are the probabilities of Type I and Type II errors as applied to the original hypothesis, H.

If A is of the form A: $\mu \neq \mu_0$, then we are faced with the fact that we do not know which value to assign to μ in calculating
$$\beta = P_A(\text{reject A}).$$

The general question is a bit difficult. We shall confine ourselves to certain illustrations involving Bernoulli populations.

Example 11.5.1 An urn is known to contain 10 marbles which may be all red, all blue, or a mixture of these colors. Let p be the probability of drawing a red marble. If the hypothesis H: $p = 0.5$ is proposed, the alternate hypothesis is A: $p \neq 0.5$. Unfortunately, there are ten possible values of p included under A. We may have
$$p = 0, 0.1, 0.2, 0.3, 0.4, 0.6, 0.7, 0.8, 0.9, 1.$$

If we are allowed three draws, with replacement, a simple rule for testing H is "reject if all three marbles are the same color." We easily find
$$\alpha = P_{0.5}(X = 0, 3) = (0.5)^3 + (0.5)^3 = 0.25$$
(see Example 11.4.4). To calculate
$$\beta = P_A(X \neq 0, 3),$$
we would need to know which of the ten possible values of p to use in the computation.

Example 11.5.2 An urn is known to contain either 8 red and 2 blue or 3 red and 7 blue marbles. In Example 11.4.3, we proposed three possible tests of H: $p = 0.8$, and found the corresponding values of α. We may now calculate the values of β for these tests, using A: $p = 0.3$ as the alternate hypothesis.

Test 1. $\beta = P_{0.3}(X = 3) = (0.3)^3 = 0.027$

Test 2. $\beta = P_{0.3}(X \geq 2) = \binom{3}{2}(0.3)^2(0.7) + (0.3)^3 = 0.216$

Test 3. $\beta = P_{0.3}(X \geq 1) = 1 - P_{0.3}(X = 0)$
$$= 1 - (0.7)^3 = 0.657$$

Combining the results of Example 11.4.3 and Example 11.5.2 we have the table in Figure 11.3.

Figure 11.3

	Test 1	Test 2	Test 3
α	0.488	0.104	0.008
β	0.027	0.216	0.657

As a general rule, tests based on a given sample size which are designed to yield small values of α result in an increased value of β. In order to minimize the risk of rejecting H when it is true, we increase the risk of accepting H when it is false. If for a given sample size, a predetermined value of α leads to an unacceptably large value of β, we have two choices. Either we must change our value of α or we must enlarge the sample size.

Example 11.5.3 To test H: $p = 0.8$ against A: $p = 0.3$, we are allowed to make 10 independent trials. Of the several possible tests, let us consider "Reject H if $X < 5$" where X represents the number of successes. We may use Table B to calculate α and β:

$$\alpha = P_{0.8}(X < 5) = 1 - P_{0.2}(X \leq 5) \doteq 1 - 0.9936 = 0.0064$$
$$\beta = P_{0.3}(X \geq 5) = 0.1502$$

EXERCISES 11.5

Calculate β if we test H: $p = 0.1$ against A: $p = 0.6$ for the test of:

- 1. Problem 5, Exercises 11.4
- 2. Problem 6, Exercises 11.4
- 3. Problem 7, Exercises 11.4
- 4. To test H against A we could simply decide to accept H without experimenting. (See Remark on page 274.) For this "test," $\alpha = 0$. What is the value of β?
- 5. What rule would enable us to guarantee that $\beta = 0$? What is the corresponding value of α?

Determine β for the test of H: $p = 0.5$, Example 11.5.1.

- 6. A: $p = 0$
- 7. A: $p = 0.1$
- 8. A: $p = 0.2$
- 9. A: $p = 0.3$
- 10. A: $p = 0.4$
- 11. Continue the preceding problems using the other possible alternate hypotheses.

★ 12. For each of the 11 possible values of p (including $p = 0.5$) we may consider the function

$$\pi(p) = 1 - \beta = P_p(\text{reject H}).$$

Use the results of Problems 4 and 5 to graph $\pi(p)$ for the test of Example 11.5.1. Recall that for $p = 0.5$, $\pi(0.5) = P_{0.5}(\text{reject H}) = \alpha$.

The function $\pi(p)$, defined in Problem 6 is called the **power function** of the test. It displays the "power" of the test to detect that H is false (when it is false) against the alternate hypotheses.

★ 13. The results of Problem 12 show that our test is "powerful" against $p = 0$, 0.1, 0.9, 1, but not very powerful against $p = 0.4, 0.6$. Explain why this is reasonable.

★ 14. If we test $p = 0.5$ against $p \neq 0.5$ for a Bernoulli population in a situation for which it is possible for p to have any value from 0 to 1, the graph of the power function will be continuous. Sketch the general shape of such a graph for the three-trial test we have been discussing.

★ 15. Suppose that we were allowed to increase the number of trials. How do you think this would affect the shape of the graph of the power function?

★ 16. What would be the general shape of the graph if we conducted a one-tailed test of $p < 0.5$?

11.6 Decisions, Decisions

Usually, a statistical test is designed to lead to a course of action (a decision) based on the results. If a test leads to the acceptance of a hypothesis, action is taken as though the hypothesis were true even when it is realized that no statistical test can *prove* the hypothesis. Many decisions in scientific endeavors, economics, manufacturing, agriculture, and government planning are of this type.

On the other hand, we must recognize that not all decisions made in the face of uncertainty are (or should be) based solely on statistical evidence. In Section 5.6, for example, we discussed situations in which we might decide to "play a game" whose expected value is negative.

In a similar way, "statistically significant" results are sometimes overridden by other considerations. There may be moral, social, philosophic, or humanistic factors involved in making a particular decision that we are not able to measure statistically. One aspect of this is pointed out in the Remark 2 at the end of Section 10.4. A "statistically significant" difference between the means of two populations does not necessarily imply significant in the sense of being "important."

It is not too difficult to conceive of cases where a statistically significant result might be ignored. Such a case, for example, may arise in connection with international policy, medical practice, ecology, legal findings, or legislative decisions. The judicial adage that "it is better that a thousand guilty men go free than one innocent man be punished" reflects the fact that more than strict statistical considerations guide our decisions.

Actually, many decisions are extremely personal or subjective. What appears to be convincing statistical evidence to one individual may well be ignored by a second. Without necessarily formalizing the thought, one individual may be content with $\alpha = 0.05$, for example, in a situation in which another insists on $\alpha < 0.005$.

Chapter Summary

1. a. The hypothesis H: $\mu_X = \mu_Y$ can be written as H: $\mu_X - \mu_Y = 0$ and so is called the *null hypothesis*. Similarly, we may consider $\mu = \mu_0$, and so on, as null hypotheses.
   b. Other hypotheses are $\mu > \mu_0$ and $\mu < \mu_0$.
2. The null hypothesis is tested by a two-sided, or two-tailed, test.
3. To test H: $\mu > \mu_0$ or H: $\mu < \mu_0$, a one-sided, or one-tailed, test is used.
4. a. Type I error: reject H when it is, in fact, true.
   b. Type II error: accept H when it is, in fact, false.
5. a. The probability of making a Type I error is

$$\alpha = P_H(\text{reject H}).$$

   b. The probability of making a Type II error is

$$\beta = P_A(\text{accept H}).$$

Chapter Review

1. What is meant by the null hypothesis in comparing the means of two populations?
2. Give an example of an experimental situation for which each of the following hypotheses is appropriate. (μ_0 represents some constant.)
   a. $\mu = \mu_0$ b. $\mu < \mu_0$ c. $\mu > \mu_0$

3. An urn contains a mixture of red and blue marbles. It is known that the fraction of red marbles is either $\frac{3}{4}$ or $\frac{1}{3}$. Let p be the probability of drawing a red marble on a single random draw. The hypothesis H: $p = \frac{3}{4}$ is tested by the rule: Draw twice with replacement, and reject H unless both marbles drawn are red. Determine α for this test.

4. For the data of Problem 3, let A: $p = \frac{1}{3}$ be the alternate hypothesis to H. Find β for the test of Problem 3.

5. Using the hypotheses of Problems 3 and 4, find α and β for the test: Reject H if both marbles are blue.

6. A certain normal population is known to have $\sigma = 12.8$. It is decided to test H: $\mu = 60$ by choosing a random sample of size 64 and rejecting H if $\bar{x}$ differs from 60 by more than 2.4. Find α for this test.

7. If we use the test of Problem 6 but with a sample size of 256, what is the value of α?

8. Random samples of size 100 are chosen for two large, independent populations, both having $\sigma = 8$. The hypothesis H: $\mu_X - \mu_Y = 0$ is to be rejected if the sample means differ by more than 2. Find α for this test.

9. For the data of Problem 8, what should the test be if we decide to choose $\alpha = 0.05$?

Appendix

NOTE ON SUMMATION NOTATION

Mathematics makes great use of symbols as a shorthand for algebraic expressions that would otherwise require considerable writing. Here we describe one that is especially useful in statistics and probability.

Suppose that we wish to add several variables. Instead of labeling them $X, Y, Z, \ldots$, it is often convenient to use the same letter and distinguish the variables by different subscripts. For instance, we might use $X_1, X_2, \ldots, X_n$ to designate n variables. Our sum, then, is written

$$X_1 + X_2 + \cdots + X_n.$$

To abbreviate this, we use the **summation notation** and write

$$\sum_{i=1}^{n} X_i = X_1 + X_2 + \cdots + X_n.$$

We read

$$\sum_{i=1}^{n} X_i$$

as "the sum of X-sub i for i equals 1 to n."* If the number of addends, n, is agreed upon in advance, we may simply write

$$\sum X_i.$$

In this text, we make use of several properties of the summation symbol. These may be explained as follows:

1. It may be useful to consider two different sets of variables $X_1, X_2, \ldots, X_n$ and $Y_1, Y_2, \ldots, Y_n$, there being exactly n X's and n Y's. Then:

$$\sum_{i=1}^{n}(X_i + Y_i) = (X_1 + Y_1) + (X_2 + Y_2) + \cdots + (X_n + Y_n)$$
$$= (X_1 + X_2 + \cdots + X_n) + (Y_1 + Y_2 + \cdots + Y_n)$$
$$= \sum_{i=1}^{n} X_i + \sum_{i=1}^{n} Y_i$$

*The symbol Σ is the Greek capital letter "sigma," corresponding to the English "S."

2. If each of the variables $X_1, X_2, \ldots, X_n$ is to be multiplied by the same constant, a, and then the results are to be added, we have:

$$\sum_{i=1}^{n} aX_i = aX_1 + aX_2 + \cdots + aX_n$$
$$= a(X_1 + X_2 + \cdots + X_n)$$
$$= a \sum_{i=1}^{n} X_i$$

3. Suppose that we wish to add a constant, b, to each X_i and then add the results:

$$\sum_{i=1}^{n} (X_i + b) = X_1 + b + X_2 + b + \cdots + X_n + b$$
$$= X_1 + X_2 + \cdots + X_n + b + b + \cdots + b$$
$$= \sum_{i=1}^{n} X_i + nb \quad \text{(since there are } n \text{ } b\text{'s to be added)}$$

This leads us to define

$$\sum_{i=1}^{n} b = nb, \quad \text{where } b \text{ is a constant.}$$

4. Combining properties 2 and 3, we have:

$$\sum_{i=1}^{n} (aX_i + b) = a \sum_{i=1}^{n} X_i + nb$$

$$\sum_{i=1}^{n} (aX_i + b)^2 = \sum_{i=1}^{n} (a^2 X_i^2 + 2ab X_i + b^2)$$
$$= a^2 \sum_{i=1}^{n} X_i^2 + 2ab \sum_{i=1}^{n} X_i + nb^2$$

5. We shall have use for some special cases, for example:

(a) $\sum \left(\dfrac{X_i}{n} \right) = \dfrac{1}{n} \sum X_i = \dfrac{\sum X_i}{n}$

(b) $\dfrac{1}{n} \sum (X_i - \mu) = \dfrac{\sum X_i}{n} - \mu$, where μ is a constant

(c) $\dfrac{1}{n} \sum (aX_i + b)^2 = \dfrac{1}{n} \sum (a^2 X_i^2 + 2ab X_i + b^2)$
$= a^2 \dfrac{\sum X_i^2}{n} + 2ab \dfrac{\sum X_i}{n} + b^2$

6. It is possible that we have two sets of variables (as in 1, on page 281) and wish to multiply them before adding. In particular, we have:

$$\sum_{i=1}^{n} X_i Y_i = X_1 Y_1 + X_2 Y_2 + \cdots + X_n Y_n$$

If for each Y_i we have $Y_i = f(X_i)$, then:

$$\sum_{i=1}^{n} X_i f(X_i) = X_1 f(X_1) + X_2 f(X_2) + \cdots + X_n f(X_n)$$

7. Sometimes we may employ more than one generalized subscript (index). Thus:

$$X_1 + X_2 + \cdots + X_n + Y_1 + Y_2 + \cdots + Y_m = \sum_{i=1}^{n} X_i + \sum_{j=1}^{m} Y_j$$

8. One particular "double sum" is of importance for us,

$$\sum_{i=1}^{n} \sum_{j=1}^{m} X_i Y_j.$$

By this is meant the sum of all possible products using all the different X's and Y's. That is, multiply X_1 in turn by Y_1 through Y_m; then X_2 in turn by Y_1 through Y_m; and so on through X_n. Finally, add these products.

If the variables are X_1, X_2, X_3, Y_1, Y_2, we have:

$$\begin{aligned}
\sum_{i=1}^{3} \sum_{j=1}^{2} X_i Y_j &= X_1 Y_1 + X_1 Y_2 + X_2 Y_1 + X_2 Y_2 + X_3 Y_1 + X_3 Y_2 \\
&= X_1(Y_1 + Y_2) + X_2(Y_1 + Y_2) + X_3(Y_1 + Y_2) \\
&= (X_1 + X_2 + X_3)(Y_1 + Y_2) \\
&= \left(\sum_{i=1}^{3} X_i\right)\left(\sum_{j=1}^{2} Y_j\right)
\end{aligned}$$

The result is general. Thus:

$$\sum_{i=1}^{n} \sum_{j=1}^{m} X_i Y_j = \left(\sum_{i=1}^{n} X_i\right)\left(\sum_{j=1}^{m} Y_j\right)$$

These conclusions are sufficient for our purposes. We have used capital letters here. We could, of course, apply the same ideas to any desired symbols.

LIST OF FORMULAS

1. Let S be a set of possible outcomes of an experiment:
$$S = \{s_1, s_2, \ldots, s_n\}$$
Let $A = \{s_1, s_2, \ldots, s_m\}$, $m \leq n$, be a finite subset (event) of S. Then
$$P(A) = \sum_{i=1}^{m} f(s_i) = f(s_1) + f(s_2) + \cdots + f(s_m). \quad \text{(Page 27)}$$

2. $P(S) = 1$ (Page 32)

3. $P(\emptyset) = 0$ (Page 32)

4. $0 \leq P(A) \leq 1$ (Page 32)

5. $P(\sim A) = 1 - P(A)$ (Page 32)

6. If A and B are events, then
$$P(A \cup B) = P(A) + P(B) - P(A \cap B). \quad \text{(Pages 37 and 41)}$$

7. If E and G are events with $P(G) \neq 0$, then
$$P(E \mid G) = \frac{P(E \cap G)}{P(G)}. \quad \text{(Page 50)}$$

8. Bayes' formula: $P(G_i \mid E) = \dfrac{P(G_i) \cdot P(E \mid G_i)}{\sum_{j=1}^{k} P(G_i) \cdot P(E \mid G_j)}$ (Page 74)

9. $\bar{x} = \dfrac{\sum_{i=1}^{n} x_i}{n} = \dfrac{\sum_{i=1}^{m} x_i \cdot t(x_i)}{n} = \sum_{i=1}^{m} x_i \cdot \dfrac{t(x_i)}{n}$ (Page 94)

10. $s_x^2 = \dfrac{\sum_{i=1}^{n}(x_i - \bar{x})^2}{n} = \dfrac{\sum_{i=1}^{n} x_i^2}{n} - \bar{x}^2$

$= \dfrac{\sum_{i=1}^{m}(x_i - \bar{x})^2 \cdot t(x_i)}{n} = \dfrac{\sum_{i=1}^{m} x_i^2 \cdot t(x_i)}{n} - \bar{x}^2$ (Page 101)

11. $s_x = \sqrt{s_x^2}$ (Page 101)

12. If X is a random variable, taking on the distinct values $x_1, x_2, \ldots, x_n$, then
$$E(X) = \mu_X = \sum_{i=1}^{n} x_i \cdot f(x_i). \quad \text{(Page 129)}$$

13. $E(aX + b) = aE(X) + b$ or $\mu_{aX+b} = a\mu_X + b$ (Pages 140 and 145)

14. $\sigma_X^2 = \text{var}(X) = E((X - \mu_X)^2)$ (Page 147)

$\sigma_X^2 = E(X^2) - \mu_X^2 = \mu_{X^2} - \mu_X^2.$ (Page 147)

15. $\sigma_X = \sqrt{\sigma_X^2}$ (Page 147)

16. Chebyshev's Inequality: $P(|X - \mu_X| > k\sigma_X) \le \dfrac{1}{k^2}$ (Page 148)

17. $E(X + Y) = E(X) + E(Y)$ or $\mu_{X+Y} = \mu_X + \mu_Y$ (Page 152)

18. If X and Y are independent random variables, then
$$\text{var}(X + Y) = \text{var}(X) + \text{var}(Y).$$ (Page 155)

19. $\text{var}(aX + b) = a^2\text{var}(X) = a^2\sigma_X^2$ (Page 158)

20. $\sigma_{aX+b} = |a|\sigma_X$ (Page 158)

21. If X is a random variable with expected value μ_X and variance σ_X^2, then
$$Z = \frac{X - \mu_X}{\sigma_X}$$
is a random variable with expected value 0 and variance 1.   (Page 160)

22. $\text{cov}(X, Y) = E((X - \mu_X)(Y - \mu_Y))$
$= \tfrac{1}{2}[\text{var}(X + Y) - \text{var}(X) - \text{var}(Y)]$
$= E(XY) - \mu_X\mu_Y = \mu_{XY} - \mu_X\mu_Y$ (Page 163)

23. $\rho(X, Y) = \dfrac{1}{\sigma_X \cdot \sigma_Y} \cdot \text{cov}(X, Y)$ (Page 165)

24. $\binom{n}{x} = \binom{n-1}{x-1} + \binom{n-1}{x}$, $0 < x < n$ (Page 181)

25. If $X_1, X_2, \ldots, X_n$ are independent Bernoulli variables, each having
$$P(X_i = 1) = p \quad \text{and} \quad P(X_i = 0) = q,$$
then the probability function of $X = X_1 + X_2 + \cdots + X_n$ (binomial distribution) is defined by
$$f(x) = b(n, p; x) = P(X = x) = \binom{n}{x} p^x q^{n-x}, \; x \in \{0, 1, 2, \ldots, n\}.$$
$E(X) = np$, $\text{var}(X) = npq$, $\sigma_X = \sqrt{npq}$ (Page 183)

26. $n! = n \cdot (n - 1) \cdots 3 \cdot 2 \cdot 1$ (Page 185)

27. $_nP_x = \dfrac{n!}{(n - x)!}$ (Page 186)

28. $_nC_x = \dfrac{n!}{x! \cdot (n - x)!} = \dfrac{_nP_x}{x!} = \binom{n}{x}$ (Page 187)

29. $\binom{n}{x} = \binom{n}{n - x}$ (Page 190)

30. $(r + b)^n = \binom{n}{n}r^n + \binom{n}{n-1}r^{n-1}b + \cdots + \binom{n}{1}rb^{n-1} + \binom{n}{0}b^n$
  (Page 192)

31. $b(n, p; x) = \binom{n}{x} p^x q^{n-x} = \binom{n}{n-x} q^{n-x} p^x = b(n, q; n - x)$

(Page 194)

32. De Moivre's Theorem: If $X = X_1 + X_2 + \cdots + X_n$, where the X_i's are independent Bernoulli variables each having $\mu = p$, then if n is sufficiently large,

$$P\left(a \leq \frac{X - np}{\sqrt{npq}} \leq b\right) \doteq P(a \leq Z \leq b)$$

$$= P(Z \leq b) - P(Z \leq a). \quad \text{(Page 211)}$$

33. $\overline{X} = \dfrac{X_1 + X_2 + \cdots + X_n}{n}$

$E(\overline{X}) = \mu_{\overline{X}} = \mu$, $\operatorname{var}(\overline{X}) = \sigma_{\overline{X}}^2 = \dfrac{\sigma^2}{n}$, $\sigma_{\overline{X}} = \dfrac{\sigma}{\sqrt{n}}$ (Page 227)

34. Central Limit Theorem: If $X_1, X_2, \ldots, X_n$ are identically distributed random variables each having expected value μ and variance σ^2, then the probability distribution of

$$Z = \frac{\overline{X} - \mu}{\dfrac{\sigma}{\sqrt{n}}}$$

is approximately standard normal if n is sufficiently large. (Page 233)

SELECTED SOLUTIONS AND ANSWERS

Solutions to all exercises marked with ● are given in this section, together with the answers to all odd-numbered problems not so marked.

Exercises 1.1 (Pages 3–4)
1. Game; fair.
3. Game; not fair. Only 3 and 6 are divisible by 3.   1, 2, 4, and 5 are not.
5. Not a "game." If 6 is thrown, the rules allow both to win.
6. Not a "game." If 2 is thrown, both "win"; if 1 is thrown, neither "wins," since the rules do not say what happens then.
7. Game; fair. Three numbers are prime (2, 3, 5); three (1, 4, 6) are not.
9. Game; not fair. There are 9 possibilities for winning, 27 for losing.
11. Game; not fair. There are 13 spades, 39 other cards.   13. Game; fair.
15. Game; not fair. There are 18 possibilities to win, 20 to lose.

Exercises 1.2 (Page 7)
There is no way of knowing which number will come next in any of these sequences. Our suggested reasoning for Problem 1, for example, might be typical.
  1. **a.** 1   **b.** 1   **c.** 2 or 1   **d.** 4 or 5
Possible Reasoning: **a.** "The numbers show an increasing pattern; we feel that it is quite likely (1) that the next number is positive. **b.** Since the pattern shows increases

from one number to the next, our belief (1) is that the next number will be larger than 32. **c.** The numbers are successive powers of 2. It is a reasonable guess (2 or 1) that the next number is $2^6 = 64$. **d.** All the given numbers are whole numbers and they increase rapidly. It is unlikely (4 or 5) that the experiment will suddenly produce a number such as 33.89."

3. **a.** 1 or 2   **b.** 4 or 5   **c.** 4 or 5   **d.** 1 or 2
5. **a.** 5   **b.** 2 or 3   **c.** 1 or 2   **d.** 4 or 5
6. **a.** 1 or 2   **b.** 4 or 5   **c.** 5   **d.** 2 or 3

Note: Based on the data presented, one's inclination would favor belief in a population increase. While the 1960 census showed the expected increase in total population, there was an actual *decrease* in population in many major cities. If this trend continues, the population of San Francisco after twenty years might turn out to be less than 775,000 (in 1960, San Francisco had a population of 740,316). Preliminary reports from the 1970 census tend to confirm this trend. This points out the need to be aware of the uncertainty in projecting information from data.

Exercises 1.3   (Pages 9–10)

1. One suggestion is to toss the two coins simultaneously. If they match, one player wins; if they fail to match, the second player wins. Since we assume that Coin 1 is fair, a match will occur on about half of all tosses even if Coin 2 always comes up heads.

3. We could retrieve the data by making five additional tosses of the coin. These extra five tosses would have little effect on the ratio of heads to tails.

4. The data could not be retrieved. If the five tosses for which data are missing were in succession, then one or more "runs" may have been lost. On the other hand, if the missing data are scattered, notice that we would not need to retrieve the missing data unless one (or more) of the missing results occurred in a potential run of 4 heads. In hth–t we do not need to retrieve the missing toss.

5. We could not retrieve the missing data if we insist on testing during the third week. Once again it would probably not be essential to worry about the missing scores unless we feel that the flu had attacked a very special group (either good or poor) of arithmetic students.

7. The details cannot be duplicated. However, if the remaining data show a clear pattern in tracking, then it might still be possible to draw conclusions that could be verified by future observations, since the conditions are more likely to be consequences of physical forces than merely accidental occurrences.

Exercises 1.4   (Pages 11–12)

1. Reordering in increasing order—1, 2, 4, 8, 16, 32, 64, 128—seems to give a pattern of successive powers of 2.

2. Reordering in increasing order—95, 95, 96, 97, 98, 99, 100, 102, 104—does not seem to give any pattern for the numbers except that there seems to be more "packing" in the lower end and a slight tendency to have "gaps" at the upper end.

3. Reordering might give a better picture of the way the heights differ. A "clustering" near 5′ 5″ might be noted: 4′ 11″, 5′ 3″, 5′ 3″, 5′ 5″, 5′ 7″, 5′ 8″, 5′ 11″, 6′ 0″, 6′ 1″.

4. These data are already in chronological order, and for the purpose, are more valuable in this form, since it shows the hourly trend of the pressure. Note that these numbers are exactly the same as those in Problem 2.
5. Reordering from lowest to highest might give an indication that, except for Denver, the annual average is higher for a location that is closer to the equator. (The lower temperature in Denver can be accounted for by its altitude.)
6. For example, opinion surveys implicitly assume that the opinions recorded are representative of the general population or a certain segment of the population. Even a routine matter such as clothing size incorporates the idea of a representative measure. On the basis of many measurements made at one time, for example, an assumption might be made that a "medium" size for shirts might mean measurements in the neighborhood of a 15″ collar, 32″ sleeve length, 38″ chest, and so forth.

Exercises 1.5   (Page 13)
2. Drawing five marbles from the box simultaneously is the same as drawing marbles one at a time repeatedly without replacement. There is no difference in the results of selecting the five marbles one at a time while inside the bag and finally removing the fistful. In each case, the marble selected is enclosed in the fist and kept isolated from the remaining marbles.

Exercises 1.6   (Pages 13–14)
Since we are asking you to draw conclusions by intuition (or by intelligent guessing), there are no "right" answers. We hope that you decided as follows:
1. a. I and II are equally likely.   b. I   c. I and II are equally likely.   d. II
2. a. I. (We expect to get "*about* 50 heads in 100 tosses." It turns out that getting *exactly* 50 heads is not very likely.)
   b. II. (We expect "about 50% of the tosses to be heads." In each experiment we are considering that from 40% to 60% of the tosses are heads. The more tosses we make, the less likely it is that the percentage of heads will vary very much from 50%.)
   c. II. (See 2b above.)   d. I. (See 2b above.)
3. a. II   b. II   c. I   d. I and II are equally likely.

Exercises 2.1   (Pages 16–17)
1. a. 2,000
   b. The first decision is to stop production and investigate. One reaction might be to question whether every item in the entire lot of 100,000 is defective. Another line of action might be to ask about testing procedures, especially about the "randomness" of selecting items to be tested. If the items were produced close together in sequence, this might indicate a period during which some machine was not functioning properly.
   c. The results seem to indicate lack of randomness; this, as in part b, can lead to distortion in interpretations.
   d. The regularity in selection makes it too easy to miss items which might be defective due to some periodic malfunctioning of a machine. For example, a smudge on a rotating drum of a copying machine might produce a blur on every fourth copy. Checking every 1000th might miss every defective item or it might include every defective item produced this way.
2. a. A lack of randomness might be introduced this way. Different characteristics might be missed or disproportionately emphasized, hence not representative of the state's population.

b. If the idea is to use the results to judge how the state is performing in general, then some steps might be taken to make the group being tested more representative of the general population. For example, centers having common features (say, urban or rural characteristics, geographical locations, and so on) might be grouped together before selection is made from the various groups. This kind of pregrouping before selection to preserve characteristics is a procedure in "stratified" sampling (see Section 9.2).

Exercises 2.2 (Pages 20–22)

1. a. {heart, spade} b. {ace, not ace}
   c. The set of all eight outcomes: $\{A_H, 8_H, A_S, 6_S, 5_S, 4_S, 3_S, 2_S\}$
   d. For the set of part c.
2. {2 heads, 1 head, 0 heads}, {hh, ht, th, tt}, {match, do not match}, and so on.
3. a. {3, 2, 1, 0} b. {hhh, hht, hth, htt, thh, tht, tth, ttt}
4. a. $32 = 2^5$ b. 2^n

5.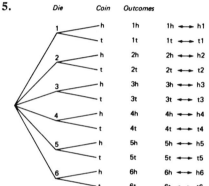

6. Either the 12 outcomes of Problem 5 or the 12 outcomes indicated in Figure 2.3. (Order is not significant here.)

7. a. match: {2, 0};
      do not match: {1}
   b. match: {rr, bb};
      do not match: {rb, br}
   c. 2: {rr}; 1: {rb, br}; 0: {bb}
   d. If we know which outcome of S_3 has occurred, we know automatically which outcome of S_1 and S_2 occurred. On the other hand, even if we know "match" occurred, we do not know whether rr or bb occurred.

8. a.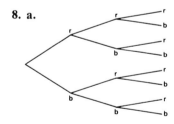

9. a.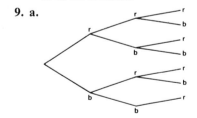

   b. 8 outcomes; {rrr, rrb, rbr, rbb, brr, brb, bbr, bbb}
   c. $2 \cdot 2 \cdot 2 = 8$ so that the multiplication principle applies.

   b. 7 outcomes; {rrr, rrb, rbr, rbb, brr, brb, bbr}
   c. No, since it requires that each of the Step 2 outcomes lead to the *same* number of Step 3 outcomes.

11. a. Outcomes are equally likely:
    {ppp, ppn, ppd, pnp, pnn, pnd, pdp, pdn, pdd, npp, npn, npd, nnp, nnn, nnd, ndp, ndn, ndd, dpp, dpn, dpd, dnp, dnn, dnd, ddp, ddn, ddd}   b. 27
    c. For example, ppp → 0.03, ppn → 0.07, pnn → 0.11, etc. The set of possible outcomes is {0.03, 0.07, 0.11, 0.12, 0.15, 0.16, 0.20, 0.21, 0.25, 0.30}. These are not equally likely. For example, 0.30 can occur in only one way while 0.25 can occur in two ways.

13. a. m^2 b. m^3 c. m^n
15. a. 11: $\{0, 1, 2, 3, 4, 5, 6, 7, 8, 9, 10\}$
    b. 101: $\{0, 0.1, 0.2, \ldots, 9.9, 10\}$ c. 1001 d. $10^{n+1} + 1$
17. a. $\{2, 1, 0\}$ b. $\{rr, rb, rg, br, bb, bg, gr, gb, gg\}$

Exercises 2.3 (Pages 24–25)

1. $\frac{1}{2}$ 2. $\frac{1}{6}$; $\frac{1}{6}$ 3. $\frac{1}{12}$ 4. $\frac{1}{k}$ 5. $\frac{1}{52}$ 6. $\frac{1}{1200}$; $\frac{1}{700}$ 7. $\frac{1}{8}$ 9. $\frac{1}{26}$

11. a. The argument assumes that every state has the same number of residents.
    b. To determine this probability, we should compare the total population of residents of the M states with the total population of the N states (against the U.S. population. As it turned out, there is not much difference here. The M states have about 15.12% of the U.S. population (1966 estimate) and the N states have about 17.49%.

13. $\frac{1}{20}$

Exercises 2.4 (Pages 28–30)

1. a. $E = \{\text{when, the, course, events}\}$, $P(E) = \frac{4}{28} + \frac{3}{28} + \frac{6}{28} + \frac{6}{28} = \frac{19}{28}$
   b. $W = \{\text{when}\}$, $P(W) = \frac{4}{28} = \frac{1}{7}$ c. $V = \{\text{the, course}\}$, $P(V) = \frac{9}{28}$
   d. $H = \{\text{when, the, human}\}$, $P(H) = \frac{12}{28} = \frac{3}{7}$
   e. $T = \{\text{in, of}\}$, $P(T) = \frac{4}{28} = \frac{1}{7}$
2. a. $\frac{1}{6}$ b. The event is $E = \{2, 4, 6\}$. $P(E) = \frac{3}{6} = \frac{1}{2}$
   c. The event is $\{2, 3, 5\}$. $P(2, 3, 5) = \frac{1}{2}$
   d. The event is $\{3, 6\}$. $P(3, 6) = \frac{2}{6} = \frac{1}{3}$
   e. The event is $\{1, 2, 3, 4\}$. $P(1, 2, 3, 4) = \frac{4}{6} = \frac{2}{3}$
   f. The event is $\{5, 6\}$. $P(5, 6) = \frac{2}{6} = \frac{1}{3}$
3. a. $\frac{1}{52}$ b. $\frac{4}{52} = \frac{1}{13}$ c. $\frac{13}{52} = \frac{1}{4}$ d. $\frac{20}{52} = \frac{5}{13}$ e. $\frac{26}{52} = \frac{1}{2}$
4. a.
   (1, 1) (2, 1) (3, 1) (4, 1) (5, 1) (6, 1)
   (1, 2) (2, 2) (3, 2) (4, 2) (5, 2) (6, 2)
   (1, 3) (2, 3) (3, 3) (4, 3) (5, 3) (6, 3)
   (1, 4) (2, 4) (3, 4) (4, 4) (5, 4) (6, 4)
   (1, 5) (2, 5) (3, 5) (4, 5) (5, 5) (6, 5)
   (1, 6) (2, 6) (3, 6) (4, 6) (5, 6) (6, 6)

   b.
Sum	2	3	4	5	6	7	8	9	10	11	12
Number of outcomes	1	2	3	4	5	6	5	4	3	2	1

   c.
X	2	3	4	5	6	7	8	9	10	11	12
Probability of X	$\frac{1}{36}$	$\frac{2}{36}$	$\frac{3}{36}$	$\frac{4}{36}$	$\frac{5}{36}$	$\frac{6}{36}$	$\frac{5}{36}$	$\frac{4}{36}$	$\frac{3}{36}$	$\frac{2}{36}$	$\frac{1}{36}$

   d. $\frac{6}{36} + \frac{2}{36} = \frac{8}{36} = \frac{2}{9}$ e. $\frac{11}{36}$
5. a. $\frac{1}{3}, \frac{2}{3}$ b. $\frac{5}{9}, \frac{4}{9}$ c. $\frac{2}{3}, \frac{1}{3}$ d. $\frac{1}{9}, \frac{8}{9}$
6. No. For either spinner the probability of red (or green) is $\frac{1}{2}$. We note that whichever color Dave chooses, we can use the same arguments he gives for the opposite color. We are assuming that both spinners are balanced. If Spinner A is "biased" towards the left or right semicircle, this would raise a more difficult question.
7. a. Not correct; the percentage is not an indication of the duration nor quantity of rainfall.

b. Not correct.   c. This is a plausible explanation. (See Section 1.2.)
d. This is a plausible attitude; it is an extension of the interpretation in (c) and is closer to the attitude adopted in practice.
e. Not correct. This explanation relies too much on guesswork; the forecaster, of course, tries to establish his predictions more substantially.

8. a. Requirements met; these events are mutually exclusive and exhaustive (see Section 2.2).
b. No, not exhaustive; outcomes with 1 head, 1 tail are excluded.
c. No; "not both heads" is three times as likely as "both heads."
d. $P(\text{both heads}) = \frac{1}{4}$. $P(\text{not both heads}) = \frac{3}{4}$.

9. $\frac{1}{4}$. (This parallels the analysis for heads and tails in coin tossing.)

11. a. $\frac{14}{42} = \frac{1}{3}$ b. $\frac{12}{42} = \frac{2}{7}$ c. $\frac{32}{42} = \frac{16}{21}$ 13. a. $\frac{18}{36} = \frac{1}{2}$ b. $\frac{1}{2}$ c. $\frac{15}{36} = \frac{5}{12}$

Exercises 2.5   (Pages 32–34)

1. a. $P(\sim E) = 1 - P(E) = 1 - \frac{19}{28} = \frac{9}{28}$
b. $1 - \frac{12}{28} = \frac{4}{7}$ c. $G = \emptyset$, and so $P(G) = 0$ d. 1

2. a. The answer depends on personal preference. It is perhaps easier to find $\sim F$, since the first 5 letters of the alphabet contain the two frequently used vowels, a and e; so there are fewer words to account for. $P(F) = \frac{6}{7}$
b. $P(\sim L) = 1$, and so $P(L) = 0$.

3. a. $\sim G$ is the event "the number on the red die is greater than or equal to the number on the green die."
b. Referring to the solution of Problem 4, Exercises 2.4, we see that $P(G) = \frac{15}{36} = \frac{5}{12}$. $P(\sim G) = \frac{21}{36} = \frac{7}{12}$.
c. $0, \frac{35}{36}, 1$ d. 0

4. a. $\frac{9}{36} = \frac{1}{4}$ b. $\frac{9}{36} = \frac{1}{4}$
c. A and B are not exhaustive. There are 18 outcomes in which one number is odd and one is even.

5. $\frac{726 + 255}{651 + 726 + 255} = \frac{981}{1632} \doteq 0.6$ 7. $P(D) = \frac{13}{52} = \frac{1}{4}; P(\sim D) = \frac{3}{4}$

9. 1:1 (We sometimes hear the peculiar sounding expression "the odds are even.")

11. $\frac{1}{101}$ 13. $\frac{a}{a+b}$

15. $\frac{n}{2n} = \frac{1}{2}$. The probability is the same for the complement.

Exercises 2.6   (Pages 38–39)

1. a. $\frac{200 + 200}{800 + 700} = \frac{400}{1500} = \frac{4}{15}$ b. $\frac{800}{1500} = \frac{8}{15}$
c. $\frac{200}{1500} = \frac{2}{15}$ d. $\frac{4}{15} + \frac{8}{15} - \frac{2}{15} = \frac{10}{15} = \frac{2}{3}$ e. $\frac{375}{1500} = \frac{1}{4}$

2. $\frac{200}{450} = \frac{4}{9}$ 3. a. $\frac{4}{20} = \frac{1}{5}$ b. $1 - \frac{1}{5} = \frac{4}{5}$ c. $\frac{7}{20} + \frac{3}{20} = \frac{10}{20} = \frac{1}{2}$
d. $\frac{7}{20} + \frac{3}{20} + \frac{6}{20} = \frac{16}{20} = \frac{4}{5}$ (Compare with part b.)

4. Recall the solution of Problem 4, Exercises 2.4.   a. $\frac{6}{36} = \frac{1}{6}$ b. $\frac{3}{36} = \frac{1}{12}$
c. $\frac{1}{36}$. The appropriate event is $\{(4, 6)\}$.   d. $\frac{6}{36} + \frac{3}{36} - \frac{1}{36} = \frac{8}{36} = \frac{2}{9}$

5. a. $\frac{6}{36} = \frac{1}{6}$ b. $\frac{1}{36}$. The appropriate event is $\{(6, 1)\}$.   c. 0
d. $P(\text{sum is 5}) = \frac{4}{36}$. $P(\text{red die shows 6}) = \frac{1}{6}$.
So $\frac{4}{36} + \frac{6}{36} = \frac{10}{36} = \frac{5}{18}$, since the two events are mutually exclusive.

6. a. $P(A \cup B) = 0.2 + 0.7 - 0.3 = 0.6$
   b. $0.8 = 0.4 + 0.5 - P(A \cap B)$; so $P(A \cap B) = 0.1$.
   c. $0.6 = 0.15 + 0.45 - P(A \cap B)$; so $P(A \cap B) = 0$.
   d. $\frac{2}{3} = \frac{1}{3} + P(B) - \frac{1}{6}$; so $P(B) = \frac{1}{2}$.

7. a. $P(A \cup B) = \frac{1}{10} + \frac{1}{25} - \frac{1}{50} = \frac{3}{25}$
   b. $A \cup B = \{1, 4, 9, 16, 25, 36, 49, 64, 81, 100, 8, 27\}$, $P(A \cup B) = \frac{12}{100} = \frac{3}{25}$

9. a. $P(B \cup R) = 0.94 + 0.91 - 0.87 = 0.98$ **b.** $1 - 0.98 = 0.02$

11. a. $\frac{4}{52} = \frac{1}{13}$ **b.** $\frac{13}{52} = \frac{1}{4}$ **c.** $\frac{1}{52}$ **d.** $\frac{4}{52} + \frac{13}{52} - \frac{1}{52} = \frac{16}{52} = \frac{4}{13}$

13. $\frac{26}{100} + \frac{48}{100} - \frac{50}{100} = \frac{24}{100} = \frac{6}{25}$

Exercises 2.7   (Page 42)

1. Proof of T_2: $A \cap {\sim}A = \emptyset$ and $A \cup {\sim}A = S$.
   Hence: $P(S) = P(A \cup {\sim}A) = P(A) + P({\sim}A)$
   $1 = p + P({\sim}A)$
   $P({\sim}A) = 1 - p$

Exercises 2.8   (Pages 43–44)

1. a and b. Either statement is a reasonable conclusion.

2. Many people are misled by this false reasoning. After three reds, the probability of black remains the same as it would have been before any reds occurred.

3. a. Yes. Each has probability $\frac{1}{1024}$.   **b.** Yes. $P(\text{hththththt}) = \frac{1}{1024}$.
   c. Again, yes. $P(\text{httthththt}) = \frac{1}{1024}$.
   d. Any *particular* sequence of heads and tails has probability $\frac{1}{1024}$.

4. No. The probability of obtaining *exactly* 50 heads in 100 tosses is small; much smaller than the probability of exactly 5 heads in 10 tosses. Nevertheless, the probability of obtaining *about* 50 heads in 100 tails is greater than that of obtaining *about* 5 heads in 10 tosses. The long-run interpretation of the law of averages is that the *ratio* $\frac{\text{heads}}{\text{number of tosses}}$ will become close to $\frac{1}{2}$ if a sufficient number of tosses are made.

5. In part b, the ratio of heads to total number of tosses is from 0.4 to 0.6 in both cases. The long-run interpretation of the law of averages is that it is *more* probable to obtain from 40% to 60% heads if the number of tosses is increased. In part c, the ratio of heads to number of tosses is from 0.45 to 0.65. It is *more* probable to obtain a ratio in this range if we make more tosses.

6. To obtain 80% or more heads in 10 tosses is unusual (has small probability, about 0.055). To obtain 80% or more heads in 100 tosses is *very* unusual (the probability is 0.000 to three decimals). The law of averages asserts that the ratio of heads to number of tosses will be "close" to 50%.

7. No. Unusual events *do* happen, and their occurrence *is* often quite surprising. Many scientists favor the second theory—that the formation of the earth is likely in view of the large number of potential earths. This is similar to the question about 9 heads in a row. The occurrence may be rare, but in a large number of tosses, can be expected. Other, equally thoughtful people maintain the first point of view—the exact circumstances have such an extremely small probability of happening by "chance" that it is not reasonable to accept it as a matter of "accident." (In a slightly different context, Professor Einstein stated that he did not believe God played dice with the cosmos.) The firm convictions held by each school of thought illustrate that here is a case where "degree of belief" is a very personal matter.

Appendix / 293

Chapter Review   (Pages 45-46)
1. a. The number of heads or the number of tails in one toss of two coins.
   b. Total number of heads (or tails) in two tosses of two coins.
   c. Sum of the numbers in two tosses of the die.
   d. {odd, even}, {greater than 4, otherwise}, and so on.
   The above are only suggested answers; there may be many other equally natural sets.

3.    a. $\frac{5}{8}$ b. $\frac{7}{8}$

5. a. 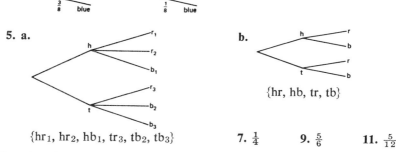   b.

   {hr$_1$, hr$_2$, hb$_1$, tr$_3$, tb$_2$, tb$_3$}   {hr, hb, tr, tb}

7. $\frac{1}{4}$ 9. $\frac{5}{6}$ 11. $\frac{5}{12}$

Exercises 3.1   (Pages 51-53)
1. Our original set of outcomes is {hhh, hht, hth, thh, htt, tht, tth, ttt} where, for example, hth represents "penny, heads; nickel, tails; dime, heads."
   a. $P(\{hhh\}) = \frac{1}{8}$
   b. Since the penny shows heads, we have the reduced set of outcomes {hhh, hht, hth, htt}. Again these are equally likely; so $P(hhh \mid \text{penny heads}) = \frac{1}{4}$.
   c. Here our reduced set is {hhh, hht, hth, thh, htt, tht, tth}. (The only outcome not included is ttt.) $P(hhh \mid \text{at least one head}) = \frac{1}{7}$.
   d. Our reduced set is {hhh, hht}. $P(hhh \mid \text{penny and nickel heads}) = \frac{1}{2}$.
   e. The *a priori* $P(\text{dime heads}) = \frac{1}{2}$. In part d we are given that the penny and nickel show heads. To obtain 3 heads we need only to have the dime show heads, and we found $P(hhh \mid \text{penny and nickel heads}) = \frac{1}{2}$. Thus, the conditional probability that the dime show heads, given that the other coins show heads, equals the *a priori* probability that the dime shows heads.

2. a. $P(A \mid R) = \dfrac{P(A \cap R)}{P(R)} = \dfrac{\frac{2}{12}}{\frac{5}{12}} = \dfrac{2}{5}$ b. $P(B \mid R) = \dfrac{P(B \cap R)}{P(R)} = \dfrac{\frac{3}{12}}{\frac{5}{12}} = \dfrac{3}{5}$

3. a. Let D be the event "a diamond was turned." $P(A \mid D) = \dfrac{P(A \cap D)}{P(D)} = \dfrac{\frac{1}{6}}{\frac{2}{6}} = \dfrac{1}{2}$

   b. Let C be the event "a club was turned." $P(A \mid C) = \dfrac{P(A \cap C)}{P(C)} = \dfrac{0}{\frac{1}{6}} = 0$

   c. $P(H \mid A) = \dfrac{P(H \cap A)}{P(A)} = \dfrac{\frac{1}{6}}{\frac{2}{6}} = \dfrac{1}{2}$.

   d. Let K be the event "a king was turned." $P(C \mid K) = \dfrac{P(C \cap K)}{P(K)} = \dfrac{\frac{1}{6}}{\frac{3}{6}} = \dfrac{1}{3}$.

   e. $P(A \mid C) = P(C \mid A) = 0$ because $P(A \cap C) = P(C \cap A) = 0$, but $P(C) \neq 0$ and $P(A) \neq 0$.

   Notice that we really do not need these "formal" calculations. We may argue directly. For part c, for example: The only two aces are clubs and hearts; since one of these has been turned, the probability that it is a heart is $\frac{1}{2}$.

4. See Solution of Problem 4, Exercises 2.4.
   a. $P(A) = \frac{6}{36}$, $P(B) = \frac{3}{36}$, $P(C) = \frac{6}{36}$, $P(A \cap C) = \frac{1}{36}$, $P(B \cap C) = \frac{1}{36}$.
   b. $P(A \mid C) = \frac{\frac{1}{36}}{\frac{6}{36}} = \frac{1}{6}$ c. $P(B \mid C) = \frac{\frac{1}{36}}{\frac{6}{36}} = \frac{1}{6}$
   d. $P(C \mid A) = \frac{\frac{1}{36}}{\frac{6}{36}} = \frac{1}{6}$ e. $P(C \mid B) = \frac{\frac{1}{36}}{\frac{3}{36}} = \frac{1}{3}$

5. a. $P(E \mid F) = \frac{0.2}{0.4} = \frac{1}{2}$, $P(F \mid E) = \frac{0.2}{0.3} = \frac{2}{3}$, $P(E \cup F) = 0.3 + 0.4 - 0.2 = 0.5$
   b. $P(A \cap B) = P(B) \cdot P(A \mid B) = (0.6)(0.5) = 0.3$
   c. $P(R \mid T) = \frac{P(R \cap T)}{P(T)}$; so $P(T) = \frac{P(R \cap T)}{P(R \mid T)} = \frac{0.3}{0.4} = \frac{3}{4}$.

6. $P(A \mid B) = \frac{P(A \cap B)}{P(B)}$; so $\frac{P(A \cap B)}{P(A \mid B)} = P(B) \leq 1$.
   Hence $P(A \cap B) \leq P(A \mid B)$. Equality holds when $\frac{P(A \cap B)}{P(A \mid B)} = 1$; this implies that $P(B) = 1$, or $B = S$.

7. We use the probabilities listed on page 26.
   a. $E = \{$when, the, course, events$\}$, $P(E) = \frac{19}{28}$
   $H \cap E = \{$when, the$\}$, $P(H \cap E) = \frac{7}{28}$
   $P(H \mid E) = \frac{7}{19}$
   $\frac{P(H \cap E)}{P(H \mid E)} = \frac{\frac{7}{28}}{\frac{7}{19}} = \frac{19}{28} = P(E)$
   b. $P(V \cap T) = 0$, $P(V \mid T) = 0$

9. a. $P(B \mid H) = \frac{1}{2}$. b. $P(B \mid T) = \frac{1}{3}$.
   c. $P(B) = P(B \cap H) + P(B \cap T) = \frac{1}{2} \cdot \frac{1}{2} + \frac{1}{2} \cdot \frac{1}{3} = \frac{5}{12}$.
   $P(H \mid B) = \frac{\frac{1}{4}}{\frac{5}{12}} = \frac{3}{5}$ d. No.

11. a. $P(E \mid E) = \frac{P(E \cap E)}{P(E)} = \frac{P(E)}{P(E)} = 1$ if $P(E) \neq 0$.
    b. $P(\emptyset \mid G) = \frac{P(\emptyset \cap G)}{P(G)} = 0$ if $P(G) \neq 0$.
    c. Since $E \subset F$, let $F = E \cup A$, where $E \cap A = \emptyset$. (A may or may not be $\emptyset$.)
    Then $F \cap G = (E \cup A) \cap G = (E \cap G) \cup (A \cap G)$.
    $P(F \cap G) = P(E \cap G) + P(A \cap G) - P(E \cap A \cap G)$.
    But $P(E \cap A \cap G) = 0$, since $E \cap A = \emptyset$.
    So $P(F \cap G) = P(E \cap G) + P(A \cap G)$. $0 \leq P(A \cap G) \leq 1$.
    Hence, $P(E \cap G) \leq P(F \cap G)$, and so $P(E \mid G) \leq P(F \mid G)$ if $P(G) \neq 0$.
    d. $P(F \mid E) = \frac{P(F \cap E)}{P(E)} = \frac{P(E)}{P(E)} = 1$ if $P(E) \neq 0$.
    e. From $P(E \mid G) + P(\sim E \mid G) = 1$, we have $P(\sim E \mid G) = 1 - P(E \mid G)$.
    f. Since $E \subset S$, we have $P(E \mid S) = P(E)$.

13. a. $\frac{1}{4}$ b. $\frac{1}{4}$ c. $\frac{1}{2}$
15. a. $\frac{3}{8}$ b. 1

Appendix / 295

Exercises 3.2   (Pages 59–60)
1. 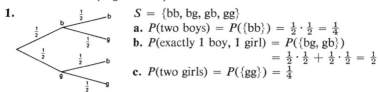 $S = \{bb, bg, gb, gg\}$
   a. $P(\text{two boys}) = P(\{bb\}) = \frac{1}{2} \cdot \frac{1}{2} = \frac{1}{4}$
   b. $P(\text{exactly 1 boy, 1 girl}) = P(\{bg, gb\})$
      $= \frac{1}{2} \cdot \frac{1}{2} + \frac{1}{2} \cdot \frac{1}{2} = \frac{1}{2}$
   c. $P(\text{two girls}) = P(\{gg\}) = \frac{1}{4}$

2. a. We deal with the reduced set of outcomes $\{bb, bg, gb\}$. Each of these is equally likely. Hence,

$$P(\{bb\}) = \frac{1}{3} \quad \text{or} \quad P(\{bb\} \mid \{bb, bg, gb\}) = \frac{P(\{bb\} \cap \{bb, bg, gb\})}{P\{bb, bg, gb\}}$$

$$= \frac{P(\{bb\})}{1 - P\{gg\}} = \frac{\frac{1}{4}}{1 - \frac{1}{4}} = \frac{1}{3}.$$

   b. Our reduced set of outcomes is $\{bb, bg\}$. Each outcome is equally likely. Hence,

$$P(\{bb\}) = \frac{1}{2} \quad \text{or} \quad P(\{bb\} \mid \{bb, bg\}) = \frac{P(\{bb\} \cap \{bb, bg\})}{P(\{bb, bg\})}$$

$$= \frac{P(\{bb\})}{P\{bb\} + P\{bg\}} = \frac{\frac{1}{4}}{\frac{1}{4} + \frac{1}{4}} = \frac{1}{2}.$$

(Problems 1 and 2 may be interpreted as follows. We meet a couple and ask how many children they have. If the answer is "Two," the *a priori* probability of two boys is $\frac{1}{4}$. If we discover that one child is a boy, the probability of two boys is $\frac{1}{3}$. If we determine further that the first born is a boy, then the probability of two boys increases to $\frac{1}{2}$!! Question: Instead of information about the first born, suppose that we discover that the second-born is a boy?)

3.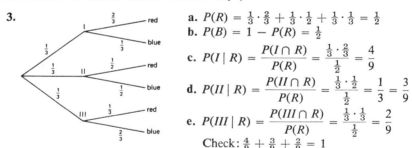
   a. $P(R) = \frac{1}{3} \cdot \frac{2}{3} + \frac{1}{3} \cdot \frac{1}{2} + \frac{1}{3} \cdot \frac{1}{3} = \frac{1}{2}$
   b. $P(B) = 1 - P(R) = \frac{1}{2}$
   c. $P(I \mid R) = \frac{P(I \cap R)}{P(R)} = \frac{\frac{1}{3} \cdot \frac{2}{3}}{\frac{1}{2}} = \frac{4}{9}$
   d. $P(II \mid R) = \frac{P(II \cap R)}{P(R)} = \frac{\frac{1}{3} \cdot \frac{1}{2}}{\frac{1}{2}} = \frac{1}{3} = \frac{3}{9}$
   e. $P(III \mid R) = \frac{P(III \cap R)}{P(R)} = \frac{\frac{1}{3} \cdot \frac{1}{3}}{\frac{1}{2}} = \frac{2}{9}$
   Check: $\frac{4}{9} + \frac{3}{9} + \frac{2}{9} = 1$

4. Refer to the solution of Problem 4, Exercises 2.4.
   a. We consider a set of 36 equally likely outcomes. There are 11 outcomes in A. $P(A) = \frac{11}{36}$. Of these, $\{(6, 4), (4, 6)\}$ are also in T. $P(T \cap A) = \frac{2}{36}$. Hence, $P(T \mid A) = \frac{2}{11}$.
   b. There are 10 outcomes in B, two in $T \cap B$. $P(T \mid B) = \frac{2}{10} = \frac{1}{5}$.
   c. There are 6 outcomes in R, one in $T \cap R$. $P(T \mid R) = \frac{1}{6}$.
   Finally, $\frac{1}{6} < \frac{2}{11} < \frac{1}{5}$. So the information in part b is most favorable to T.

5. a. 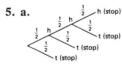 $S = \{t, ht, hht, hhh\}$
   b. $P(\text{tails}) = P(t) + P(ht) + P(hht) = \frac{1}{2} + \frac{1}{4} + \frac{1}{8}$
      $= \frac{7}{8}$,
      or $P(\text{tails}) = 1 - P(hhh) = 1 - \frac{1}{8} = \frac{7}{8}$
   c. $P(X = 0) = P(t) = \frac{1}{2}$, $\quad P(X = 1) = P(ht) = \frac{1}{4}$,
      $P(X = 2) = P(hht) = \frac{1}{8}$, $\quad P(X = 3) = P(hhh) = \frac{1}{8}$ d. No

6. The 10 equally likely outcomes are:

1 and 2 sum: odd	2 and 5 sum: odd	1 and 5 sum: even
1 and 4 odd	3 and 5 even	2 and 4 even
2 and 3 odd	1 and 3 even	3 and 4 odd
		4 and 5 odd

a. $P(\text{sum even}) = \frac{4}{10}$
b. $P(\text{both odd}) = \frac{3}{10}$
c. $P(\text{both odd} \mid \text{sum even}) = \frac{3}{4}$
d. $P(\text{sum even} \mid \text{both odd}) = 1$
e. $P(\text{sum odd} \mid \text{both even}) = 0$

7. a. 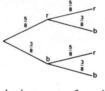 **b.** $\frac{2}{3}$ **c.** $\frac{5}{6}$

9. a. **b.** $\frac{85}{144}$ **c.** $\frac{5}{18}$

Exercises 3.3 (Pages 62–64)

1. a.

Notice that this is precisely the diagram of Example 3.3.1 with $p = \frac{5}{8}$ and "h,t" replaced by "red, blue."

b. Assignment of probabilities would be the same for the two cases with replacement.

2. a. For the problem of drawing without replacement, knowing the ratio of the colors is not enough; the number of marbles of each color is needed, since removing a marble changes the ratio.

b. In the case of the urn with 5 red, 3 blue marbles, drawing a red leaves 4 red and 3 blue in the urn; drawing a blue, leaves 5 red, 2 blue. In the case of the urn with 10 red, 6 blue marbles, drawing a red leaves 9 red, 6 blue; drawing a blue leaves 10 red, 5 blue.

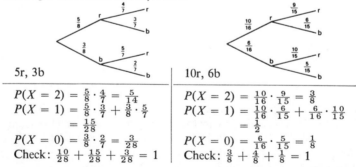

5r, 3b

$P(X = 2) = \frac{5}{8} \cdot \frac{4}{7} = \frac{5}{14}$
$P(X = 1) = \frac{5}{8} \cdot \frac{3}{7} + \frac{3}{8} \cdot \frac{5}{7} = \frac{15}{28}$
$P(X = 0) = \frac{3}{8} \cdot \frac{2}{7} = \frac{3}{28}$
Check: $\frac{10}{28} + \frac{15}{28} + \frac{3}{28} = 1$

10r, 6b

$P(X = 2) = \frac{10}{16} \cdot \frac{9}{15} = \frac{3}{8}$
$P(X = 1) = \frac{10}{16} \cdot \frac{6}{15} + \frac{6}{16} \cdot \frac{10}{15} = \frac{1}{2}$
$P(X = 0) = \frac{6}{16} \cdot \frac{5}{15} = \frac{1}{8}$
Check: $\frac{3}{8} + \frac{4}{8} + \frac{1}{8} = 1$

c. Drawing two marbles simultaneously is the same as drawing without replacement; the first marble kept in the fist parallels the first marble drawn from the urn.

3.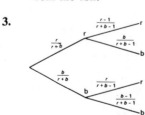

a. $P(B \mid R) = \dfrac{b}{r + b - 1};$

$P(R \mid B) = \dfrac{r}{r + b - 1} \neq \dfrac{b}{r + b - 1}$ if $r \neq b$.

b. $P(B, R) = \dfrac{b}{r + b} \cdot \dfrac{r}{r + b - 1};$

$P(R, B) = \dfrac{r}{r + b} \cdot \dfrac{b}{r + b - 1} = P(B, R)$

4. 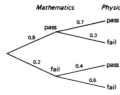 $P(\text{pass physics}) = (0.8)(0.7) + (0.2)(0.4) = 0.64$

5. $P(\text{pass math} \mid \text{fail physics})$
$= \dfrac{P(\text{pass math and fail physics})}{P(\text{fail physics})} = \dfrac{(0.8)(0.3)}{0.36}$
$= \tfrac{2}{3}$,
since $P(\text{fail physics}) = 1 - P(\text{pass physics})$ from Problem 4.

6. Let A be the event "A hits the bull's-eye," and B be the event "B hits the bull's-eye."

a. $P(\text{both hit}) = P(A \cap B)$
$= 0.48$
b. $P(\text{exactly one hits})$
$= P(A \cap \sim B) + P(\sim A \cap B)$
$= (0.6)(0.2) + (0.4)(0.8)$
$= 0.44$

c. $P(\text{at least one hits}) = 1 - P(\text{no hits}) = 1 - (0.4)(0.2) = 0.92$
or $P(\text{at least one hits}) = P(\text{both hit}) + P(\text{exactly one hits})$
$= 0.48 + 0.44 = 0.92$

7. a. $\tfrac{1}{4}$ b. $\tfrac{1}{4}$ c. $\tfrac{1}{4}$ d. $\tfrac{1}{4}$ 9. $\tfrac{4}{9}$

10. We think of two steps. First, the person is either healthy or ill. Second, the test shows positive or negative.

$S = \{$healthy and positive, healthy and negative, ill and positive, ill and negative$\}$
a. $P(\text{healthy}) = \tfrac{995}{1000} = 0.995$
b. $P(\text{positive})$
$= P(\text{healthy and positive}) + P(\text{ill and positive})$
$= (0.995)(0.01) + (0.005)(0.94) = 0.01465$

c. $P(\text{healthy} \mid \text{positive}) = \dfrac{P(\text{healthy and positive})}{P(\text{positive})} = \dfrac{0.00995}{0.01465} \doteq 0.68$

About 68% of all "positive" test results in this example came from healthy persons. While these probabilities are fictitious, the problem illustrates a real difficulty in medical diagnosis.

11. a. $\tfrac{11}{20} = 0.55$ or 55% b. $\tfrac{4}{11}$ or approximately 36%

Exercises 3.4   (Pages 68–69)

1. a. $P(A) = \tfrac{1}{13}$, $P(D) = \tfrac{1}{4}$, $P(A \cap D) = \tfrac{1}{52}$. (There is exactly one ace of diamonds.) Therefore, $P(A \cap D) = P(A) \cdot P(D)$, and the events are independent.
b. $P(A) = \tfrac{1}{13}$, $P(H) = \tfrac{20}{52} = \tfrac{5}{13}$, $P(A \cap H) = \tfrac{4}{52} = \tfrac{1}{13}$. (There are 4 aces and all the aces are honors.) So $P(A \cap H) \neq P(A) \cdot P(H)$, and the events are *not* independent.
c. $P(D) = \tfrac{1}{4}$, $P(H) = \tfrac{5}{13}$, $(PD \cap H) = \tfrac{5}{52}$. (There are five diamonds which are honor cards.) So $P(D \cap H) = P(D) \cdot P(H)$, and the events are independent.

2. a. $P(A) = \tfrac{1}{3}$, $P(H) = \tfrac{1}{2}$, $P(A \cap H) = \tfrac{1}{6}$
$P(A \cap H) = P(A) \cdot P(H)$, and so A and H are independent events.
b. $P(A) = \tfrac{1}{3}$, $P(C) = \tfrac{1}{6}$, $P(A \cap C) = 0$.
$P(A \cap C) \neq P(A) \cdot P(C)$, and so A and H are not independent events.

3. a. $P(A) = \frac{1}{6}$, $P(B) = \frac{1}{12}$. $P(A \cap B) = 0 \neq P(A) \cdot P(B)$, and so A and B are not independent events.

b. $P(A) = \frac{1}{6}$, $P(C) = \frac{1}{6}$, $P(A \cap C) = \frac{1}{36}$. (There is one way for the dice to total 7 if the red die shows a 4.)
$P(A \cap C) = P(A) \cdot P(C)$, and so A and C are independent events.

c. $P(B) = \frac{1}{12}$, $P(C) = \frac{1}{6}$. $P(B \cap C) = \frac{1}{36} \neq P(B) \cdot P(C)$, and so B and C are not independent events.

4. a. If E and G are mutually exclusive, then $P(E \cap G) = 0$. Since $P(E) \neq 0$, $P(G) \neq 0$, it follows that $P(E \cap G) \neq P(E) \cdot P(G)$, and so E and G are *not* independent events.

b. If E and G are independent, then $P(E \cap G) = P(E) \cdot P(G)$. Now $P(E) \neq 0$ and $P(G) \neq 0$ imply that $P(E \cap G) \neq 0$. Hence, E and G are *not* mutually exclusive.

5. $E \cap {\sim}E = \emptyset$, and so $P(E \cap {\sim}E) = 0$. Since $P(E) \neq 1$, we have $P({\sim}E) \neq 0$. Further, it is given that $P(E) \neq 0$, and so $P(E \cap {\sim}E) \neq P(E) \cdot P({\sim}E)$. The events are not independent. (See also Problem 4.)

6. $P(\text{fail math}) = \frac{15}{120}$, $P(\text{fail chem}) = \frac{16}{120}$, $P(\text{fail both}) = \frac{6}{120}$, $\frac{6}{120} \neq \frac{15}{120} \cdot \frac{16}{120}$, and so the events are not independent.

7. No   **9.** No   **11.** Yes

Exercises 3.5   (Pages 72–73)

1.

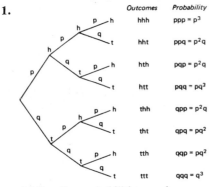

Adding the probabilities, we have
$$p^3 + p^2q + p^2q + pq^2 + p^2q + pq^2 + pq^2 + q^3$$
$$= p^2(p+q) + pq(p+q) + pq(p+q) + q^2(p+q)$$
$$= p^2 + pq + pq + q^2 = p(p+q) + q(p+q) = p+q = 1.$$

It is worth noting that the sum of the probabilities may be written as:
$$p^3 + 3p^2q + 3pq^2 + q^3.$$

2. a. $S = \{rr, rb, br, bb\}$, $E = \{rb, br, bb\}$, $G = \{rb, br\}$; $E \cap G = \{rb, br\}$, $P(E \cap G) = \frac{1}{2}$, $P(E) = \frac{3}{4}$, $P(G) = \frac{1}{2}$. $P(E \cap G) \neq P(E) \cdot P(G)$, and so E and G are *not* independent events.

b. $S = \{rrr, rrb, rbr, brr, rbb, brb, bbr, bbb\}$,
$E = \{rbb, brb, bbr, bbb\}$, $G = \{rrb, rbr, brr, rbb, brb, bbr\}$;
$E \cap G = \{rbb, brb, bbr\}$, $P(E \cap G) = \frac{3}{8}$, $P(E) = \frac{1}{2}$,
$P(G) = \frac{3}{4}$. $P(E \cap G) = P(E) \cdot P(G)$, and so E and G are independent events.

Remark: This problem illustrates that our intuition may sometimes be wrong. It seems that E and G should be independent in both cases or not independent in both. See Problems 6 and 7 for other such examples.

3. a. *Afflicted patient* $P(\text{diagnosis}) = 0.8 + (0.2)(0.8) = 0.96$
or $P(\text{diagnosis}) = 1 - P(\text{no diagnosis})$
$= 1 - (0.2)(0.2) = 1 - (0.2)^2$
$= 0.96$

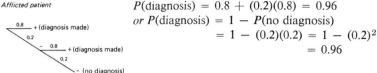

b. For three tests, $P(\text{diagnosis}) = 1 - P(\text{no diagnosis})$
$= 1 - (0.2)(0.2)(0.2) = 1 - (0.2)^3 = 0.992.$
The probability of diagnosis increases from 0.960 to 0.992.
c. For 4 tests, $P(\text{diagnosis}) = 1 - (0.2)^4 = 0.9984$
The increase is now from 0.9920 to 0.9984. Situations of this kind reveal that successive testing reaches a point of "diminishing returns." Some decision as to when to stop testing must be made.

5. $\left(\dfrac{2}{13}\right)^4 = \dfrac{16}{13^4} = \dfrac{16}{28561} \doteq 0.00056$ (The numbered cards are 2, 3, 4, ..., 10.)

6. a. $H = \{bg, gb\}$, $A = \{bb, bg, gb\}$, $H \cap A = \{bg, gb\}$.
$P(H) = \frac{2}{4} = \frac{1}{2}$, $P(A) = \frac{3}{4}$. $P(H \cap A) = \frac{1}{2} \neq P(H) \cdot P(A)$, and so H and A are not independent events.
b. $H = \{bbg, bgb, gbb, bgg, gbg, ggb\}$, $P(H) = \frac{6}{8} = \frac{3}{4}$,
$A = \{bbb, bbg, bgb, gbb\}$, $P(A) = \frac{4}{8} = \frac{1}{2}$, $H \cap A = \{bbg, bgb, gbb\}$.
$P(H \cap A) = \frac{3}{8} = P(H) \cdot P(A)$, and so H and A are independent events.

7. a. $\sim H = \{bb, gg\}$, $\sim A = \{gg\}$, $\sim H \cap \sim A = \{gg\}$.
$P(\sim H) = \frac{1}{2}$, $P(\sim A) = \frac{1}{4}$. $P(\sim H \cap \sim A) = \frac{1}{4} \neq P(\sim H) \cdot P(\sim A)$, and so $\sim H$ and $\sim A$ are not independent events.
b. $\sim H = \{bbb, ggg\}$, $\sim A = \{bgg, gbg, ggb, ggg\}$, $\sim H \cap \sim A = \{ggg\}$,
$P(\sim H) = \frac{1}{4}$, $P(\sim A) = \frac{1}{2}$. $P(\sim H \cap \sim A) = \frac{1}{8} = P(\sim H) \cdot P(\sim A)$, and so in this case $\sim H$ and $\sim A$ are independent events.

9. $75 for A, $25 for B.

11.

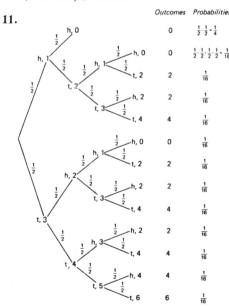

a. $P(X = 0) = \frac{6}{16}$
b. $P(X = 1) = 0$
c. $P(X = 2) = \frac{5}{16}$
d. $P(X = 3) = 0$
e. $P(X = 4) = \frac{4}{16}$
f. $P(X = 5) = 0$
g. $P(X = 6) = \frac{1}{16}$
Check: $\frac{6}{16} + \frac{5}{16} + \frac{4}{16} + \frac{1}{16} = 1$
h. $P(X = 0 \mid \text{heads on first toss})$
$= \frac{5}{8}$

Exercises 3.6 (Pages 77–78)

1. $P(II \mid D) = \dfrac{(0.75)(0.03)}{(0.25)(0.01) + (0.75)(0.03)} = \dfrac{0.0225}{0.0250} = 0.90$

3. $P(A \mid \text{boy}) = \dfrac{P(A \cap \text{boy})}{P(\text{boy})} = \dfrac{\frac{1}{3}}{\frac{1}{2}} = \dfrac{2}{3}$

5. **a.** $P(\text{second red} \mid \text{first red}) = \frac{7}{10}$
   b. $P(\text{first red} \mid \text{second red}) = \frac{7}{10}$

7. $P(\text{second red}) = \dfrac{r}{r+b} \cdot \dfrac{r+t}{r+b+t} + \dfrac{b}{r+b} \cdot \dfrac{r}{r+b+t}$

   $= \dfrac{r(r+t) + br}{(r+b)(r+b+t)} = \dfrac{r(r+b+t)}{(r+b)(r+b+t)} = \dfrac{r}{r+b}$

   $= P(\text{first red})$

Chapter Review (Pages 78–79)

1. $\{(h, 4), (t, 4)\}$

3. $P(A \cap B) = P(A) \cdot P(B \mid A) = P(B) \cdot P(A \mid B)$
   $(0.6)(0.5) = (0.4) \cdot P(A \mid B)$
   $P(A \mid B) = \dfrac{0.3}{0.4} = 0.75$

5. **a.**

   [tree diagram with branches labeled 1/2, 2/3, 1/3 leading to: d 1/3 (defective found on 3d draw); d 1/2, g 1/3 (defective found on 2d draw); d 1, g 1/3 (defective found on 1st draw)]

   b. $P(\text{defective bulb found in one of the first two tests})$
   $= 1 - P(\text{defective bulb found in the third test}) = 1 - \frac{1}{3} = \frac{2}{3}$

7. **a.** $P(\text{heads, red}) = \frac{1}{2} \cdot \frac{2}{3} = \frac{1}{3}$ **b.** $P(\text{tails}) = \frac{1}{2}$
   c. $P(\text{red}) = P(\text{heads, red}) + P(\text{tails, red}) = \frac{1}{3} + \frac{1}{6} = \frac{1}{2}$
   d. $P(\text{blue}) = 1 - P(\text{red}) = \frac{1}{2}$ **e.** $P(\text{heads} \mid \text{red}) = \dfrac{\frac{1}{3}}{\frac{1}{2}} = \dfrac{2}{3}$

9. $P(\text{failed mathematics}) = \frac{15}{120}$, $P(\text{failed chemistry}) = \frac{16}{120}$. To be independent, $P(\text{failed both}) = P(\text{failed math}) \cdot P(\text{failed chemistry}) = \frac{15}{120} \cdot \frac{16}{120} = \frac{2}{120}$. Two students failed both.

Exercises 4.1 (Pages 86–87)

1.
Al 10.41	Cr 19.87	Y 32.95
Si 11.29	Mn 20.75	Zr 33.83
Cl 13.82	Fe 21.62	Cb 34.72
K 15.55	Co 22.48	Mo 35.59
Ca 16.41	Ni 23.34	Ru 37.32
Ti 18.14	Cu 24.20	Pd 39.06
V 19.00	Zn 25.07	Ag 39.83

It is of historical interest to note that by studying this rank-order listing and by finding the differences between the changes listed, Moseley was able to predict correctly the number of chemical elements that were missing between those listed here. For example, phosphorus and sulfur come between silicon and chlorine, and argon comes between chlorine and potassium. (The name of columbium was later changed to niobium.)

3. 7 //
   6 ///
   5 ////
   4 //// /
   3 ////
   2 ///
   1 /

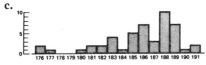

5. a. 191 //
   190 /
   189 //// //
   188 //// ////
   187 ///
   186 //// //
   185 ////
   184 /
   183 ////
   182 //
   181 //
   180 /
   179
   178
   177 /
   176 //

b.

x_i	$t(x_i)$	Rel. Freq.
191	2	$\frac{2}{48}$
190	1	$\frac{1}{48}$
189	7	$\frac{7}{48}$
188	10	$\frac{10}{48}$
187	3	$\frac{3}{48}$
186	7	$\frac{7}{48}$
185	5	$\frac{5}{48}$
184	1	$\frac{1}{48}$
183	4	$\frac{4}{48}$
182	2	$\frac{2}{48}$
181	2	$\frac{2}{48}$
180	1	$\frac{1}{48}$
179	0	$\frac{0}{48}$
178	0	$\frac{0}{48}$
177	1	$\frac{1}{48}$
176	2	$\frac{2}{48}$
	48	

c.

d. Between 190.5 and 191.5

7. a.

x_i	$t(x_i)$	Rel. Freq.
21	4	$\frac{4}{20}$
20	1	$\frac{1}{20}$
19	2	$\frac{2}{20}$
18	1	$\frac{1}{20}$
17	6	$\frac{6}{20}$
16	1	$\frac{1}{20}$
15	2	$\frac{2}{20}$
14	2	$\frac{2}{20}$
13	0	$\frac{0}{20}$
12	1	$\frac{1}{20}$
	20	

b.

9. a.

x_i	$t(x_i)$	Rel. Freq.
100	2	$\frac{2}{20}$
95	3	$\frac{3}{20}$
90	6	$\frac{6}{20}$
85	7	$\frac{7}{20}$
80	2	$\frac{2}{20}$
	20	

b.

Exercises 4.2 (Pages 91–92)

1. a, b.

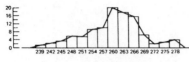

c.

2. a, b.

x_i	$t(x_i)$	Rel. Freq.
210	1	0.04
200	1	0.04
190	1	0.04
180	4	0.16
170	3	0.12
160	5	0.20
150	5	0.20
140	2	0.08
130	2	0.08
120	1	0.04
$m = 25$		1.00

c.

3. a.

x	$t(x_i)$
164.5	///
156.5	/
148.5	////
140.5	
132.5	//
124.5	𝍫 ////
116.5	𝍫 𝍫 /
108.5	𝍫 ///
100.5	𝍫 //
92.5	///
84.5	////
76.5	/
68.5	
60.5	//
52.5	/
44.5	//
36.5	//

b.

x	$t(x_i)$
165.5	///
155.5	//
145.5	///
135.5	//
125.5	𝍫 ////
115.5	𝍫 𝍫 𝍫
105.5	𝍫 //
95.5	𝍫 //
85.5	////
75.5	/
65.5	/
55.5	//
45.5	//
35.5	//

c, d.

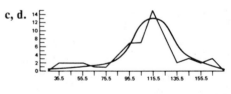

5. a.

x_i	$t(x_i)$
27.5	4
22.5	7
17.5	7
12.5	8
7.5	4
	30

b.

Exercises 4.3 (Page 96)

1. $\bar{x} = \dfrac{1(1) + 2(3) + 3(5) + 4(6) + 5(5) + 6(3) + 7(2)}{25} = \dfrac{103}{25} = 4.12$

3. $\bar{x} = \dfrac{191(2) + 190(1) + 189(7) + \cdots + 176(2)}{48} = \dfrac{8914}{48} \doteq 185.71$

5.

y_i	$t(y_i)$	$y_i \cdot t(y_i)$
6	2	12
5	1	5
4	7	28
3	10	30
2	3	6
1	7	7
0	5	0
−1	1	−1
−2	4	−8
−3	2	−6
−4	2	−8
−5	1	−5
−6	0	0
−7	0	0
−8	1	−8
−9	2	−18
		34

$\bar{y} = \frac{34}{48} \doteq 0.71$

$\bar{x} = \bar{y} + 185 \doteq 185.71$

This is an application of the simple property of $\bar{x}$ mentioned on page 95.

7.

x_i	Rel. Freq.	$x_i \cdot$ (R.F.)
278	0.03	8.34
275	0.02	5.50
272	0.01	2.72
269	0.06	16.14
266	0.15	39.90
263	0.17	44.71
260	0.20	52.00
257	0.10	25.70
254	0.09	22.86
251	0.05	12.55
248	0.06	14.88
245	0.03	7.35
242	0.02	4.84
239	0.01	2.39
	1.00	$\bar{x} = 259.88$

9.

x_i	$t(x_i)$	$x_i \cdot t(x_i)$
165.5	3	496.5
155.5	2	311.0
145.5	3	436.5
135.5	2	271.0
125.5	9	1129.5
115.5	15	1732.5
105.5	7	738.5
95.5	7	668.5
85.5	4	342.0
75.5	1	75.5
65.5	1	65.5
55.5	2	111.0
45.5	2	91.0
35.5	2	71.0
	60	6540.0

$\bar{x} = \frac{6540}{60} = 109$

11. a. $74 = \frac{\sum x_i}{50}$; $\sum x_i = (50)(74) = 3700$

b. $68 = \frac{\sum y_i}{30}$; $\sum y_i = (30)(68) = 2040$

c. The combined mean is

$$\frac{3700 + 2040}{80} = \frac{5740}{80} = 71.75.$$

12. a. $\dfrac{n\bar{x} + m\bar{y}}{n + m}$

b. $\dfrac{n\bar{x} + m\bar{y}}{n + m} = \dfrac{\bar{x} + \bar{y}}{2}$ if $\bar{x} = \bar{y}$ or if $n = m$.

13. $\dfrac{n\bar{x} + m\bar{y} + k\bar{z}}{n + m + k}$

Exercises 4.4 (Pages 102–103)

1. We first find that $\bar{x} = 5.25$.

First Method

x_i	$x_i - \bar{x}$	$(x_i - \bar{x})^2$
8	2.75	7.5625
8	2.75	7.5625
7	1.75	3.0625
6	.75	0.5625
5	−.25	0.0625
4	−1.25	1.5625
3	−2.25	5.6250
1	−4.25	18.0625
		43.5000

$$s_x^2 = \frac{43.5}{8} = 5.4375$$

Second Method

x_i	x_i^2
8	64
8	64
7	49
6	36
5	25
4	16
3	9
1	1
	264

$$s_x^2 = \frac{264}{8} - (5.25)^2 = 33 - 27.5625 = 5.4375$$

3. We hope you used the formula

$$s_x^2 = \frac{\sum x_i^2 \cdot t(x_i)}{n} - \bar{x}^2 \ !!!$$

x_i	x_i^2	$t(x_i)$	$x_i^2 \cdot t(x_i)$
10	100	1	100
9	81	3	243
8	64	8	512
7	49	10	490
6	36	6	216
5	25	11	275
4	16	2	32
3	9	6	54
2	4	2	8
1	1	0	0
0	0	1	0
		50	1930

$$s_x^2 = \frac{1930}{50} - (5.84)^2 = 4.4944$$

$$s_x \doteq 2.12$$

5.

y_i	y_i^2
31.7	1004.89
35.8	1281.64
43.3	1874.89
55.7	3102.49
65.6	4303.36
75.9	5760.81
81.5	6642.25
79.8	6368.04
71.3	5083.69
60.2	3624.04
44.6	1989.16
35.8	1281.64
681.2	42316.90

$$\bar{y} = \frac{681.2}{12} \doteq 56.8$$

$$s_y^2 = \frac{42316.90}{12} - (56.8)^2 \doteq 300.17$$

$$s_y \doteq 17.32$$

Although the mean temperature for Kansas City is approximately the same as that for San Francisco, the standard deviation for the temperature distribution for Kansas City is more than 4 times that of San Francisco. This agrees with the rough estimates provided by the ranges: $81.5 - 31.7 = 49.8$ for Kansas City against $62.0 - 50.7 = 11.3$ for San Francisco. We have cautioned, however, against too much reliance on the use of the range as a measure of spread.

Strictly, we might argue that the averages for each month should not be given equal "weight" because the number of days are not always the same. However, averages obtained by "weighting" each month according to the number of days it has will not differ very much from the simple averages.

7. $s_x \doteq 29.9$

8. We shall use the formula

$$s_x^2 = \frac{\sum (x_i - \bar{x})^2 \cdot t(x_i)}{n};$$

$\bar{x} = \frac{30}{6} = 5$ and $\bar{y} = \frac{45}{9} = 5$. From Problem 12, Exercises 4.3, $\bar{z} = \bar{x} = \bar{y}$. We are given that $x_i = y_i = z_i$ and $t(z_i) = t(x_i) + t(y_i)$. Also, if the distribution for the 6 numbers x_i has variance s_x^2, then

$$\sum (x_i - \bar{x})^2 \cdot t(x_i) = 6s_x^2.$$

Likewise, since there are 9 numbers y_i,

$$\sum (y_i - \bar{y})^2 \cdot t(y_i) = 9s_y^2.$$

Since $x_i = y_i = z_i$, $\bar{x} = \bar{y} = \bar{z}$, and $t(z_i) = t(x_i) + t(y_i)$, we have:

$$\frac{\sum(z_i - \bar{z})^2 \cdot t(z_i)}{m+n} = \frac{6s_x^2 + 9s_y^2}{6+9} = \frac{6(6.67) + 9(5.33)}{15} \doteq 5.87.$$

The relationship here, with $\bar{x} = \bar{y}$, is: $s_z^2 = \frac{ms_x^2 + ns_y^2}{m+n}$.

9. **a.** $\bar{z} = \frac{100}{24} = \frac{25}{6} \doteq 4.17$. $s_z^2 = \frac{568}{24} - (\frac{25}{6})^2 \doteq 23.67 - 17.36 = 6.31$

   b. $\frac{k\bar{w} + n\bar{y}}{k+n} = \frac{15(\frac{11}{3}) + 9(5)}{15+9} = \frac{100}{24} = \frac{25}{6}$.

   This equals $\bar{z}$ (compare Problem 12, Exercises 4.3).

   c. $\frac{ks_w^2 + ns_y^2}{k+n} \doteq \frac{15(6.22) + 9(5.33)}{15+9} \doteq 5.89 \neq 6.31$

   The reason that this formula is not applicable is that the two distributions do not have the same mean, and s^2 is based on the deviations from the mean.

Exercises 4.5a (Page 104)

(Remember, these suggested answers are only reasonable guesses.)

1. **a.** IV. There must be some scores which deviate from 70 by many points. It is unlikely that all these deviating points will be below the mean.
   b. Again, we cannot be sure. The best guess is II, followed by IV or III.
2. **a.** V. (We are more sure of this conclusion.)
   b. Probably IV, although we cannot be sure.
3. **a.** II. (We are not sure, however.)   **b.** IV.
4. III. (In fact the mean of the entire 500 is precisely 60.)

Exercises 4.5b (Pages 105–106)

1. $\bar{x} = 7$, median $= 5$, mode $= 4$.
2. There is no real answer to this. Some authors choose 12 (50% of the scores are at or below 12); some choose 14 (50% of the scores are at or above 14); still others 13 (halfway between 12 and 14.)
3. There are many such practical situations. One occurs when testing some physical characteristic of a human population. A frequency curve for strength of grip among 1000 males and 1000 females might look like that shown at the right.

   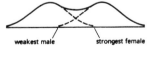
4. The mean is drastically changed by including extreme scores. For a certain business the salaries (in thousands of dollars) of the 7 employees might be:

   $$8, 8, 8, 9, 10, 10, 26.$$

   The mean ($11,000) and the mode ($8000) are both misleading. There is really no completely satisfactory reporting number to use as the "average," but the median ($9000) is surely the best of the three.

5. a. Between the 25th and 75th percentiles.
   b. Using q_1 and q_3 for the first and third quartiles, the semi-interquartile range is
$$\frac{q_3 - q_1}{2}.$$
   c. No. Of 100 scores, for example, 99 lie below the largest score.
6. a. The mode is 5 (although this distribution is nearly "bi-modal").
   b. The modal interval is the interval 67.5–72.5.
7. 68.3

Chapter Review   (Pages 107–108)

1. a.

No. of days	Frequency
12	1
11	2
10	1
9	5
8	2
7	1
	$n = 12$

b.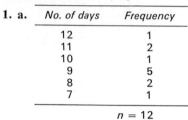

c. $\bar{x} = \dfrac{12 + 22 + 10 + 45 + 16 + 7}{12} \doteq 9.33$

x_i	$t(x_i)$	$x_i \cdot t(x_i)$	x_i^2	$x_i^2 \cdot t(x_i)$
12	1	12	144	144
11	2	22	121	242
10	1	10	100	100
9	5	45	81	405
8	2	16	64	128
7	1	7	49	49
Total	12	112		1068

$\bar{x} = \frac{112}{12} \doteq 9.33$, $s_x^2 = \frac{1068}{12} - (9.33)^2 \doteq 2$

3. a.

Score Interval	Midpoint	Frequency of Interval
85–95	90	2
75–85	80	3
65–75	70	6
55–65	60	6
45–55	50	5
35–45	40	3
		$n = 25$

b.

Midpoint (x_i)	$t(x_i)$	$x \cdot t(x_i)$	x_i^2	$x_i^2 \cdot t(x_i)$
90	2	180	8100	16200
80	3	240	6400	19200
70	6	420	4900	29400
60	6	360	3600	21600
50	5	250	2500	12500
40	3	120	1600	4800
		1570		103700

$\bar{x} = \frac{1570}{25} = 62.8$, $s_x^2 = \frac{103700}{25} - (62.8)^2 \doteq 204$, $s_x \doteq 14.3$

Exercises 5.1   (Pages 112–113)

1. Let X represent the total number of heads (or tails) obtained in the four tosses. The possible values of X are 0, 1, 2, 3, 4. (Other answers are acceptable.)
2. 0, 1, 3. (It is not possible to get exactly two matches; if two match, the third must also match.)
3. 2, 4, 6, 8, 10, 12. $Z = 2X$
4. $-2\frac{1}{2}, -1\frac{1}{2}, -\frac{1}{2}, \frac{1}{2}, 1\frac{1}{2}, 2\frac{1}{2}$. $Z = X - 3\frac{1}{2}$
5. 1, 4, 9, 16, 25, 36. $Z = X^2$
6. $\frac{1}{4}, 2\frac{1}{4}, 6\frac{1}{4}$. $Z = (X - 3\frac{1}{2})^2$
7. We must multiply $1 \times 1, 1 \times 2, \ldots, 6 \times 6$. The possible values are 1, 2, 3, 4, 5, 6, 8, 9, 10, 12, 15, 16, 18, 20, 24, 25, 30, 36. (If X and Y are random variables corresponding to the numbers on the individual dice, then $Z = X \cdot Y$.)

8.

Value of Z	1	2	3	4	5	6	8	9	10	12	15	16
Number of outcomes	1	2	2	3	2	4	2	1	2	4	2	1

Value of Z	18	20	24	25	30	36
Number of outcomes	2	2	2	1	2	1

Total, 36 outcomes

For example, $Z = 12$ is obtained from the outcomes $(3, 4), (4, 3), (2, 6), (6, 2)$; $Z = 4$ from the outcomes $(1, 4), (4, 1), (2, 2)$.

9. a. $-\frac{10}{8}, -\frac{9}{8}, \ldots, 0, \frac{1}{8}, \frac{2}{8}, \ldots, \frac{10}{8}$.
   b. The price quoted when the market opened does not matter. At issue is the range from the highest to lowest quotation.
   c. X takes on the single value 0 (no change).

11. In the tree diagram below, the number along each path indicates the number spun; the number at each vertex shows the square on which the player lands. (For convenience, the trees are drawn with branches spreading downwards.)

    First spin

    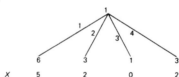

    After the first spin, X can take on values 0, 2, 5.

    Second spin

    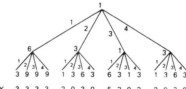

    After the second spin, X can take on values $-3, -2, 0, 2, 3, 5$. No new number (except 9) appears on the secondary vertices; all possible values of X have been listed. (For square 9, the game ends; possible value of X on landing here is 0.)

    Possible values of X are $-3, -2, 0, 2, 3, 5$.

Exercises 5.2   (Pages 118–120)

1. a. 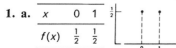   b.

c.

x	0	1	2	3
f(x)	$\frac{1}{8}$	$\frac{3}{8}$	$\frac{3}{8}$	$\frac{1}{8}$

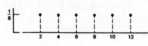

2.

z	2	4	6	8	10	12
f(z)	$\frac{1}{6}$	$\frac{1}{6}$	$\frac{1}{6}$	$\frac{1}{6}$	$\frac{1}{6}$	$\frac{1}{6}$

3.

z	$-\frac{5}{2}$	$-\frac{3}{2}$	$-\frac{1}{2}$	$\frac{1}{2}$	$\frac{3}{2}$	$\frac{5}{2}$
f(z)	$\frac{1}{6}$	$\frac{1}{6}$	$\frac{1}{6}$	$\frac{1}{6}$	$\frac{1}{6}$	$\frac{1}{6}$

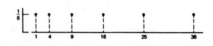

4.

z	1	4	9	16	25	36
f(z)	$\frac{1}{6}$	$\frac{1}{6}$	$\frac{1}{6}$	$\frac{1}{6}$	$\frac{1}{6}$	$\frac{1}{6}$

(Notice that these distributions are all different. Further, they all differ from the distribution of Example 5.2.3.)

5.

z	$\frac{1}{4}$	$\frac{9}{4}$	$\frac{25}{4}$
f(z)	$\frac{1}{3}$	$\frac{1}{3}$	$\frac{1}{3}$

$Z = (X - 3\frac{1}{2})^2$. Two different values of X lead to the same value of Z. $P(Z = \frac{1}{4}) = P(X = 3) + P(X = 4) = \frac{1}{6} + \frac{1}{6} = \frac{1}{3}$.

6. From the results of Problems 7 and 8 of Exercises 5.1 we have

z	1	2	3	4	5	6	8	9	10	12	15	16	18	20	24
f(z)	$\frac{1}{36}$	$\frac{2}{36}$	$\frac{2}{36}$	$\frac{3}{36}$	$\frac{2}{36}$	$\frac{4}{36}$	$\frac{2}{36}$	$\frac{1}{36}$	$\frac{2}{36}$	$\frac{4}{36}$	$\frac{2}{36}$	$\frac{1}{36}$	$\frac{2}{36}$	$\frac{2}{36}$	$\frac{2}{36}$

z	25	30	36
f(z)	$\frac{1}{36}$	$\frac{2}{36}$	$\frac{1}{36}$

7.

x	1	2	3	4
f(x)	$\frac{5}{12}$	$\frac{7}{24}$	$\frac{1}{6}$	$\frac{1}{8}$

or

x	1	2	3	4
f(x)	$\frac{10}{24}$	$\frac{7}{24}$	$\frac{4}{24}$	$\frac{3}{24}$

9.

10.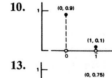

11.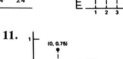

12.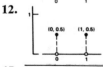

13.

14.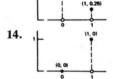

15.

x	0	1	2
f(x)	$\frac{3}{7}$	$\frac{2}{7}$	$\frac{2}{7}$

17. a. See tree diagrams in solutions to Problem 11, Exercises 5.1.
b. Possible values of X: 1, 3, 6, 9
c. $P(X = 9$ after first spin$) = 0$

Exercises 5.3 (Page 124)
1. $f(x) = \frac{1}{10}$ for $2 \le x \le 12$, $f(x) = 0$ for all other x.
   a. $P(X \le 5) = \frac{1}{10}(5 - 2) = \frac{3}{10}$
   b. $P(X < 5) = \frac{1}{10}(5 - 2) = \frac{3}{10}$. Remember that $P(X = 5) = 0$.
   c. $P(X \ge 5) = 1 - \frac{3}{10} = \frac{7}{10}$ d. $P(4 \le X \le 8) = \frac{1}{10}(8 - 4) = \frac{2}{5}$
   e. $P(2 \le X \le 3 \text{ or } 9 \le X \le 11)$
      $= P(2 \le X \le 3) + P(9 \le X \le 11) = \frac{1}{10}(3 - 2) + \frac{1}{10}(11 - 9) = \frac{3}{10}$
   f. $P(4 \le X \le 9 \text{ or } 6 \le X \le 10) = P(4 \le X \le 10) = \frac{1}{10}(10 - 4) = \frac{3}{5}$
2. a. $\frac{1}{10}(x - 2) = 0.5$; $x = 7$ b. $\frac{1}{10}(x - 2) = 0.25$; $x = 4.5$
   c. $\frac{1}{10}(x - 3) = 0.5$; $x = 8$
3. The sets $\{X < c\}$ and $\{X = c\}$ are mutually exclusive and $\{X \le c\} = \{X < c\} \cup \{X = c\}$. Hence, $P(X \le c) = P(X < c) + P(X = c)$. But $P(X = c) = 0$, and the result follows.
4. a. $P(X \le 4)$ = area of triangle bounded by

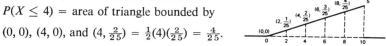

   $(0, 0)$, $(4, 0)$, and $(4, \frac{2}{25}) = \frac{1}{2}(4)(\frac{2}{25}) = \frac{4}{25}$.
   b. $P(X \ge 8) = 1 - P(X \le 4) - P(4 \le X \le 8) = 1 - \frac{4}{25} - \frac{12}{25} = \frac{9}{25}$
   c. $P(2 \le X \le 6)$
      = area of trapezoid bounded by $(2, 0)$, $(2, \frac{1}{25})$, $(6, \frac{3}{25})$, $(6, 0)$
      $= \frac{1}{2}(6 - 2)(\frac{3}{25} + \frac{1}{25}) = \frac{8}{25}$
   d. $P(X \le x) = 0.5$
      $\frac{1}{2}(x - 0)\left(\frac{x}{50}\right) = 0.5$; so $\frac{x^2}{100} = 0.5$, and we take the positive root,
      $x = \sqrt{50} \doteq 7.07$.
5. a. $f(x) = \begin{cases} 0 \text{ for } x < 0 \\ 1 \text{ for } 0 \le x \le 1 \\ 0 \text{ for } x > 1 \end{cases}$ b. $f(x) = \begin{cases} 0 \text{ for } x < -5 \\ \frac{1}{10} \text{ for } -5 \le x \le 5 \\ 0 \text{ for } x > 5 \end{cases}$
   c. $f(x) = \begin{cases} 0 \quad \text{ for } x < a \\ \dfrac{1}{b - a} \text{ for } a \le x \le b \\ 0 \quad \text{ for } x > b \end{cases}$
7. $P(c \le X \le e \text{ or } d \le X \le f) = P(\{c \le X \le e\} \cup \{d \le X \le f\})$.
   We apply our formula:
   $$P(A \cup B) = P(A) + P(B) - P(A \cap B)$$
   $$P(\{c \le X \le e\} \cap \{d \le X \le f\}) = P(d \le X \le e)$$
   Therefore, $P(c \le X \le e \text{ or } d \le X \le f)$
   $= P(c \le X \le e) + P(d \le X \le f) - P(d \le X \le e)$.

Exercises 5.4 (Page 128)
1. a. $P(Z \le 0.50) = 0.6915$
   b. $P(|Z| \le 0.50) = P(Z \le 0.50) - P(Z \le -0.50)$
      $= 0.6915 - 0.3085 = 0.3830$
2. a. $P(Z \ge 0.50) = 1 - P(Z \le 0.50) = 1 - 0.6915 = 0.3085$
   b. $P(|Z| \ge 0.50) = 1 - P(|Z| \le 0.50) = 1 - 0.3830 = 0.6170$
      or $P(|Z| \ge 0.50) = P(Z \ge 0.50) + P(Z \le -0.50)$
      $= 2P(Z \le -0.50) = 0.6170$
3. a. $P(Z \le 1.32) = 0.9066$ b. $P(Z \le 0.92) = 0.8212$

310 / Appendix

4. a. $P(-1.00 \leq Z \leq 1.00) = 0.8413 - 0.1587 = 0.6826$
   b. $P(-2.00 \leq Z \leq 2.00) = 0.9773 - 0.0227 = 0.9546$
5. a. $P(0.92 \leq Z \leq 1.32) = 0.9066 - 0.8212 = 0.0854$
   b. $P(-1.07 \leq Z \leq -0.36) = 0.3594 - 0.1423 = 0.2171$
6. a. $P(-1.25 \leq Z \leq 2.25) = 0.9878 - 0.1056 = 0.8822$ 7. 0
   b. $P(-0.74 \leq Z \leq 1.30) = 0.9032 - 0.2296 = 0.6736$ 9. 1.27
11. $P(-1.00 \leq Z \leq z) = P(Z \leq z) - P(Z \leq 1.00) = P(Z \leq z) - 0.1587$.
    So $P(Z \leq z) - 0.1587 = 0.6408$;
    $P(Z \leq z) = 0.6408 + 0.1587 = 0.7995$. Hence, z is 0.84.
13. a. $\{Z < z\}$ and $\{Z \geq z\}$ are mutually exclusive and exhaustive; so
    $P(Z < z) + P(Z \geq z) = 1$. Hence $P(Z \geq z) = 1 - P(Z < z)$.
    $P(Z = z) = 0$; so $P(Z < z) = P(Z \leq z)$ and $P(Z \geq z) = 1 - P(Z \leq z)$.
    In words, the area under the normal curve to the right of $Z = z$ is equal to the difference of the total area (total area is 1) and the area to the left of $Z = z$.
    b. $P(|Z| \geq z) = P(Z \geq z) + P(Z \leq -z)$. That is, it is the area under the normal curve to the right of $Z = z$ plus the area to the left of $Z = -z$.
    By symmetry, the two areas are equal. So $P(|Z| \geq z) = 2P(Z \geq z)$.
    But from (a) above, $P(Z \geq z) = 1 - P(Z \leq z)$.
    Hence $P(|Z| \geq z) = 2[1 - P(Z \leq z)]$.
    c. $P(|Z| \leq z) = P(-z \leq Z \leq z)$. That is, it is the area under the normal curve from $Z = -z$ to $Z = z$. Since
    $$P(Z \leq -z) + P(-z \leq Z \leq z) + P(Z \geq z) = 1,$$
    we have $P(Z \geq z) + P(Z \leq -z) + P(-z \leq Z \leq z) = 1$. The first two terms are precisely $P(|Z| \geq z)$ and the third term is $P(|Z| \leq z)$; so
    $P(|Z| \geq z) + P(|Z| \leq z) = 1$.
    Hence $P(|Z| \leq z) = 1 - P(|Z| \geq z)$
    $= 1 - 2[1 - P(Z \leq z)] = 2P(Z \leq z) - 1$.

Exercises 5.5   (Pages 131–133)

1. $\frac{1}{6}(2 + 4 + 6 + 8 + 10 + 12) = 2 \cdot \frac{1}{6}(1 + 2 + 3 + 4 + 5 + 6) = 2 \cdot \frac{7}{2} = 7$. Notice that this is twice the $E(X)$ in Example 5.5.2.
2. $\frac{1}{6}(-2\frac{1}{2} - 1\frac{1}{2} - \frac{1}{2} + \frac{1}{2} + 1\frac{1}{2} + 2\frac{1}{2}) = 0$. Notice that $3\frac{1}{2}$ is the $E(X)$ in Example 5.5.2, and the random variable in this Problem, say, Z, is related to the one in the Example by $Z = X - E(X)$.
3. $\frac{1}{6}(1 + 4 + 9 + 16 + 25 + 36) = \frac{91}{6}$. Notice that $Z = X^2$ but $E(Z)$ is not the square of $E(X)$, which was $\frac{7}{2}$.
4. $\frac{1}{3}(\frac{1}{4} + \frac{9}{4} + \frac{25}{4}) = \frac{35}{12}$ 5. $E(X) = 0 \cdot \frac{1}{2} + 1 \cdot \frac{1}{4} + 2 \cdot \frac{1}{8} + 3 \cdot \frac{1}{8} = \frac{7}{8}$
6. $E(X) = 1 \cdot \frac{5}{12} + 2 \cdot \frac{7}{24} + 3 \cdot \frac{1}{6} + 4 \cdot \frac{1}{8} = 2$
7. $E(Y) = \frac{1}{36}[2 + 2(3) + 3(4) + 4(5) + 5(6) + 6(7) + 5(8) + 4(9)$
   $+ 3(10) + 2(11) + 1(12)] = 7$.
   Observe that if X and Z are random variables for points on each die, then $E(X) = E(Z) = \frac{7}{2}$ (Example 5.5.2). It can be seen also that $E(Y) = E(X + Z) = 7$; so $E(X + Z) = E(X) + E(Z)$.
8. $1 \cdot \frac{1}{36} + 2 \cdot \frac{2}{36} + 3 \cdot \frac{2}{36} + 4 \cdot \frac{3}{36} + 5 \cdot \frac{2}{36} + 6 \cdot \frac{4}{36} + 8 \cdot \frac{2}{36} + 9 \cdot \frac{1}{36}$
   $+ 10 \cdot \frac{2}{36} + 12 \cdot \frac{4}{36} + 15 \cdot \frac{2}{36} + 16 \cdot \frac{1}{36} + 18 \cdot \frac{2}{36} + 20 \cdot \frac{2}{36}$
   $+ 24 \cdot \frac{2}{36} + 25 \cdot \frac{1}{36} + 30 \cdot \frac{2}{36} + 36 \cdot \frac{1}{36}$
   $= \frac{1}{36}(1 + 4 + 6 + 12 + 10 + 24 + 16 + 9 + 20 + 48$
   $+ 30 + 16 + 36 + 40 + 48 + 25 + 60 + 36) = \frac{441}{36} = \frac{49}{4} = (\frac{7}{2})^2$

Appendix / **311**

9. a. $0 \cdot \frac{1}{2} + 1 \cdot \frac{1}{2} = \frac{1}{2}$ b. $0 \cdot \frac{1}{4} + 1 \cdot \frac{2}{4} + 2 \cdot \frac{1}{4} = \frac{2}{2} = 1$
   c. $0 \cdot \frac{1}{8} + 1 \cdot \frac{3}{8} + 2 \cdot \frac{3}{8} + 3 \cdot \frac{1}{8} = \frac{3}{2}$
   d. $0 \cdot \frac{1}{16} + 1 \cdot \frac{4}{16} + 2 \cdot \frac{6}{16} + 3 \cdot \frac{4}{16} + 4 \cdot \frac{1}{16} = \frac{4}{2} = 2$

11.

a. $E(X) = 2 \cdot \frac{25}{64} + 1 \cdot \frac{15}{64} + 1 \cdot \frac{15}{64} + 0 \cdot \frac{9}{64}$
$= \frac{80}{64} = \frac{5}{4}$

b. $E(X) = 3(\frac{5}{8})^3 + 6(\frac{5}{8})^2(\frac{3}{8}) + 3(\frac{5}{8})(\frac{3}{8})^2 + 0(\frac{3}{8})^3$
$= \frac{1}{512}(375 + 450 + 135)$
$= \frac{15}{8}$

c. $E(Y) = 0 \cdot \frac{25}{64} + 1 \cdot \frac{15}{64} + 1 \cdot \frac{15}{64} + 2 \cdot \frac{9}{64} = \frac{48}{64} = \frac{3}{4}$

12. $E(X) = 3(\frac{1}{5})^3 + 6(\frac{1}{5})^2(\frac{4}{5}) + 3(\frac{1}{5})(\frac{4}{5})^2 + 0(\frac{4}{5})^3$
$= \frac{1}{125}(3 + 24 + 48) = \frac{75}{125} = \frac{3}{5}$

13. $E(X) = \frac{1}{4}(1 + 2 + 3 + 4) = \frac{5}{2}$

15. $E(X) = (35)\frac{1}{38} + (-1)\frac{37}{38} = -\frac{2}{38} = -\frac{1}{19} \doteq -0.05$

17. For each i, $y_i = x_i - k$, $P(Y = y_i) = P(Y = x_i - k) = P(X = x_i) = f(x_i)$.
Therefore,
$$E(Y) = \sum_i y_i P(Y = y_i) = \sum_i (x_i - k) P(X = x_i)$$
$$= \sum_i x_i P(X = x_i) - \sum_i k P(X = x_i) = \sum_i x_i f(x_i) - k \sum_i f(x_i)$$
Now $\sum_i f(x_i) = 1$; hence $E(Y) = E(X) - k$, or $E(X - k) = E(X) - k$.

18. $E(X - k) = 0$ if and only if $E(X) - k = 0$. Hence $k = E(X)$.
We may write our result in the form $E(X - E(X)) = 0$.

19. a. $E(X) = \frac{a+b}{2}$ b. $E(X) = 0$ c. $E(X) = 6\frac{2}{3}$

20. $E(X \cdot Y) = \frac{1}{36}(1 + 4 + 6 + 12 + 10 + 24 + 16 + 9 + 20 + 48$
$+ 30 + 16 + 36 + 40 + 48 + 25 + 60 + 36)$
$= \frac{441}{36} = (\frac{21}{6})^2 = (\frac{7}{2})^2$
Since $E(X) = E(Y) = \frac{7}{2}$, $E(X \cdot Y) = E(X) \cdot E(Y)$. (See Prob. 8, Ex. 5.1.)

b. $E(X^2) = \frac{1 + 4 + 9 + 16 + 25 + 36}{6} = \frac{91}{6}$. So $E(X^2) \neq [E(X)]^2$.

21. The random variables X and Y are independent; X and X are surely not independent.

Exercises 5.6   (Pages 134–135)

1. The student knows that the insurance company makes the premium high enough so that the *company's* expected value is positive. It must be positive in order to provide for the cost of salesmen's commissions, overhead, maintenance of reserves, taxes, and profit. This means, of course, that the student's expected value is negative. He is (essentially) betting the company that he will have an accident in which the cost will exceed the expense of the premiums. The insurers, using long-run statistics as a basis, believe that he will not. Still, most people with expensive cars *do* carry collision insurance. Why? They argue that the premiums are relatively inexpensive in relation to the possible disastrous loss if a car worth several thousand dollars is demolished. A particular individual must decide for himself which has more "utility" for him—the many small premium payments or the protection offered.

2. The argument is essentially that of Problem 1. Many people wish to protect themselves against *large* possible losses despite the fact that the "game" has negative expected value for them. (It is of interest to note that many people now carry "major medical" insurance which provides no payment for minor illnesses but which does provide coverage for disastrous long-term sickness and hospitalization.)

3. The view of the state is that the social worth of the insurance requirement overrides other considerations. Liability insurance provides payments to an innocent party who is injured by someone else's automobile. The state argues, essentially, that the owner of the automobile must pay for the protection of others.

4. $E(X) = (1000)\frac{1}{2} + (3000)\frac{1}{3} + (6000)\frac{1}{6} = 2500$
$E(Y) = (-1000)\frac{1}{2} + (0)\frac{1}{3} + (18000)\frac{1}{6} = 2500$
In either case, his "expected" profit is $2500. His decision depends on many things. One is the expected profit from his other acreage. With such a "cushion," he might be willing to take a greater risk. Another is more personal. Crop A offers a minimum profit of $1000, while Crop B faces him with a likely loss but the chance of a large profit. The need for this type of decision is quite common in business. There are no hard and fast rules to guide us to a "proper" decision.

5. From the point of view of expected value, it is *certainly unwise*. The expected loss on each ticket is
$$(-0.25)\frac{999{,}999}{1{,}000{,}000} + (100{,}000)\frac{1}{1{,}000{,}000} \doteq -0.15.$$
However, many people are willing to risk a $0.25 loss in face of the admittedly remote possibility of a huge profit.

6. The expected value is
$$(-20{,}000)\frac{9{,}999}{10{,}000} + (500{,}000{,}000)\frac{1}{10{,}000} \doteq 30{,}000.$$
The decision depends on how important the $20,000 is to us. If $20,000 is unimportant, perhaps one could justify the risk. Most people, however, would refuse to risk such a large sum for such a remote chance of a large fortune.

7. The gains and losses to the patient are:

		Help		
		major	minor	none
Side Effects	major	5	−3	−5
	minor	8	0	−2
	none	10	2	0

Expected value to patient:
$(5)(0.2) + (8)(0.05) + (10)(0.05)$
$+ (-3)(0.10) + (0)(0.20) + (2)(0.10)$
$+ (-5)(0.10) + (-2)(0.05) + (0)(0.15) = 2.1 - 0.9 = 1.2$

Since $E(X)$ is greater than 1, treatment begins.

8. Expected value to patient:
$(5)(0.10) + (8)(0.10) + (10)(0.05)$
$+ (-3)(0.30) + (0)(0.05) + (2)(0)$
$+ (-5)(0.20) + (-2)(0.10) + (0)(0.10) = 1.8 - 2.1 = -0.3$
$E(X) < 1$; no treatment.

Chapter Review   (Pages 136–137)

1. a.   **b.** 0

3. a.

y, the value of Y	−4	6
f(y), probability of y	3/5	2/5

b.

c. 0   **d.** No charge should be made.

5. a.

z \ x	1	2	3	4	5	6
1			3	4	5	6	7
2		3		5	6	7	8
3		4	5		7	8	9
4		5	6	7		9	10
5		6	7	8	9		11
6		7	8	9	10	11	

The entries in the body of this table are the values, y, of $Y = X + Z$.

b. 7   **c.** $\frac{7}{2}$

7.

x	0	1
f(x)	2/3	1/3

9. a. $h = \frac{1}{6}$
b. $\frac{1}{8}$
c. This is the area of the trapezoid between 7 and 13. $P(7 < X < 13) = \frac{1}{2}$.

11. a. $z = -3$ **b.** $z = 2$ **c.** $z = 1$

Exercises 6.1   (Pages 145–147)

1. a. $\mu_X = -1(\frac{1}{3}) + 0(\frac{1}{6}) + 4(\frac{1}{2}) = \frac{5}{3}$
b. $\mu_{6X+2} = 6(\frac{5}{3}) + 2 = 12$
c. $\mu_{-6X+2} = -6(\frac{5}{3}) + 2 = -8$
d. $\mu_{6X-2} = 6(\frac{5}{3}) - 2 = 8$
e. $E(X^2) = (-1)^2 \cdot f(-1) + (0)^2 \cdot f(0) + (4)^2 \cdot f(4)$
$= 1(\frac{1}{3}) + 0(\frac{1}{6}) + 16(\frac{1}{2}) = \frac{25}{3}$

3. a. $\mu_X = 0 \cdot \frac{1}{3} + 1 \cdot \frac{2}{3} = \frac{2}{3}$
b. $E(X - \frac{2}{3}) = (-\frac{2}{3})(\frac{1}{3}) + (\frac{1}{3})(\frac{2}{3}) = 0$ **c.** $E(X^2) = 0^2(\frac{1}{3}) + 1^2(\frac{2}{3}) = \frac{2}{3}$
d. $E((X - \frac{2}{3})^2) = (0 - \frac{2}{3})^2(\frac{1}{3}) + (1 - \frac{2}{3})^2(\frac{2}{3}) = \frac{4}{9} \cdot \frac{1}{3} + \frac{1}{9} \cdot \frac{2}{3} = \frac{2}{9}$
e. $E(X^2) - \mu_X^2 = \frac{2}{3} - (\frac{2}{3})^2 = \frac{2}{9} = E((X - \frac{2}{3})^2)$

5. a. $\mu_X = 0 \cdot q + 1 \cdot p = p$
(Recall that for the Bernoulli distribution, $\mu_X = p$.)
b. $E(X - p) = (-p)q + (1 - p)p = -pq + qp = 0$
c. $E(X^2) = 0^2 \cdot q + 1^2 \cdot p = p$
d. $E((X - p)^2) = (-p)^2 q + (1 - p)^2 p$
$= p^2 q + q^2 p = pq(p + q) = pq \cdot 1 = pq$
e. $E(X^2) - \mu_X^2 = p - p^2 = p(1 - p) = pq = E((X - p)^2)$

6.

x	0	1	2
f(x)	1/4	1/2	1/4

$\mu_X = 1$
$E(X^2) = 0^2 \cdot \frac{1}{4} + 1^2 \cdot \frac{1}{2} + 2^2 \cdot \frac{1}{4} = \frac{3}{2}$

$E((X - \mu_X)^2) = E((X - 1)^2) = (0 - 1)^2 \cdot \frac{1}{4} + (1 - 1)^2 \cdot \frac{1}{2} + (2 - 1)^2 \cdot \frac{1}{4}$
$= 1 \cdot \frac{1}{4} + 0 \cdot \frac{1}{2} + 1 \cdot \frac{1}{4} = \frac{1}{2}$
$E(X^2) - \mu_X^2 = \frac{3}{2} - 1^2 = \frac{1}{2} = E(X - \mu_X)^2$

314 / Appendix

7.

x	0	1	2	3
$f(x)$	$\frac{1}{8}$	$\frac{3}{8}$	$\frac{3}{8}$	$\frac{1}{8}$

$\mu_X = 0 \cdot \frac{1}{8} + 1 \cdot \frac{3}{8} + 2 \cdot \frac{3}{8} + 3 \cdot \frac{1}{8} = \frac{3}{2}$

$E(X^2) = 0^2 \cdot \frac{1}{8} + 1^2 \cdot \frac{3}{8} + 2^2 \cdot \frac{3}{8} + 3^2 \cdot \frac{3}{8}$
$ = 3$

$E((X - \mu_X)^2) = E((X - \frac{3}{2})^2)$
$ = (0 - \frac{3}{2})^2 \cdot \frac{1}{8} + (1 - \frac{3}{2})^2 \cdot \frac{3}{8} + (2 - \frac{3}{2})^2 \cdot \frac{3}{8} + (3 - \frac{3}{2})^2 \cdot \frac{1}{8}$
$ = \frac{9}{4} \cdot \frac{1}{8} + \frac{1}{4} \cdot \frac{3}{8} + \frac{1}{4} \cdot \frac{3}{8} + \frac{9}{4} \cdot \frac{1}{8} = \frac{3}{4}$

$E(X^2) - \mu_X^2 = 3 - (\frac{3}{2})^2 = 3 - \frac{9}{4} = \frac{3}{4} = E((X - \mu_X)^2)$

8. a. $E(X^2) = \frac{91}{6}$ (See Problem 3, Exercises 5.5.)
   b. $E(X - \frac{7}{2})^2 = (-\frac{5}{2})^2 \cdot \frac{1}{6} + (-\frac{3}{2})^2 \cdot \frac{1}{6} + (-\frac{1}{2})^2 \cdot \frac{1}{6} + (\frac{1}{2})^2 \cdot \frac{1}{6}$
   $\phantom{E(X - \frac{7}{2})^2} + (\frac{3}{2})^2 \cdot \frac{1}{6} + (\frac{5}{2})^2 \cdot \frac{1}{6} = \frac{35}{12}$ c. $\frac{35}{12} = \frac{91}{6} - (\frac{7}{2})^2$

9. By definition, $E(U) = \sum u_i \cdot f(u_i)$
   $ = c \cdot 1 + \text{(terms for which } f(u_i) = 0) = c$.

10.

x	-2	0	2	5
$f(x)$	0.1	0.3	0.4	0.2
z	-8	-2	4	13
x^2	4	0	4	25
z^2	64	4	16	169

$Z = 3X - 2$
$E(X) = 1.6$ (Example 6.1.1)
$E(Z) = 2.8$ (Example 6.1.1)
$E(X^2) = 7$ (Example 6.1.3)
$E(Z^2) = 47.8$ (Example 6.1.5)

11. a. A reasonable guess would be $E((aX + b)^2) = a^2 E(X^2) + 2ab E(X) + b^2$.
    b. By the formula in part a:
    $E((3X - 2)^2) = 9E(X^2) + 2(3)(-2)E(X) + 4$
    $ = 9(7) - 12(1.6) + 4 = 47.8 = E(Z^2)$
    c. $E((aX + b)^2) = E(a^2 X^2 + 2ab X + b^2)$
    $ = E(a^2 X^2) + E(2ab X) + E(b^2)$
    $ = a^2 E(X^2) + 2ab E(X) + b^2$

Exercises 6.2 (Pages 151–152)

1. a.

x	$f(x)$	$xf(x)$	$x - \mu$	$(x - \mu)^2$	$(x - \mu)^2 f(x)$
-3	$\frac{3}{12}$	$-\frac{9}{12}$	-5	25	$\frac{75}{12}$
1	$\frac{1}{12}$	$\frac{1}{12}$	-1	1	$\frac{1}{12}$
2	$\frac{4}{12}$	$\frac{8}{12}$	0	0	0
6	$\frac{4}{12}$	$\frac{24}{12}$	4	16	$\frac{64}{12}$
sum	1	$2 = \mu$			$\frac{140}{12}$

$\sigma_X^2 = \frac{140}{12} = \frac{35}{3}$; $\sigma_X \doteq 3.42$

b. $E(X^2) = (-3)^2 \frac{3}{12} + 1^2 \frac{1}{12} + 2^2 \frac{4}{12} + 6^2 \frac{4}{12}$
$ = \frac{27}{12} + \frac{1}{12} + \frac{16}{12} + \frac{144}{12} = \frac{188}{12}$
$\sigma_X^2 = E(X^2) - \mu_X^2 = \frac{188}{12} - 2^2 = \frac{140}{12} = \frac{35}{3}$

3. $\mu_X = \frac{7}{2}$ (See Example 5.5.2.) $E(X^2) = \frac{91}{6}$ (See Problem 3, Exercises 5.5.)
$\sigma_X^2 = \frac{91}{6} - (\frac{7}{2})^2 = \frac{91}{6} - \frac{49}{4} = \frac{35}{12}$

5. We can use the "center of mass" principle, or we may compute directly to show that $\mu_X = 7$.
$E(X^2) = 2^2 \cdot \frac{1}{36} + 3^2 \cdot \frac{2}{36} + 4^2 \cdot \frac{3}{36} + 5^2 \cdot \frac{4}{36} + 6^2 \cdot \frac{5}{36} + 7^2 \cdot \frac{6}{36}$
$ + 8^2 \cdot \frac{5}{36} + 9^2 \cdot \frac{4}{36} + 10^2 \cdot \frac{3}{36} + 11^2 \cdot \frac{2}{36} + 12^2 \cdot \frac{1}{36}$
$ = \frac{1}{36}(4 + 18 + 48 + 100 + 180 + 294 + 320 + 324$
$ + 300 + 242 + 144)$
$ = \frac{1974}{36}$ $\sigma_X^2 = \frac{1974}{36} - 7^2 = \frac{35}{6}$; $\sigma_X = \sqrt{\frac{35}{6}} \doteq 2.42$

7. The values of Z are: $-9, 3, 6, 18$. $\mu_Z = 3\mu_X = 3 \cdot 2 = 6$
   $E(Z^2) = (-9)^2 \cdot \frac{3}{12} + 3^2 \cdot \frac{1}{12} + 6^2 \cdot \frac{4}{12} + 18^2 \cdot \frac{4}{12} = 141$
   $\sigma_Z^2 = 141 - 6^2 = 105 = 3\sigma_X^2$.

8. The values of Z are: $-7, 5, 8, 20$. $\mu_Z = 3\mu_X + 2 = 8$
   $E(Z^2) = 49 \cdot \frac{3}{12} + 25 \cdot \frac{1}{12} + 64 \cdot \frac{4}{12} + 400 \cdot \frac{4}{12} = 169$
   $\sigma_Z^2 = 169 - 8^2 = 105 = 3\sigma_X^2$.

9. We observe that:
   $\sigma_{3X}^2 = 9\sigma_X^2$ (Multiplying each value of X by 3 has the effect of multiplying the variance by $3^2 = 9$.)
   $\sigma_{3X+2}^2 = \sigma_{3X}^2$ (Adding 2 does not affect the variance.)

10. No. $\sigma_X^2 = \sum (x_i - \mu)^2 f(x_i)$. For each i, $(x_i - \mu)^2 \geq 0$, and $f(x_i) \geq 0$. Hence, each term is nonnegative and therefore the sum is nonnegative.

11. 0. If we take the view that we can consider a random variable X having $P(X = k) = 1$, then $\mu_X = k$ and $\sigma_X^2 = (k - k)^2 \cdot 1 = 0$.

12. a. $\mu_X = \frac{1}{2}$; $E(X^2) = 0^2 \cdot \frac{1}{2} + 1^2 \cdot \frac{1}{2} = \frac{1}{2}$; $\text{var}(X) = \frac{1}{2} - (\frac{1}{2})^2 = \frac{1}{4}$
    (Or we may apply Example 6.2.3. X has a Bernoulli distribution with $p = \frac{1}{2}$.)
    b. $\mu_X = 1$; $E(X^2) = 0^2 \cdot \frac{1}{4} + 1^2 \cdot \frac{1}{2} + 2^2 \cdot \frac{1}{4} = \frac{3}{2}$
    $\text{var}(X) = \frac{3}{2} - 1^2 = \frac{1}{2} = \frac{2}{4}$
    c. $\mu_X = \frac{3}{2}$; $E(X^2) = 0^2 \cdot \frac{1}{8} + 1^2 \cdot \frac{3}{8} + 2^2 \cdot \frac{3}{8} + 3^2 \cdot \frac{1}{8} = 3$
    $\text{var}(X) = 3 - (\frac{3}{2})^2 = \frac{3}{4}$

13. $\frac{n}{4}$

15. $\text{var}(X) = \frac{5}{4}$

16. a. $\mu = 7$, $\sigma^2 = \frac{35}{6}$. We are interested in $P(4 \leq X \leq 10)$ or $P(|X - 7| \leq 3)$. By Chebyshev's Inequality,
    $$P(|X - 7| \leq k\sigma) \geq 1 - \frac{1}{k^2}.$$
    Using $k\sigma = 3$, we find $\frac{1}{k^2} = \frac{\sigma^2}{9} = \frac{35}{54}$. Thus,
    $$P(|X - 7| \leq 3) \geq 1 - \frac{35}{54} = \frac{19}{54}.$$
    b. From Problem 4b, Exercises 2.4, we find that only 6 of the 36 outcomes for the sum are not within the specified range. So $P(4 \leq X \leq 10) = \frac{30}{36} = \frac{5}{6} = \frac{45}{54}$. By comparison with the results obtained in part a, we can see that here Chebyshev's Inequality is a very conservative estimate for the probability.

Exercises 6.3 (Page 157)

1. a.

	1	2	3	4	5	6	7	8
	$\frac{1}{8}$	$\frac{1}{8}$	$\frac{1}{8}$	$\frac{1}{8}$	$\frac{1}{8}$	$\frac{1}{8}$	$\frac{1}{8}$	$\frac{1}{8}$

$P(\text{even}) = \frac{4}{8} = \frac{1}{2}$ $P(\text{div. by 3}) = \frac{2}{8} = \frac{1}{4}$

x	0	2
$f(x)$	$\frac{1}{2}$	$\frac{1}{2}$

y	0	3
$g(y)$	$\frac{3}{4}$	$\frac{1}{4}$

b. $\mu_X = 1$; $\mu_Y = \frac{3}{4}$
$\sigma_X^2 = E(X^2) - \mu_X^2 = 2 - 1 = 1$
$\sigma_Y^2 = E(Y^2) - \mu_Y^2 = \frac{9}{4} - \frac{9}{16} = \frac{27}{16}$

c. One of the eight numbers (6) is divisible by both 2 and 3; here $z = 5$ and $h(z) = \frac{1}{8}$.
One of these eight numbers (3) is divisible by 3 but not by 2; here $z = 3$ and $h(z) = \frac{1}{8}$.

Three (2, 4, 8) are divisible by 2 but not by 3; here $z = 2$ and $h(z) = \frac{3}{8}$.
Three (1, 5, 7) are divisible neither by 2 nor by 3; here $z = 0$ and $h(z) = \frac{3}{8}$.

z	0	2	3	5
h(z)	$\frac{3}{8}$	$\frac{3}{8}$	$\frac{1}{8}$	$\frac{1}{8}$

d. $\mu_Z = (0)\frac{3}{8} + (2)\frac{3}{8} + (3)\frac{1}{8} + (5)\frac{1}{8} = \frac{7}{4}$
$\sigma_Z^2 = E(Z^2) - \mu_Z^2 = 0^2 \cdot \frac{3}{8} + 2^2 \cdot \frac{3}{8} + 3^2 \cdot \frac{1}{8} + 5^2 \cdot \frac{1}{8} - (\frac{7}{4})^2$
$= \frac{46}{8} - \frac{49}{16} = \frac{43}{16}$

e. Yes. We can check the table for Z and notice that
$h(2) = P(Z = 2) = P(X = 2 \text{ and } Y = 0) = f(2) \cdot g(0) = \frac{1}{2} \cdot \frac{3}{4} = \frac{3}{8}$.
Similarly for other values of $h(z)$. We also observe that $\sigma_Z^2 = \sigma_X^2 + \sigma_Y^2$.
Intuitively, X and Y are not independent, since the outcomes for X and Y have overlapping membership. This example shows again that the definition for random variables to be independent does not necessarily agree with our intuitive use of the word "independent."

2. a. Prime numbers are 2, 3, 5, 7. $P(\text{prime}) = \frac{4}{8} = \frac{1}{2}$

x	0	2
f(x)	$\frac{1}{2}$	$\frac{1}{2}$

y	0	3
g(y)	$\frac{1}{2}$	$\frac{1}{2}$

b. $\mu_X = 1, \sigma_X^2 = 1$ $\mu_Y = \frac{3}{2}, \sigma_Y^2 = \frac{9}{2} - \frac{9}{4} = \frac{9}{4}$

c. 2 is prime and divisible by 2; here $z = 5$ and $h(z) = \frac{1}{8}$.
1 is neither prime nor divisible by 2; here $z = 0$ and $h(z) = \frac{1}{8}$.
2, 4, and 6 are divisible by 2; 3, 5, and 7 are odd primes.

z	0	2	3	5
h(z)	$\frac{1}{8}$	$\frac{3}{8}$	$\frac{3}{8}$	$\frac{1}{8}$

d. $\mu_Z = (0)\frac{1}{8} + (2)\frac{3}{8} + (3)\frac{3}{8} + (5)\frac{1}{8} = \frac{5}{2}$
$\sigma_Z^2 = 0^2 \cdot \frac{1}{8} + 2^2 \cdot \frac{3}{8} + 3^2 \cdot \frac{3}{8} + 5^2 \cdot \frac{1}{8} - (\frac{5}{2})^2 = \frac{64}{8} - \frac{25}{4} = \frac{7}{4}$

e. No. $P(Z = 0) = P(X = 0 \text{ and } Y = 0) = \frac{1}{8}$, but
$P(X = 0) \cdot P(Y = 0) = f(0) \cdot g(0) = \frac{1}{4} \neq \frac{1}{8}$.
We also observe that $\sigma_Z^2 \neq \sigma_X^2 + \sigma_Y^2$. The concept of being "not independent" in this case agrees with our intuition.

3. a. We have five independent Bernoulli trials with $p = \frac{3}{4}$. The total number of red marbles obtained is represented by the random variable
$$X = X_1 + X_2 + X_3 + X_4 + X_5.$$
$E(X) = 5p = \frac{15}{4}$, $\text{var}(X) = 5pq = 5(\frac{3}{4})(\frac{1}{4}) = \frac{15}{16}$

b. Now $p = \frac{1}{4}$, $E(X) = 5p = \frac{5}{4}$, and $\text{var}(X) = \frac{15}{16}$.

4. a. We note first that $\mu_X = 0$, $\text{var}(X) = 1$. $Y = X^2$

y	0	4
g(y)	$\frac{3}{4}$	$\frac{1}{4}$

b. $P(Z = 0) = P(X = 0 \text{ and } Y = 0)$
$= P(X = 0 \text{ and } X^2 = 0)$
$= P(X = 0) = \frac{3}{4}$
$P(Z = 2) = P(X = -2 \text{ and } X^2 = 4) = P(X = -2) = \frac{1}{8}$
$P(Z = 6) = P(X = 2 \text{ and } X^2 = 4) = P(X = 2) = \frac{1}{8}$
$Z = X + X^2$
But $P(X = 0) \cdot P(Y = 0) = \frac{3}{4} \cdot \frac{3}{4} = \frac{9}{16}$
$\neq P(Z = 0)$.

z	0	2	6
h(z)	$\frac{3}{4}$	$\frac{1}{8}$	$\frac{1}{8}$

So X and Y are not independent random variables.
(Is it "obvious" that the values of X^2 "depend" on the values of X?)

c. $\mu_Y = 1$; var$(Y) = \frac{16}{4} - 1 = 3$.
  $\mu_Z = 1$; var$(Z) = (0^2 \cdot \frac{3}{4} + 2^2 \cdot \frac{1}{8} + 6^2 \cdot \frac{1}{8}) - (1)^2$
  $= 5 - 1 = 4 = 1 + 3 = $ var$(X) + $ var(Y)

5. a. $E(X) = \frac{3}{4}$, $E(Y) = \frac{3}{4}$, $E(Z) = \frac{3}{2}$
   b. var$(X) = \frac{3}{16}$, var$(Y) = \frac{3}{16}$, var$(Z) = \frac{1}{4}$

Exercises 6.4   (Page 162)

1. a. 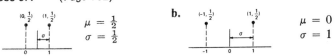   b. $\mu = \frac{1}{2}$ $\sigma = \frac{1}{2}$ $\mu = 0$ $\sigma = 1$

3. $\mu_X = \frac{3}{2}$, $\sigma_X = \frac{\sqrt{3}}{2}$ (Problem 12c, Exercises 6.2.) The standardized variable is

$Z = \dfrac{X - \frac{3}{2}}{\frac{\sqrt{3}}{2}}$. Values for z corresponding to $x = 0, 1, 2, 3$ are

$-\sqrt{3}, -\frac{1}{3}\sqrt{3}, \frac{1}{3}\sqrt{3}, \sqrt{3}$.

z	$-\sqrt{3}$	$-\frac{\sqrt{3}}{3}$	$\frac{\sqrt{3}}{3}$	$\sqrt{3}$
$g(z)$	$\frac{1}{8}$	$\frac{3}{8}$	$\frac{3}{8}$	$\frac{1}{8}$

5. Choose $Z = \dfrac{X + 5}{100}$.
   ($a = 0.01$, $b = 0.05$)

z	-3	-1	1	2	4
$g(z)$	0.15	0.25	0.10	0.20	0.30

$\mu_Z = -3(0.15) + (-1)(0.25) + 1(0.10) + 2(0.20) + 4(0.30) = 1$
$\sigma_Z^2 = 9(0.15) + 1(0.25) + 1(0.10) + 4(0.20) + 16(0.30) - (1)^2 = 6.3$
$\sigma_Z \doteq 2.5$
Since $X = 100Z - 5$, $\mu_X = 100(1) - 5 = 95$. $\sigma_X \doteq 100(2.5) = 250$.
Alternately, we could choose $Z = \dfrac{X - 95}{100}$, in which case, $a = 0.01$,
$b = -0.95$. The arithmetic is simpler on this choice.

6. a. $\mu_Y = 10 + 4 = 14$, $\sigma_Y^2 = 1^2 \cdot 9 = 9$, $\sigma_Y = 3$.
   b. $\mu_Y = -2(10) + 3 = -17$, $\sigma_Y^2 = 4 \cdot 9 = 36$, $\sigma_Y = 6$ (NOT -6 !!)
   c. $\mu_Y = \frac{1}{4}(10) + 1 = 3.5$, $\sigma_Y^2 = \frac{1}{16} \cdot 9 = \frac{9}{16}$, $\sigma_Y = \frac{3}{4}$.
   d. $\mu_Y = 0$, $\sigma_Y^2 = 1$, $\sigma_Y = 1$. Clearly no computation is required here; this transformation standardizes the random variable, X.

7.    a.   b.

   c.   d.

Exercises 6.5a   (Pages 163–164)

1. a. cov$(X, Y) = \frac{1}{2}$
   b. cov$(X, Y) = -\frac{1}{2}$
   c. cov$(X, Y) = \frac{3}{4}$

Exercises 6.5b   (Page 164)

1. a. $\text{cov}(Y, X) = E(YX) - \mu_Y\mu_X = E(XY) - \mu_X\mu_Y = \text{cov}(X, Y)$
   b. $\text{cov}(aX, bY) = E(aX \cdot bY) - \mu_{aX}\mu_{bY}$
   $= E(abXY) - (a\mu_X)(b\mu_Y) = ab[E(XY) - \mu_X\mu_Y]$
   $= ab\,\text{cov}(X, Y)$
   c. $\text{cov}(X - \mu_X, Y - \mu_Y) = E[(X - \mu_X)(Y - \mu_Y)] - \mu_{X-\mu_X} \cdot \mu_{Y-\mu_Y}$
   $= \text{cov}(X, Y) - E(X - \mu_X) \cdot E(Y - \mu_Y)$
   But $E(X - \mu_X) = E(Y - \mu_Y) = 0$, and the desired result follows.

3. a. $\text{cov}(X, X) = \sigma_X^2$
   b. For this problem, $Y = 2X$ and $\mu_X = \frac{3}{2}$, $\sigma_X^2 = \frac{1}{4}$. Thus,
   $\text{cov}(X, 2X) = 2\,\text{cov}(X, X) = 2\sigma_X^2 = \frac{1}{2}$.

Exercises 6.5c   (Page 165)

1. $a = \dfrac{1}{\sigma_X}$

3. This follows directly from Problem 2, Exercises 6.5b, with $a = \dfrac{1}{\sigma_X}$, $b = \dfrac{1}{\sigma_Y}$.

5. Consider first $Y = aX + b$, $X + Y = (a + 1)X + b$.
   $\text{cov}(X, Y) = \frac{1}{2}[\text{var}(X + Y) - \text{var}(X) - \text{var}(Y)]$
   Note that $\text{var}(X + Y) = \text{var}((a + 1)X + b)$
   $= (a + 1)^2\sigma_X^2 = (a^2 + 2a + 1)\sigma_X^2$
   $\text{var}(X) = \sigma_X^2$; $\text{var}(Y) = a^2\sigma_X^2$
   Hence, $\text{cov}(X, Y) = \frac{1}{2}(a^2 + 2a + 1 - 1 - a^2)\sigma_X^2 = a\sigma_X^2$.

   $$\rho(X, Y) = \frac{1}{\sigma_X\sigma_Y} a \cdot \sigma_X^2 = \frac{1}{\sigma_X \cdot |a| \cdot \sigma_X} \cdot a\sigma_X^2 = \frac{a}{|a|}$$

   Hence, $\rho(X, Y) = \begin{cases} 1 & \text{if } a > 0, \\ -1 & \text{if } a < 0. \end{cases}$

Chapter Review   (Pages 175–176)

1. a. $E(X) = \frac{7}{2}$, $Y = 4X + 3$, $E(Y) = 4E(X) + 3 = 14 + 3 = 17$
   Expected winning $= 17 - 20$; or $-3¢$
   b. $100(-3) = -300$; expected loss is \$3.

3. a.

x	−1	0	3	5
f(x)	0.3	0.2	0.1	0.4
z	1	3	9	13
x²	1	0	9	25
z²	1	9	81	169

   b. $E(X) = -0.3 + 0.3 + 2 = 2$
   $E(Z) = 2(2) + 3 = 7$
   c. $E(X^2) = 0.3 + 0.9 + 10 = 11.2$
   d. $E(Z^2) = 0.3 + 1.8 + 8.1 + 67.6 = 77.8$

5. a.

Outcomes	Probabilities
rr	$\frac{1}{2}$
rb	$\frac{1}{4}$
br	$\frac{1}{4}$

   b.

x	0	1
f(x)	$\frac{1}{4}$	$\frac{3}{4}$

y	0	1
g(y)	$\frac{1}{4}$	$\frac{3}{4}$

z	1	2
h(z)	$\frac{1}{2}$	$\frac{1}{2}$

   c. $\mu_Z = \frac{3}{2}$; $\mu_X = \frac{3}{4}$; $\mu_Y = \frac{3}{4}$. $\mu_Z = \mu_X + \mu_Y$
   d. $\text{var}(Z) = E(Z^2) - [E(Z)]^2 = \frac{5}{2} - \frac{9}{4} = \frac{1}{4}$
   $\text{var}(X) = \frac{3}{4} - \frac{9}{16} = \frac{3}{16} = \text{var}(Y)$
   $\frac{1}{4} \neq \frac{3}{16} + \frac{3}{16}$, and so $\text{var}(Z) \neq \text{var}(X) + \text{var}(Y)$.

7. a. $Y = aX + b$
$\mu_Y = a\mu_X + b$

$1 = 2a + b$
$-5 = 5a + b$
$-6 = 3a;$ $a = -2, b = 1 - 2a = 5$

b. $\sigma_Y = |a|\sigma_Y = 2 \cdot 3 = 6$

9. $Z = X - \mu = X - 12$

Exercises 7.1   (Pages 184–185)

1. a. $P(X = 0) = \binom{5}{0}\left(\frac{1}{2}\right)^0\left(\frac{1}{2}\right)^5 = \frac{1}{32}$, $P(X = 1) = \binom{5}{1}\left(\frac{1}{2}\right)^1\left(\frac{1}{2}\right)^4 = \frac{5}{32}$

and so on

x	0	1	2	3	4	5
$f(x)$	$\frac{1}{32}$	$\frac{5}{32}$	$\frac{10}{32}$	$\frac{10}{32}$	$\frac{5}{32}$	$\frac{1}{32}$

b.

x	0	1	2	3	4	5	6	7	8	9	10
$f(x)$	$\frac{1}{1024}$	$\frac{10}{1024}$	$\frac{45}{1024}$	$\frac{120}{1024}$	$\frac{210}{1024}$	$\frac{252}{1024}$	$\frac{210}{1024}$	$\frac{120}{1024}$	$\frac{45}{1024}$	$\frac{10}{1024}$	$\frac{1}{1024}$

Notice that $P(5 \text{ heads, } 10 \text{ tosses}) = \frac{252}{1024} \doteq 0.25$ as mentioned in Problem 4, Exercises 2.8.

2. a. $P(X = 0) = \binom{3}{0}\left(\frac{1}{6}\right)^0\left(\frac{5}{6}\right)^3 = \frac{125}{216}$,

$P(X = 1) = \binom{3}{1}\left(\frac{1}{6}\right)^1\left(\frac{5}{6}\right)^2 = \frac{75}{216}$, and so on

x	0	1	2	3
$f(x)$	$\frac{125}{216}$	$\frac{75}{216}$	$\frac{15}{216}$	$\frac{1}{216}$

b. $E(X) = 0 \cdot \frac{125}{216} + 1 \cdot \frac{75}{216} + 2 \cdot \frac{15}{216} + 3 \cdot \frac{1}{216} = \frac{108}{216} = \frac{1}{2}$

$\text{var}(X) = 0^2 \cdot \frac{125}{216} + 1^2 \cdot \frac{75}{216} + 2^2 \cdot \frac{15}{216} + 3^2 \cdot \frac{1}{216} - \left(\frac{1}{2}\right)^2$

$= \frac{144}{216} - \frac{1}{4} = \frac{5}{12} = \frac{15}{36} = 3 \cdot \frac{1}{6} \cdot \frac{5}{6}$

3. a. $P(X = 0) = \binom{4}{0}\left(\frac{3}{4}\right)^0\left(\frac{1}{4}\right)^4 = \frac{1}{256}$,

$P(X = 1) = \binom{4}{1}\left(\frac{3}{4}\right)^1\left(\frac{1}{4}\right)^3 = \frac{12}{256}$, and so on

x	0	1	2	3	4
$f(x)$	$\frac{1}{256}$	$\frac{12}{256}$	$\frac{54}{256}$	$\frac{108}{256}$	$\frac{81}{256}$

b. $E(X) = \frac{1}{256}(0 + 12 + 108 + 324 + 324) = 3$

c. $\sigma_X^2 = \frac{1}{256}(0 + 12 + 216 + 972 + 1296) - 3^2 = \frac{192}{256} = \frac{3}{4}$

$\sigma_X = \frac{\sqrt{3}}{2}$

4. $P(Y = 0) = \binom{4}{0}\left(\frac{1}{4}\right)^0\left(\frac{3}{4}\right)^4 = \frac{81}{256} = P(X = 4)$

$P(Y = 1) = \binom{4}{1}\left(\frac{1}{4}\right)^1\left(\frac{3}{4}\right)^3 = \frac{108}{256} = P(X = 3)$, and so on

y	0	1	2	3	4
g(y)	$\frac{81}{256}$	$\frac{108}{256}$	$\frac{54}{256}$	$\frac{12}{256}$	$\frac{1}{256}$

Notice that this table is the reverse of the one for Problem 3a. $g(y) = f(4 - y)$. This is reasonable, since drawing y blue marbles means drawing $4 - y$ red marbles.

5. $P(X = 0) = \binom{4}{0}(0.3)^0(0.7)^4 = 0.2401$

   $P(X = 1) = \binom{4}{1}(0.3)^1(0.7)^3 = 0.4116$

   $P(X = 2) = \binom{4}{2}(0.3)^2(0.7)^2 = 0.2646$

   $P(X = 3) = \binom{4}{3}(0.3)^3(0.7)^1 = 0.0756$

   $P(X = 4) = \binom{4}{4}(0.3)^4(0.7)^0 = 0.0081$

   $\phantom{P(X = 4) = \binom{4}{4}(0.3)^4(0.7)^0 = }\overline{1.0000}$

6. **a.** This gives the same distribution as in Example 7.1.2.
   b. $P(X \geq 4) = 0.0154 + 0.0015 + 0.0001 = 0.0170$

7. $\sum_{x=0}^{6} \binom{6}{x} = \binom{6}{0} + \binom{6}{1} + \binom{6}{2} + \binom{6}{3} + \binom{6}{4} + \binom{6}{5} + \binom{6}{6}$
   $= 1 + 6 + 15 + 20 + 15 + 6 + 1 = 64 = 2^6$

9. $\frac{216}{625}$, or 0.3453   **11. a.** $\frac{2048}{5000}$, or 0.4096   **b.** $\frac{7184}{10,000}$, or 0.7184

Exercises 7.2 (Pages 188–189)

1. $_7P_2 = 7 \cdot 6 = 42$, $_7P_5 = 7 \cdot 6 \cdot 5 \cdot 4 \cdot 3 = 2520$

   $_7C_2 = \frac{7!}{2! \cdot 5!} = \frac{7 \cdot 6}{2} = 21$, $_7C_5 = \frac{7!}{5! \cdot 2!} = 21$

3. **a.** $_{52}P_3 = 52 \cdot 51 \cdot 50 = 132{,}600$ **b.** $_{52}C_3 = \frac{_{52}P_3}{3!} = \frac{132{,}600}{6} = 22{,}100$

5. $_{12}C_2 = \frac{12!}{2! \cdot 10!} = \frac{12 \cdot 11}{1 \cdot 2} = 66$

7. There are $_4P_4 = 24$ possible orderings of the four volumes. Only one of these is the order I, II, III, IV. $P(\text{proper order}) = \frac{1}{24}$

8. We may select the 5 articles in $_{10}C_5 = 252$ ways.
   a. Here we must select all 5 from the 7 nondefective articles. Thus:
   $$P(0 \text{ defective}) = \frac{_7C_5}{252} = \frac{21}{252} = \frac{1}{12}.$$
   b. $P(1 \text{ defective}) = \frac{_7C_4 \cdot _3C_1}{252} = \frac{105}{252} = \frac{5}{12}$

   c. $P(2 \text{ defective}) = \frac{_7C_3 \cdot _3C_2}{252} = \frac{105}{252} = \frac{5}{12}$

   d. $P(3 \text{ defective}) = \frac{_7C_2 \cdot _3C_3}{252} = \frac{21}{252} = \frac{1}{12}$

9. **a.** $P(3 \text{ Republicans}) = \frac{_5C_3}{220} = \frac{10}{220}$

b. $P(\text{1 Republican, 2 Democrats}) = \dfrac{{}_5C_1 \cdot {}_7C_2}{220} = \dfrac{105}{220}$

c. $P(\text{3 Democrats}) = \dfrac{{}_7C_3}{220} = \dfrac{35}{220}$

Notice that these probabilities, together with the $\frac{70}{220}$ of Example 7.2.4, total 1.

10. $P(\text{4-flush}) = \dfrac{{}_4C_1 \cdot {}_{13}C_4 \cdot {}_{39}C_1}{{}_{52}C_5} = \dfrac{4 \cdot 715 \cdot 39}{52 \cdot 51 \cdot 20 \cdot 49} = \dfrac{143}{3332} \doteq 0.04$. It is about 20 times as probable to receive a 4-flush as a flush (Compare Example 7.2.5).

11. a. He needs an 8 or a king from any suit. $P(\text{straight}) = \frac{8}{47}$
  b. $\frac{4}{47}$ **c.** $\frac{45}{1081}$ **d.** $\frac{55}{1081}$ **e.** $\frac{55}{45}$, or $\frac{11}{9}$

13. $P(\text{enough}) = \dfrac{{}_6C_2 + {}_3C_1}{{}_7C_3} = \dfrac{18}{35}$; $P(\text{not enough}) = 1 - \frac{18}{35} = \frac{17}{35}$

Exercises 7.3   (Pages 192–193)

1.

n \ x	0	1	2	3	4	5	6	7	8	9	10	11	12
11	1	11	55	165	330	462	462	330	165	55	11	1	
12	1	12	66	220	495	792	924	792	495	220	66	12	1

2. $\dbinom{11}{6} = \dfrac{11!}{6! \cdot 5!} = \dfrac{11 \cdot 10 \cdot 9 \cdot 8 \cdot 7}{5 \cdot 4 \cdot 3 \cdot 2 \cdot 1} = 462$

3. $\dbinom{100}{2} = \dfrac{100!}{2! \cdot 98!} = \dfrac{100 \cdot 99}{2 \cdot 1} = 4950, \ \dbinom{100}{98} = \dfrac{100!}{98! \cdot 2!} = 4950$

$\dbinom{52}{3} = \dfrac{52!}{3! \cdot 49!} = \dfrac{52 \cdot 51 \cdot 50}{3 \cdot 2 \cdot 1} = 22{,}100$

4. $\dbinom{n}{x} = \dfrac{n!}{x! \cdot (n-x)!}, \ \dbinom{n}{n-x} = \dfrac{n!}{(n-x)! \cdot [n-(n-x)]!} = \dfrac{n!}{(n-x)! \cdot x!}$

5. $\dbinom{25}{0} = \dfrac{25!}{0! \cdot 25!} = 1, \ \dbinom{50}{50} = \dfrac{50!}{50! \cdot 0!} = 1$

6. $\dbinom{n}{0} = \dfrac{n!}{0! \cdot (n-0)!} = \dfrac{n!}{n!} = 1 = \dbinom{n}{n}$

7. a–b. $\dfrac{7!}{x! \cdot (7-x)!} = \dfrac{5!}{x! \cdot (5-x)!}, \ \dfrac{7 \cdot 6 \cdot 5!}{5!} = \dfrac{(7-x)(6-x)[(5-x)!]}{(5-x)!}$

Hence, $(7-x)(6-x) = 42$ or $x(x-13) = 0$.
The acceptable solution is 0; the other solution, 13, cannot be accepted since $13 > 5$ and $13 > 7$; $\dbinom{5}{13}$ and $\dbinom{7}{13}$ have no meaning.

8. $2^n = (1+1)^n = \dbinom{n}{n}1^n + \dbinom{n}{n-1}1^{n-1} \cdot 1 + \cdots + \dbinom{n}{0}1^0 \cdot 1^n$

$= \dbinom{n}{n} + \dbinom{n}{n-1} + \cdots + \dbinom{n}{0} = \displaystyle\sum_{x=0}^{n} \dbinom{n}{x}$

9. $(w+r)^5 = w^5 + 5w^4r + 10w^3r^2 + 10w^2r^3 + 5wr^4 + r^5$

10. $(3r - 2b)^4 = (3r)^4 + 4(3r)^3(-2b) + 6(3r)^2(-2b)^2 + 4(3r)(-2b)^3 + (-2b)^4$
$= 81r^4 - 216r^3b + 216r^2b^2 - 9rb^3 + 16b^4$

11. $(\frac{1}{3} + \frac{2}{3})^6 = (\frac{1}{3})^6 + 6(\frac{1}{3})^5(\frac{2}{3}) + 15(\frac{1}{3})^4(\frac{2}{3})^2 + 20(\frac{1}{3})^3(\frac{2}{3})^3$
$\qquad + 15(\frac{1}{3})^2(\frac{2}{3})^4 + 6(\frac{1}{3})(\frac{2}{3})^5 + (\frac{2}{3})^6$
$\qquad = \frac{1}{729}(1 + 12 + 60 + 160 + 240 + 192 + 64) = 1$ (of course!)

12. If $(p + q) = 1$, then $1 = (p + q)^n = \sum_{x=0}^{n} \binom{n}{x} p^x q^{n-x}$.

Exercises 7.4 (Page 199)

1. a. b. c. d. See Figure 7.4.

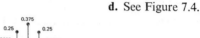

e.
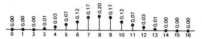

2. a. $\mu_X = 0.5$, $\sigma_X = \sqrt{(1)(0.5)(0.5)} = 0.5$
b. $\mu_X = 1$, $\sigma_X = \sqrt{0.5} \doteq 0.71$ c. $\mu_X = 2$, $\sigma_X = 1$
d. $\mu_X = 4$, $\sigma_X = \sqrt{2} \doteq 1.4$ e. $\mu_X = 8$, $\sigma_X = 2$

3. a. b. c. d.

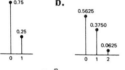
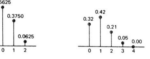
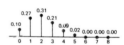

e.

5. $n = 4$ $n = 16$

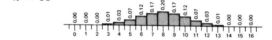

6. a. $\bar{x} = \frac{\sum x_i \cdot t(x_i)}{n}$ (page 94)

$= \dfrac{0 \cdot 4 + 1 \cdot 31 + 2 \cdot 109 + 3 \cdot 219 + 4 \cdot 274 + 5 \cdot 219 + 6 \cdot 109 + 7 \cdot 31 + 8 \cdot 4}{1000}$

$= \dfrac{8 \cdot 4 + 8 \cdot 31 + 8 \cdot 109 + 8 \cdot 219 + 4 \cdot 274}{1000} = \dfrac{4000}{1000} = 4$

b. $s_x^2 = \dfrac{\sum (x_i - \bar{x})^2 \cdot t(x_i)}{n}$ (page 101)

$= \dfrac{(-4)^2(4) + (-3)^2(31) + \cdots + (3)^2(31) + (4)^2 \cdot 4}{1000}$

$= \dfrac{2 \cdot 16 \cdot 4 + 2 \cdot 9 \cdot 31 + 2 \cdot 4 \cdot 109 + 2 \cdot 1 \cdot 219}{1000} = \dfrac{1996}{1000} \doteq 2$

(The rounded values do not give exactly 2. If fractions had been used we would have

$s_x^2 = \dfrac{(2 \cdot 16 \cdot \frac{1}{256} + 2 \cdot 9 \cdot \frac{8}{256} + 2 \cdot 4 \cdot \frac{28}{256} + 2 \cdot 1 \cdot \frac{56}{256})1000}{1000} = 2.)$

$s_x = \sqrt{2} \doteq 1.4$. If we treat the theoretical distribution as though it were numerical data, we observe that $\bar{x}$ agrees with μ and s_x with σ.

7. No. Unless each coin has the same probability for heads, the conditions for a binomial distribution are not fulfilled.
8. No. We *would* expect that the actual distribution would not differ greatly from the theoretical distribution.
9. a. $n = 6$, $p = \frac{1}{3}$

x	0	1	2	3	4	5	6
$b(6, \frac{1}{3}; x)$	$\frac{64}{729}$	$\frac{192}{729}$	$\frac{240}{729}$	$\frac{160}{729}$	$\frac{60}{729}$	$\frac{12}{729}$	$\frac{1}{729}$

Compare this result with that of Problem 11, Exercises 7.3.

b. $n = 3$, $p = \frac{1}{6}$

x	0	1	2	3
$b(3, \frac{1}{6}; x)$	$\frac{125}{216}$	$\frac{75}{216}$	$\frac{15}{216}$	$\frac{1}{216}$

11. 2 (See Chapter 4, "mode", page 105.)

Exercises 7.5a (Page 200)
1. a. Several arguments might be reasonable.
   (1) He first misses badly, then he adjusts the gunsights, which increases the value of p for the next shot.
   (2) With some initial success he might shoot with increased confidence, thus increasing p.
   b. Again, there might be several answers.
   (1) After each shot, he gets tired and is less able to hold the gun steady, thus decreasing his accuracy.
   (2) Some misses at the beginning of the firing might make the marksman nervous or uncertain, thus decreasing p.
2. Many factors affect a batter's probability of batting successfully. Most important is the ability of the opposing pitcher. A batting "average" is a composite of results obtained in a variety of situations. The probability of getting a hit is not a "constant" 0.300.
3. a. The trials are not independent. (a) and (b) are satisfied, but not (c).
   b. The value of p is not constant; (b) is not satisfied. For another example of this kind of situation, see Problem 7, Exercises 7.4.
   c. These trials are not Bernoulli; the random variable can assume more than two values.

Exercises 7.5b (Pages 200–201)
1. a. $\frac{55{,}000}{100{,}000} \cdot \frac{54{,}999}{99{,}999} \cdot \frac{54{,}998}{99{,}998}$
   b. Yes. $(0.55)^3$ differs very little from the answer to 1a. If you have access to a computer, you may work it out.
   c. $P(X = 2) = \binom{3}{2}(0.55)^2(0.45) = 0.408375$
2. a. $P(X_2 = 1) = \frac{1}{2} \cdot \frac{4}{9} + \frac{1}{2} \cdot \frac{5}{9} = \frac{1}{2}$
   b. $P(X_2 = 1 \mid X_1 = 1) = \frac{4}{9} \neq P(X_2 = 1)$
   c. The two events are not independent.
3. The events are independent. If an even number is selected first, then ($X_1 = 1$). This does not reduce the available supply of even numbers, and $P(X_2 = 1) = P(X_2 = 1 \mid X_1 = 1)$.

Chapter Review   (Page 202)

1. a. $E(X) = np = 7 \cdot \frac{3}{13} = \frac{21}{13}$ **b.** $E(Y) = nq = \frac{70}{13} = 7 - E(X)$
   c. $\sigma_X^2 = npq;\ \sigma_Y^2 = nqp = \sigma_X^2$

3. a. $np = \frac{15}{13}$ **b.** $\dfrac{\binom{5}{3}}{\binom{13}{3}} = \dfrac{60}{1716}$ **c.** $P(X = 3) = \frac{60}{1716}$

5. $\dfrac{13 \cdot 12 \cdot \dfrac{4!}{3! \cdot 1!} \cdot \dfrac{4!}{2! \cdot 2!}}{\dfrac{52!}{47! \cdot 5!}}$

Exercises 8.1   (Pages 205–206)

1. Standardizing, we have $Z = \dfrac{X - 100}{15}$.
   a. $P(X \leq 122.5) = P(Z \leq 1.5) \doteq 0.9332$
   b. $P(X \geq 90) \doteq P(Z \geq -0.67) = P(Z \leq 0.67) \doteq 0.7486$
   c. $P(85 \leq X \leq 125) \doteq P(-1 \leq Z \leq 1.33) \doteq 0.9082 - 0.1587 = 0.7495$
   d. $P(|X - 100| \leq 12) = P(|Z| \leq 0.8) = P(Z \leq 0.8) - P(Z \leq -0.8)$
     $\doteq 0.7881 - 0.2119 = 0.5762$

3. $P(X \leq 150) \doteq P(Z < 3.33)$, which is close to 1.

5. $X = 15Z + 100$
   a. $0.5000 = P(Z \leq 0)$; so $P(X \leq 100) = 0.5;\ k = 100$.
   b. $0.8997 = P(Z \leq 1.28)$; so $P(X \leq 119.2) = 0.8997;\ k = 119.2$.
   c. $0.3108 = P(|Z| \leq 0.4)$; so $P(|X - 100| \leq 6) = 0.3108;\ k = 6$.
   d. $0.5000 \doteq P(-0.67 \leq Z \leq 0.67)$; so $P(-3.35 \leq X - 100 \leq 3.35)$
     $\doteq 0.5000$; so $k \doteq 3.35$.

7. a. $P(25.8 \leq X \leq 28.3) = P(1 \leq Z \leq 2) \doteq 0.9773 - 0.8413 = 0.1360$
   b. $P(X \geq 28.3) = P(Z \geq 2) = P(Z \leq -2) \doteq 0.0227$
   c. $P(X \geq 25.8) = P(Z \geq 1) \doteq 0.1587$. By the multiplication principle, assuming independence, we have P(two cars exceeding 25.8 mpg) $= (0.1587)^2 = 0.0252$.

9. $P(X \leq 18) = P\left(Z \leq -\dfrac{5.3}{2.5}\right) = P(Z \leq -2.12) \doteq 0.0170$

Exercises 8.2   (Pages 211–212)

1. $b(100, \frac{1}{2}; 50)$: $Z \doteq Y = \dfrac{X - 50}{5}$.
   $P(X = 50) = P(49.5 \leq X \leq 50.5) \doteq P(|Z| \leq 0.1) = 0.0796$

2. $b(20, \frac{1}{2}; 16)$: $np = 10,\ npq = 5;\ Y = \dfrac{X - 10}{\sqrt{5}} = \left(\dfrac{X - 10}{5}\right)\sqrt{5}$
   $P(X = 16) = P(15.5 \leq X \leq 16.5) \doteq P(2.46 \leq Z \leq 2.91)$
     $= 0.9981 - 0.9931 = 0.0050$
   $P(X = 10) = P(9.5 \leq X \leq 10.5)$
     $\doteq P(-0.22 \leq Z \leq 0.22) \doteq 0.5871 - 0.4129 = 0.1742$
   $\dfrac{0.1742}{0.0051} \doteq 34.1$; the probability of getting 10 heads in 20 tosses is about 34 times that of getting 16 heads in 20 tosses.

3. $\dfrac{0.1742}{0.0796} \doteq 2.2$; the probability of getting 10 heads in 20 tosses is more than twice the probability of getting 50 heads in 100 tosses.

4. For 10 tosses (Table B):
$$P(4 \leq X \leq 6) = P(X = 4) + P(X = 5) + P(X = 6)$$
$$= 0.2051 + 0.2461 + 0.2051 = 0.6563$$

For 100 tosses (Example 8.2.2): $P(40 \leq X \leq 60) = 0.9642$
Even though $b(100, \frac{1}{2}; 50) = 0.0796$ is less than $\frac{1}{3}$ of $b(10, \frac{1}{2}; 5) = 0.2461$ (Problem 1), the probability of getting *about* 50 out of 100 (0.9642) is much greater than the probability of getting *about* 5 out of 10 (0.6563).

5. $Z \doteq Y = \dfrac{X - 50}{5}$; $P(44.5 \leq X \leq 55.5) \doteq P(|Z| \leq 1.1) \doteq 0.7286$

$$Z \doteq \frac{Y - 500}{5\sqrt{10}} = \left(\frac{Y - 500}{50}\right)\sqrt{10}$$

$P(449.5 \leq X \leq 550.5) \doteq P(|Z| \leq 3.19)$, which is close to 1.

6. $P(X \geq 8) = P(X = 8) + P(X = 9) + P(X = 10)$
$= 0.0439 + 0.0098 + 0.0010 = 0.547$
$P(X \geq 80) = P(X \geq 79.5) \doteq P(Z \leq -4.9) \doteq 0.0000$

7.

Binomial X	Standardized Y	Interval of values	Standardized binomial P(Y in interval)	Standard normal P(Z in interval)
0	−2	−2.33, −1.67	0.0317	0.0376
1	−1.33	−1.67, −1	0.1267	0.1112
2	−0.67	−1, −0.33	0.2323	0.2120
3	0	−0.33, 0.33	0.2581	0.2586
4	0.67	0.33, 1	0.1936	0.2120
5	1.33	1, 1.67	0.1032	0.1112
6	2	1.67, 2.33	0.0401	0.0376
7	2.67	2.33, 3	0.0115	0.0086
8	3.33	3, 3.67	0.0024	0.0012
9	4	3.67, 4.33	0.0004	0.0000
10	4.67	4.33, 5	0.0000	0.0000
11	5.33	5, 5.67	0.0000	0.0000
12	6	5.67, 6.33	0.0000	0.0000

Exercises 8.3 (Page 214)

1. **a.** Poisson, since $np = 2$ and $2 < 5$.
   b. Standard normal, since $np = 20$, and $20 > 5$.
   c. This can be computed directly or found from Table B.

$$b(10, 0.5; 5) = \binom{10}{5}(0.5)^5(0.5)^5 = \binom{10}{5}\left(\frac{1}{2}\right)^{10} = \frac{252}{1024} \doteq 0.2461$$

   From Table B, $b(10, 0.5; 5) \doteq 0.2461$.
   d. $np = 5$. Our rule of thumb permits us to use the standard normal approximation. Notice, however, that Table P includes entries for $\lambda = 5$. (See Problem 2, below.)

2. For the standard normal, we associate with X, the random variable
$$Y = \frac{X - 5}{\sqrt{100(0.05)(0.95)}} = \frac{X - 5}{\sqrt{4.95}}.$$

$P(2.5 \leq X \leq 3.5) \doteq P(-1.12 \leq Z \leq -0.67) \doteq 0.2514 - 0.1314 = 0.1200$
Using the Poisson distribution we have, for $\lambda = 5$, $x = 3$, the probability 0.1404. Using logarithms, we find $b(100, 0.05; 3) \doteq 0.1388$.
Thus, for this particular case, the Poisson approximation is more accurate than the standard normal.

Exercises 8.4 (Pages 215–216)

1. **a.** (1, 2) (2, 3) (3, 4) (4, 5) (5, 6) $\frac{3}{2}, \frac{5}{2}, \frac{7}{2}, \frac{9}{2}, \frac{11}{2}$
   (1, 3) (2, 4) (3, 5) (4, 6)   2, 3, 4, 5
   (1, 4) (2, 5) (3, 6) $\frac{5}{2}, \frac{7}{2}, \frac{9}{2}$
   (1, 5) (2, 6)   3, 4
   (1, 6) $\frac{7}{2}$

   b.

$\bar{x}$	$\frac{3}{2}$	2	$\frac{5}{2}$	3	$\frac{7}{2}$	4	$\frac{9}{2}$	5	$\frac{11}{2}$
$f(\bar{x})$	$\frac{1}{15}$	$\frac{1}{15}$	$\frac{2}{15}$	$\frac{2}{15}$	$\frac{3}{15}$	$\frac{2}{15}$	$\frac{2}{15}$	$\frac{1}{15}$	$\frac{1}{15}$

   c. $\mu_{\bar{X}} = \frac{7}{2} = \mu_X$ (by computation or by symmetry)
   d. $\mu^2_{\bar{X}} = \frac{1}{15}[\frac{9}{4} + 4 + 2(\frac{25}{4}) + (2)9 + 3(\frac{49}{4}) + 2(16) + 2(\frac{81}{4}) + 25 + \frac{121}{4}]$
        $- (\frac{7}{2})^2$
        $= \frac{7}{6} = \frac{4}{10} \cdot \frac{35}{12} = \frac{4}{10} \cdot \sigma^2_X$

2. **a.** $\binom{6}{3} = 20$, each with probability $\frac{1}{20}$.
   b. The smallest value of $\bar{X}$ arises from the draw 1, 2, 3. Hence the smallest value is 2.
   c. The largest value is $\frac{4+5+6}{3} = 5$.
   d.
	Smallest value	Largest value
X	1	6
$\bar{X}$, 2-draws	$\frac{3}{2} = 1\frac{1}{2}$	$\frac{11}{2} = 5\frac{1}{2}$
$\bar{X}$, 3-draws	2	5

   e. $\sigma^2_{\bar{X}}$ for drawing 3 at a time will be less than $\sigma^2_{\bar{X}}$ for 2 at a time.

3. **a.** $\mu_{\bar{X}} = \frac{7}{2} = \mu_X$ (by computation or by symmetry)
   b. At least two generalizations: All three random variables have the same expected value; the spread of the distributions is successively smaller as the number of marbles drawn increases from one to two to three.

4. **a.** $\mu_{\bar{X}} = \frac{7}{2} = \mu_X$
   b. $\sigma^2_{\bar{X}}$ for four at a time is less than for three at a time.
   c. Some guesses:
   $\mu_{\bar{X}} = \mu_X = \frac{1001}{2}$ (Do you see an easy way to find μ_X?)
   $\sigma^2_{\bar{X}}$ will be much smaller than σ^2_X.
   The distribution of $\bar{X}$ for 30 at a time will be approximately normal. (This is a "wild" guess based on our experience with the binomial case.)

5. **a.** $\bar{X}$ for drawing one at a time may be identified with the original random variable X.
   b. $\bar{X}$ for six at a time takes on only one value,
   $$\frac{1+2+3+4+5+6}{6} = \frac{7}{2}$$
   c. $\mu_{\bar{X}} = \frac{7}{2} = \mu_X$, $\sigma^2_{\bar{X}} = \sigma^2_X = \frac{35}{12}$ **d.** $\mu_{\bar{X}} = \frac{7}{2} = \mu_X$, $\sigma^2_{\bar{X}} = 0$

6. a. $\sigma_{\bar{X}}^2 = \dfrac{7}{6} = \dfrac{35}{12} \cdot \dfrac{4}{10} = \dfrac{35}{12} \cdot \dfrac{6-2}{2(6-1)} = \sigma_X^2 \cdot \dfrac{6-2}{2(6-1)}$

b. $n = 3;\ \sigma_{\bar{X}}^2 = \sigma_X^2 \cdot \dfrac{6-3}{3(6-1)} = \dfrac{35}{12} \cdot \dfrac{1}{5} = \dfrac{7}{12}$

$n = 4;\ \sigma_{\bar{X}}^2 = \sigma_X^2 \cdot \dfrac{6-4}{4(6-1)} = \dfrac{35}{12} \cdot \dfrac{1}{10} = \dfrac{7}{24}$

$n = 5;\ \sigma_{\bar{X}}^2 = \sigma_X^2 \cdot \dfrac{6-5}{5(6-1)} = \dfrac{35}{12} \cdot \dfrac{1}{25} = \dfrac{7}{60}$

$n = 6;\ \sigma_{\bar{X}}^2 = \sigma_X^2 \cdot \dfrac{6-6}{6(6-1)} = \dfrac{35}{12} \cdot 0 = 0$

c. We know that $\sigma_{\bar{X}}^2 = \sigma_X^2$ for $n = 1$. Our formula gives $\sigma_{\bar{X}}^2 = \sigma_X^2 \cdot \dfrac{6-1}{1(6-1)} = \sigma_X^2$; so the formula holds for $n = 1$.

Chapter Review (Pages 217–218)

1. $Z = \dfrac{X - 100}{15}$

3. $Z = \dfrac{X - \frac{7}{2}}{\sqrt{\frac{35}{12}}}$

5. a. $P(X < 575) \doteq P(Z < -1.77) \doteq 0.03836$
b. $x = 600$
c. $x \doteq 9.6$

Exercises 9.2 (Pages 225–226)

1. a. $n = 60$ **b.** $\bar{x} = \dfrac{15}{60} = \dfrac{1}{4}$
c. A (theoretically infinite) Bernoulli population with $P(\text{five}) = \frac{1}{6}$ (*a priori* probability of obtaining 5 on a single throw).
d. 10

2. Method (a) would "bias" the sample in the sense that it would over-represent the rural and suburban groups. Method (b) provides a stratified sample. We would wish to use stratification if it were felt that the arithmetic performance of children from different types of school systems might differ. Method (c) *might* be satisfactory but the method is awkward since the number of third graders from school to school would vary greatly. Furthermore, it is not obvious that a random selection of schools will lead to a random sample of third graders.

3. a. Toss the coin. Choose Urn I if the coin lands heads, Urn II if the coin lands tails.

b. Associate each member of the population with "a on red die, b on green," where $1 \leq a \leq 6$, $1 \leq b \leq 6$. If the population size is less than 36 then some throws would have to be ignored and the dice thrown again.

c. Associate each of the twelve objects with exactly one of the twelve equally likely outcomes of the experiment "toss the coin and throw the die."

4. Many methods are possible. For a class size of 36 or less, we could modify the method of Problem 3b.

5. a. (1) 400 (2) 190   **b.** (1) 10,000 (2) 4900
c. (1) 1000 (2) 120   **d.** (1) 1,000,000 (2) 161,700

7. Yes, samples having 1 blue marble and 2 red marbles.

Exercises 9.3 (Page 230)

1. a. $\mu_{\bar{X}} = 100;\ \sigma_{\bar{X}} = \dfrac{15}{1} = 15$ **b.** $\mu_{\bar{X}} = 100;\ \sigma_{\bar{X}} = \dfrac{15}{\sqrt{100}} = 1.5$

c. $\mu_{\bar{X}} = 100$; $\sigma_{\bar{X}} = \dfrac{15}{\sqrt{10{,}000}} = 0.15$

2. $\mu_{\bar{X}} = \frac{2}{3}$; $\sigma_{\bar{X}} = \sqrt{\dfrac{\frac{2}{3} \cdot \frac{1}{3}}{72}} = \dfrac{1}{18}$

3. a. $\mu_{\bar{X}} = \frac{1}{6}$; $\sigma_{\bar{X}} = \sqrt{\dfrac{\frac{1}{6} \cdot \frac{5}{6}}{320}} = \dfrac{1}{48}$

b. We have a binomial situation. The expected value is $\frac{1}{6} \cdot 320 = 53\frac{1}{3}$.

c. Standard deviation $= \sqrt{npq} = \sqrt{(320) \cdot \frac{1}{6} \cdot \frac{5}{6}} = \frac{40}{6} = 6\frac{2}{3}$ or standard deviation $= \frac{1}{48}(320) = \frac{20}{3} = 6\frac{2}{3}$.

5. $\dfrac{10{,}000 - 100}{100(10{,}000 - 1)} = \dfrac{9{,}900}{100(9{,}999)} = \dfrac{99}{9999} = 0.0099$

$\left(\text{which, of course, is very close to } \dfrac{1}{n} = 0.01\right)$

7. a. $\mu = 4$, $\sigma^2 = \frac{8}{3}$ **b.** $\mu_{\bar{X}} = \frac{36}{9} = 4 = \mu$, $\sigma_{\bar{X}}^2 = \dfrac{12}{9} = \dfrac{4}{3} = \dfrac{\sigma^2}{n}$

c. $\mu_{\bar{X}} = \frac{12}{3} = 4 = \mu$, $\sigma_{\bar{X}}^2 = \frac{2}{3} = \frac{1}{2}\left(\frac{1}{2}\right)\frac{8}{3} = \frac{1}{n}\left(\dfrac{N-n}{N-1}\right)\sigma^2$

Exercises 9.4 (Page 232)

1. Thinking of red as success, $\mu = p = 0.6$. $0.5 = 0.6 - 0.1$ and $0.7 = 0.6 + 0.1$. Thus, we seek the probability that $|\bar{X} - 0.6| > 0.1$ to be less than 0.05; that is,

$$P(|\bar{X} - 0.6| > 0.1) < 0.05.$$

The alternate form of Chebyshev's Inequality is $\quad P(|\bar{X} - \mu| > c) < \dfrac{\sigma^2}{nc^2}.$

Now $\mu = 0.6$, $c = 0.1$, $\sigma^2 = pq = (0.6)(0.4) = 0.24$. We have

$$\dfrac{0.24}{n(0.1)^2} = 0.05; \quad n = \dfrac{0.24}{(0.05)(0.01)} = \dfrac{24}{0.05} = 480$$

[We shall see in Chapter 10 (Problem 15, Exercises 10.4) that the number of draws required can be reduced considerably by using a refinement of the Chebyshev Inequality.]

3. For any c, $P(|\bar{X} - \mu| < c) \geq 1 - \dfrac{\sigma^2}{nc^2}$. We have $\mu = 100$, $\sigma = 15$, $c = 5$.

Therefore we wish to determine n so that

$$1 - \dfrac{225}{25n} \geq 0.99, \quad \dfrac{9}{n} \leq 0.01, \text{ and finally } n \geq 900.$$

4. In 10 tosses, we "expect" about 5 heads. The excess of heads on the first 10 tosses will eventually be "swamped," or lose its significance, when many more tosses are made. Eventually, after several hundred tosses, the 5 "extra" heads would not be noticed. After 10, 100, 200, 300 tosses the law of large numbers assures us that the *ratio* of heads to the total number of tosses approaches $\frac{1}{2}$. We might have, typically;

n	Number of heads	Ratio
10	10	1
100	62	0.62
200	108	0.54
300	152	0.507

Exercises 9.5 (Pages 234–235)

1. $\mu_{\bar{X}} = 100$, $\sigma_{\bar{X}} = \dfrac{15}{\sqrt{400}} = 0.75$, $Z = \dfrac{\bar{X} - 100}{0.75}$

   a. $P(\bar{X} < 101) = P\left(\dfrac{\bar{X} - 100}{0.75} < \dfrac{1}{0.75}\right) \doteq P(Z < 1.33) \doteq 0.9082$

   b. $P(|\bar{X} - 100| < 1) = P(|Z| < 1.33) \doteq 0.9082 - 0.0918 = 0.8164$

   c. $P\left(\dfrac{\bar{X} - 100}{0.75} > \dfrac{2}{0.75}\right) \doteq P(Z > 2.66) = P(Z < -2.66) = 0.0039$

2. a. Since the samples are independent, it follows, using the result of Problem 1a, that
   $P(\bar{X} < 101 \text{ and } \bar{Y} < 101) = P(\bar{X} < 101) \cdot P(\bar{Y} < 101)$
   $\doteq (0.9082)^2 \doteq 0.8248$.

   b. $P(\bar{X} > 101) = 1 - P(\bar{X} < 101) \doteq 0.0918$
   $P(\bar{X} > 101 \text{ or } \bar{Y} > 101) \doteq 0.0918 + 0.0918 - 0.0084 = 0.1752$.
   We could also argue:
   $P(\bar{X} > 101 \text{ or } \bar{Y} > 101) = 1 - P(\bar{X} \leq 101 \text{ and } \bar{Y} \leq 101)$
   $= 1 - 0.8248 = 0.1752$

3. a. We may let $Z = \dfrac{\bar{X} - np}{\sqrt{npq}} = \dfrac{\bar{X} - 50}{5}$ and find $P\left(\dfrac{\bar{X} - 50}{5} > -\dfrac{2}{5}\right)$ as in Chapter 8.

   Or we may let $Z = \dfrac{\bar{X} - p}{\sqrt{\dfrac{pq}{n}}} = \dfrac{\bar{X} - \frac{1}{2}}{0.05}$ and find $P\left(\dfrac{\bar{X} - \frac{1}{2}}{0.05} > -\dfrac{0.02}{0.05}\right)$.

   Using either method, we have $P(Z > -0.4) \doteq 0.6554$, or about 65.54%.

   b. $P(|Z| \leq 1) \doteq 0.6826$, or 68.26%

4. $Z = \dfrac{\bar{X} - np}{\sqrt{npq}} = \dfrac{\bar{X} - 500}{5\sqrt{10}}$

   a. $P(\bar{X} > 480) = P(Z > -0.4\sqrt{10}) \doteq P(Z > -1.26) \doteq 0.8962$, or about 89.62%

   b. $P(450 < \bar{X} < 550) = P(-\sqrt{10} < Z < \sqrt{10}) \doteq P(|Z| < \sqrt{10}) \doteq 1$, or about 100%

5. $\mu_{\bar{X}} = 3.5$; $\sigma_{\bar{X}} = \dfrac{\sigma}{\sqrt{100}} = \dfrac{3.5}{12}$; $Z = \dfrac{\bar{X} - 3.5}{\frac{3.5}{12}}$

   $P(|\bar{X} - 3.5| < 0.05) = P\left(|Z| < \dfrac{0.05}{\frac{3.5}{12}}\right) = P\left(|Z| < \dfrac{0.60}{3.5}\right)$
   $\doteq P(|Z| < 0.17) \doteq 0.5675 - 0.4325 = 0.1350$

7. $\mu_{\bar{X}} = \mu = 40$; $\sigma_{\bar{X}} = \dfrac{\sigma}{\sqrt{36}} = \dfrac{3}{6} = \dfrac{1}{2}$; $Z = \dfrac{\bar{X} - 40}{\frac{1}{2}}$

   a. $P(\bar{X} < 36) = P(Z < -8) = 0.0000$
   b. $P(\bar{X} < 39) = P(Z < -2) = 0.0227$

Exercises 9.6a (Page 236)

1. We are certain that $p \neq 0$ and $p \neq 1$. Therefore "$p = \frac{1}{2}$" is correct.

2. We know that $p \neq 0$. We might guess that $p = 1$, although we are not very confident.

3. While we are not certain, the law of large numbers leads us to believe $p = \frac{1}{2}$ is not likely. We are very confident that $p = 1$.
4. We have no way of telling whether $p = 0$ or $p = \frac{1}{2}$. Perhaps we might choose $p = 0$. (See Problem 5.)
5. a. If $p = 0$, $P(X = 0) = 1$ b. If $p = \frac{1}{2}$, $P(X = 0) = \frac{1}{2}$
6. a. If $p = 1$, $P(X = 3) = 1$ b. If $p = \frac{1}{2}$, $P(X = 3) = \frac{1}{8}$
7. Probably
9. (c) gives the best support since 102 is the largest sample size. Of course (a) gives about the same degree of confidence.
10. For (a) $\sigma_{\bar{X}} = \dfrac{5}{\sqrt{100}} = 0.5$, for (b) $\sigma_{\bar{X}} = \dfrac{50}{\sqrt{100}} = 5$.

Thus, distribution of the possible values of $\bar{X}$ for case (a) has a much smaller standard deviation. Therefore, we conclude that it is more likely to obtain a value of $\bar{X}$ close to μ, since the possible values of $\bar{X}$ "deviate" less from the mean.

Exercises 9.6b   (Pages 237–238)

1. a. $p = 0.01$, $\mu = np = 8.5$. We "expect" eight or nine 99's per page. The large number on the first page is extremely unusual, but could occur. The lesser number of 99's on the next two pages might be an indication that "swamping" has begun.
   b. In a table of 1,000,000 triplets of digits, we expect $0.001 \times 1,000,000$, or 1,000, triplets of a specific digit. Thus, 100 would be surprising.
2. The law of large numbers does not promise that these unusual events do not happen; nor does it promise to rectify imbalance immediately. Since $P(X = 7) = \frac{6}{36} = \frac{1}{6}$, the lady probably assumed that having 22 sevens in a row has very small probability. It is true that $(\frac{1}{6})^{22}$ is small—about 7.6×10^{-18} (by logarithms). However, she forgot that a die is not supposed to have memory; thus:
$P(X = 7 \text{ on 22nd trial} \mid X = 7 \text{ on each of 21 preceding trials})$ is still equal to $\frac{1}{6}$.
3. Answers would vary. Most likely the decision whether or not to ignore the 1% (as an extreme case) would depend on other factors such as the consequences of *not* operating.
4. Again, answers vary. This is an example of a piece of "maverick" data. Without any knowledge of the particular circumstances surrounding the F grade, it seems that this is the record of a good (B+) student who had one bad classroom experience, which may be ignored in view of his over-all record.
5. Answers vary. If we were following the man's record year by year, when we encounter the jump from 140 to 180, it is highly unlikely that we would feel we could ignore this information. Such a sudden change in the blood pressure might be a symptom of a medical problem or a signal of an impending one. Another check on the blood pressure (after a reasonable interval) might be taken to confirm the 180 reading. If the reading were verified, then certain procedures might be prescribed for the patient. However, looking at whole list as presented, with subsequent readings "back to normal," the physician might take the long-term view and dismiss the 180 as an anomaly that had been controlled. However, this attitude would depend on the role that the physician is expected to take. If he were a consulting physician screening the man for insurance, his viewpoint of this anomaly might be more critical.

Chapter Review   (Pages 239–240)

1. a. $\mu = 4$, $\sigma^2 = 5$

b.

$\bar{x}$	1	2	3	4	5	6	7
t	1	2	3	4	3	2	1

$\mu_{\bar{X}} = 4 = \mu$, $\sigma_{\bar{X}}^2 = \dfrac{5}{2} = \dfrac{\sigma^2}{2}$

c.

$\bar{x}$	2	3	4	5	6
t	1	1	2	1	1

$\mu_{\bar{X}} = 4 = \mu$, $\sigma_{\bar{X}}^2 = \dfrac{5}{3} = \dfrac{1}{2}\left(\dfrac{4-2}{4-1}\right)\sigma^2$

3. A proper interpretation of the "law of averages" is what we have called the Law of Large Numbers.

5. a. Associated with every student at the college is a *number*, the total number of hours he used the reading room divided by the number of weeks in the term. The population is this collection of numbers, *not* the students themselves.

b. Any sample chosen from among library users will not contain any representative of those members of the population for which the number of hours use is 0. (There are other possible answers.)

Exercises 10.1   (Pages 242–243)

1. $\mu + 7.84$ corresponds to $z = \dfrac{7.84}{4} = 1.96$ (since $\sigma = 4$). $P(|Z| \leq 1.96) = 0.95$

2. $31.56 = 25 + 6.56$, which corresponds to $z = \dfrac{6.56}{4} = 1.64$. Similarly, 18.44 corresponds to $z = -1.64$.
$P(|Z| > 1.64) = 1 - P(|Z| < 1.64) = 1 - 0.90 = 0.10$

3. a. $\mu = 6$
b. $2\sigma \doteq 7.4$. Since the lowest card is a 2 and the highest a 10,
$$P(|X - 6| > 7.4) = 0.$$

4. The only definite conclusion is $P(X = 1) \neq 0$.

5. Using Chebyshev's Inequality with $k = 3$, we find
$$P(|X - \mu| < 3\sigma) \geq 1 - \dfrac{1}{3^2} \doteq 0.889.$$

6. a. $\tfrac{1}{4}$ **b.** $\tfrac{1}{4}$ **c.** $\tfrac{1}{2}$

7. Of the three choices, we feel that 25 is the most reasonable choice.

9. a. $\mu = 9$ **b.** $\sigma^2 = 0$

Exercises 10.3   (Pages 247–248)

1. From Section 5.4 (or Table N), we find that 0.90 of such intervals contain the true value of μ.

2. a. Verify that $P(|Z| < 2.33) \doteq 0.98$; thus, $a \doteq 2.33$.
b. $a \doteq 2.58$ **c.** $a = 1$

3. a. Lengthen. The larger the value of a, the longer the interval and the greater the confidence.
b. Decreased. The shorter the interval, the less likely it is that we have "captured" μ.

4. Increase n. This decreases the value of $a\dfrac{\sigma}{\sqrt{n}}$ for every a and hence shortens the interval without decreasing the confidence.

5. a. 0.05   **b.** $\frac{1}{2} \times 0.05 = 0.025$
c. $P(Z \geq -2.58) \doteq P(Z \leq 2.58) \doteq 0.9951$. Thus, 99.5% of all such intervals contain μ. (We are 99.5% sure that μ lies to the right of $\bar{x} - 2.58 \frac{\sigma}{\sqrt{n}}$.)

7. a. [19.04, 22.96]   **b.** [20.02, 21.98]   **c.** [20.216, 21.784]

Exercises 10.4   (Pages 252–253)

1. a. 75% of the normal distribution lies between $Z = -1.15$ and $Z = 1.15$. Hence, the desired interval is

$$50 \pm 1.15 \frac{20}{\sqrt{100}}, \text{ or } 50 \pm 2.30, \text{ or } [47.70, 52.30].$$

b. For 90% confidence, $a = 1.64$:

$$50 \pm 1.64 \frac{20}{\sqrt{100}}, \text{ or } 50 \pm 3.28, \text{ or } [46.72, 53.28]$$

c. $50 \pm 1.96 \frac{20}{\sqrt{100}}$, or 50 ± 3.92, or [46.08, 53.92]

d. $50 \pm 2.58 \frac{20}{\sqrt{100}}$, or 50 ± 5.16, or [44.84, 55.16]

2.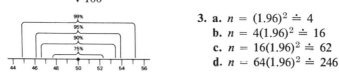

3. a. $n = (1.96)^2 \doteq 4$
b. $n = 4(1.96)^2 \doteq 16$
c. $n = 16(1.96)^2 \doteq 62$
d. $n = 64(1.96)^2 \doteq 246$

5. $32 \pm 1.64 \frac{8}{\sqrt{100}}$, or 32 ± 1.31, or [30.69, 33.31]

6. $6.2 \pm 1.96 \frac{2.1}{\sqrt{1000}}$, or 6.2 ± 0.13, or [6.07, 6.33]

7. $\bar{x} = \frac{240}{400} = 0.60$. We use $\sqrt{(0.60)(0.40)} \doteq 0.49$ for σ. [0.55, 0.65]

9. a. $1.64 \frac{\sigma}{10} = \frac{1}{2}(1.4) = 0.7; \sigma = \frac{7}{1.64} \doteq 4.27$

b. $1.64 \frac{\sigma}{\sqrt{n}} = \frac{1}{2}(0.7) = 0.35; \frac{7}{\sqrt{n}} = 0.35; \sqrt{n} = \frac{7}{0.35} = 20$

So the desired sample size is 20^2, or 400.

c. No. To construct an interval half as wide as a given interval, we need to increase the sample size by a factor of 4. That is, if $w = 2a \frac{\sigma}{\sqrt{n}}$, then $\frac{1}{2}w = 2a \frac{\sigma}{\sqrt{4n}}$.

10. For 95% confidence, we wish to have $1.96 \frac{40}{\sqrt{n}} = 10$.

Thus $\sqrt{n} = \frac{(1.96)(40)}{10} = 7.84; n \doteq 62$.

11. The width of the desired interval is $9.7 - 9.5$, or 0.2;

$$0.1 = a \frac{3.3}{\sqrt{900}}, \quad a = \frac{3}{3.3} \doteq 0.91.$$

From Table N, we find that about 64% of the standard normal distribution lies between $Z = -0.91$ and $Z = 0.91$. Hence, [9.5, 9.7] is the 64% confidence interval.

12. We wish $|\bar{x} - \mu| \leq 5$; $5 = 1.96 \dfrac{35}{\sqrt{n}}$; so $n \doteq 188$.

13. Consider the set of all real numbers to the left of $\bar{x} + a\dfrac{\sigma}{\sqrt{n}}$. From Table N, $a = P(Z \leq 0.90) \doteq 1.28$. So $5 = 1.28 \dfrac{35}{\sqrt{n}}$; $n \doteq 80$.

15. For 95% confidence we wish $1.96 \dfrac{\sqrt{(0.6)(0.4)}}{\sqrt{n}} = \dfrac{1}{2}(0.7 - 0.5) = 0.1$.

$$\sqrt{n} = \dfrac{1.96\sqrt{(0.6)(0.4)}}{0.1}, \quad n \doteq \dfrac{(3.84)(0.24)}{0.01} \doteq 92$$

In Problem 1, Exercises 9.4, using the Chebyshev Inequality, we found $n = 480$. Note how the modification using the normal approximation improves on the estimate, thus requiring a considerably smaller sample size.

Exercises 10.5   (Pages 257–258)

1. a. Yes. The 99% confidence interval is

$$\mu \pm 2.58 \dfrac{\sigma}{\sqrt{n}} = 52.3 \pm 2.58 \dfrac{32}{\sqrt{400}} = 52.3 \pm 4.13.$$

$\bar{x} = 57.5$ lies outside the interval [48.2, 56.4].

  b. Yes. The appropriate one-sided interval (see Figure 10.4) has

$$\mu + 2.33 \dfrac{\sigma}{\sqrt{n}} = 52.3 + 2.33 \dfrac{32}{\sqrt{400}} \doteq 52.3 + 3.73 \doteq 56.0$$

as its upper limit.

2. a. No. $\bar{x}$ might be either greater than $\mu + 1.96 \dfrac{\sigma}{\sqrt{n}}$ or it might be smaller than $\mu - 1.96 \dfrac{\sigma}{\sqrt{n}}$.

  b. 2.5% $\left(97.5\% \text{ of the normal distribution lies to the right of } \mu - 1.96 \dfrac{\sigma}{\sqrt{n}}.\right)$

  c. $\bar{x}$ is significantly greater than μ at the 2.5% level.

3. a. Yes   b. No   c. Yes

  Note: The preceding problems show that to determine significance levels two-sided intervals are more "conservative." For this reason, a two-sided interval is frequently used to determine whether a sample differs significantly from a population even when a one-sided interval appears to be the more natural choice. Restating: Even if we are interested in whether the sample is "better" than the population, we often are content to determine that it is (statistically) "different" from the population.

5. No. For $p = 0.022$, the confidence interval at the 2% level is

$$0.022 \pm 2.33 \sqrt{\dfrac{(0.022)(0.978)}{1000}} \doteq 0.022 \pm 0.011,$$

or [0.011, 0.033]. $\tfrac{12}{1000} = 0.012$, and this is within the interval; so the adjustment has not made any significant *difference* at the 2% level.

6. A 98% one-sided interval (Figure 10–4) has as its lower limit

$$0.022 - 2.05\sqrt{\frac{(0.022)(0.978)}{1000}} \doteq 0.022 - 0.009 \doteq 0.013$$

Since less than 2% of all samples of size 1000 with $p = 0.022$ would give $\bar{x} < 0.013$, the adjusted lathe gives significantly *better* performance at the 2% level.

7. For 68% of the tires to come within 1,200 miles of the mean, one standard deviation equals 1200. For the 1% level, we have

$$20{,}000 \pm 2.58(\tfrac{1200}{8}) \doteq 20{,}000 \pm 387$$

Since 19,600 is not within the interval [19,613, 20,387], the mean of the sample is significantly different at the 1% level. For the 5% level, we have

$$20{,}000 \pm 1.96(\tfrac{1200}{8}), \text{ or } [19{,}706,\ 20{,}294].$$

So the mean of the sample is significantly different at the 5% level.

9. A plausible generalization (which turns out to be correct) is: The value of a at the $k\%$ level of a one-sided interval is the same as the $2k\%$ level of a two-sided interval.

Exercises 10.6 (Pages 259–260)

1. For any two random variables, we know that the expected value of the sum equals the sum of the expected values (Section 6.3); so

$$\bar{D} = \bar{X} - \bar{Y} = \bar{X} + (-\bar{Y}).$$

a. $\mu_{\bar{D}} = E(\bar{D}) = E(\bar{X}) + E(-\bar{Y}) = E(\bar{X}) - E(\bar{Y}) = \mu_{\bar{X}} - \mu_{\bar{Y}}$
b. Since $\mu_{\bar{X}} = \mu_X$, $\mu_{\bar{Y}} = \mu_Y$, we have $\mu_{\bar{D}} = \mu_X - \mu_Y$.

2. a. Since $\bar{X}$ and $\bar{Y}$ are assumed to be independent, we have:

$$\sigma_{\bar{D}}^2 = \sigma_{\bar{X}+(-\bar{Y})}^2 = \sigma_{\bar{X}}^2 + \sigma_{-\bar{Y}}^2$$
$$= \sigma_{\bar{X}}^2 + (-1)^2 \sigma_{\bar{Y}}^2 = \sigma_{\bar{X}}^2 + \sigma_{\bar{Y}}^2$$

(See Section 6.4, $\sigma_{aW}^2 = a^2 \sigma_W^2$.)

b. $\sigma_{\bar{X}}^2 = \dfrac{\sigma_X^2}{n_X}$; $\sigma_{\bar{Y}}^2 = \dfrac{\sigma_Y^2}{n_Y}$. Therefore $\sigma_{\bar{D}}^2 = \dfrac{\sigma_X^2}{n_X} + \dfrac{\sigma_Y^2}{n_Y}$

3. $Z = \dfrac{\bar{D} - \mu_{\bar{D}}}{\sigma_{\bar{D}}} = \dfrac{(\bar{X} - \bar{Y}) - (\mu_X - \mu_Y)}{\sqrt{\dfrac{\sigma_X^2}{n_X} + \dfrac{\sigma_Y^2}{n_Y}}}$

4. We hope you made the (correct) guess that Z has a distribution which is approximately standard normal.

5. For large sample sizes, we would replace σ_X^2 and σ_Y^2 by the values of s_x^2 and s_y^2 obtained from the actual samples. Thus,

$$Z' = \dfrac{(\bar{X} - \bar{Y}) - (\mu_X - \mu_Y)}{\sqrt{\dfrac{s_x^2}{n_x} + \dfrac{s_y^2}{n_y}}}$$

will have an approximately standard normal distribution.

Appendix / 335

Chapter Review (Pages 261–262)
1. a. The best guess is $k = \frac{2}{3}$ b. $\frac{3}{11}$
3. (See page 247). "Confidence" expresses our degree of belief that something has happened which we would expect (*a priori*) to happen with a certain probability.
5. [23.126, 24.274] 7. [59.99, 64.81]

Exercises 11.1 (Page 266)
1. The following are statistical hypotheses:
   b. We are comparing the probabilities p_J and p_A of the players in making free throws.
   c. The hypothesis is H: $p > \frac{1}{6}$. e. H: $\mu_{men} = \mu_{women}$.
   f. For the population of voters, let $p = P$(vote for Smith). The hypothesis is H: $p > \frac{1}{2}$.
2. Our evidence leads us to be "confident" that the true value of p is "close" to 0.5. We cannot be sure that p is *exactly* 0.5.
3. We rewrite $\mu = \mu_0$ as $\mu - \mu_0 = 0$. In practice, this form arises as the result of sampling. The sample is drawn from *some* population. The question "Does this population have a mean different from that of some known population?" becomes "Shall we reject H: $\mu - \mu_0 = 0$?" where μ_0 is the mean of the known population.
4. a. H: $p = \frac{1}{4}$
   b. H: $\mu_0 > \mu_T$, where μ_T = mean recovery time for patients with the new drug, μ_0 = mean recovery time for patients without the drug.
   c. H: $\mu_X = \mu_Y$, where X and Y represent the performances of the populations of high-jumpers, with and without the special weight-lifting training.

Exercises 11.2 (Page 269)
1. With $\alpha = 0.02$ we reject H if the obtained value of $\overline{X}$ is less than $-2.33 \, \frac{6}{\sqrt{144}}$ or greater than $2.33 \, \frac{6}{\sqrt{144}}$. The rejection region is the set of values of $\overline{X}$ that lie outside the interval $[-1.165, 1.165]$.

3. If $p = 0.5$ is true, then $\frac{\sigma}{\sqrt{n}} = \sqrt{\frac{(0.5)(0.5)}{100}} = 0.05$.

We accept H if the obtained value of $\overline{X}$ falls in the interval $0.425 \leq \overline{X} \leq 0.575$ or $|\overline{X} - 0.5| \leq \frac{1}{2}(0.575 - 0.425)$. Thus;

$$\left|\frac{\overline{X} - 0.5}{0.05}\right| \leq \frac{0.075}{0.05} = 1.5 = a$$

From Table N, $P(|Z| \geq 1.5) \doteq 1 - (0.9332 - 0.0668) = 0.1336 = \alpha$. (Note that the values 43 to 57 are replaced by 42.5 and 57.5 to agree with the approximation by the standard normal, with boundaries extending a half unit beyond the corresponding boundaries.)

5. With $\alpha = 0.01$, the 99% confidence interval for $\overline{X}$ is

$$47 \pm 2.58 \, \frac{12}{\sqrt{400}} \doteq 47 \pm 1.55, \text{ or } [45.45, 48.55].$$

Since 50 lies outside this interval, we reject H: $\mu = 47$.

7. **a.** H: $\mu_X - \mu_Y = 0$
   b. With s_x and s_y as approximations for σ_X and σ_Y, the test statistic is
   $$\frac{\overline{X} - \overline{Y}}{\sqrt{\frac{15^2}{500} + \frac{16^2}{400}}} = \frac{\overline{X} - \overline{Y}}{\frac{1}{10}\sqrt{109}} = \frac{\overline{X} - \overline{Y}}{1.04}.$$

   For $\alpha = 0.01$, our rejection region is
   $$\left|\frac{\overline{X} - \overline{Y}}{1.04}\right| > 2.58 \quad \text{or} \quad |\overline{X} - \overline{Y}| > 2.42$$

   Since $|62.1 - 57.4| = 2.7$, we reject the null hypothesis.

8. $p = \frac{1}{6}$, $\sigma^2 = (\frac{1}{6})(\frac{5}{6})$, $\sigma = \frac{\sqrt{5}}{6}$, $\overline{X} = \frac{72}{600} = 0.12$
   a. For $\alpha = 0.02$, $a = 2.33$ and the confidence interval is
   $$\tfrac{1}{6} \pm 2.33 \frac{\sqrt{5}}{6\sqrt{600}} \doteq 0.1667 \pm (2.33)(0.0152),$$
   or $[0.1313, 0.2021]$.

   Since 0.12 falls outside this region, we reject H.
   b. For $\alpha = 0.01$, $a = 2.58$, and the confidence interval is
   $$0.1667 \pm (2.58)(0.0152) = [0.1265, 0.2059].$$
   0.12 still falls outside this interval (though by not much); we reject H.

Exercises 11.3   (Pages 272–273)

1. Using s_x as an approximation for σ, we may establish the appropriate 5% rejection region.
   a. $|\overline{X} - 10{,}000| > 1.96 \frac{300}{\sqrt{100}}$, or $|\overline{X} - 10{,}000| > 58.8$

   For the given value of $\overline{X}$, $|9947 - 10{,}000| < 58.8$; so we do not reject H.
   b. $\overline{X} - 10{,}000 < -1.64 \frac{300}{\sqrt{100}}$, or $\overline{X} < 9950.8$

   Since $9947 < 9950.8$, we reject H.
   c. $\overline{X} - 10{,}000 > 1.64 \frac{300}{\sqrt{100}}$, or $\overline{X} > 10{,}049.2$. The obtained value, 9947, does not lie in the rejection region; hence we accept H.

2. **a.** Rejection region is at
   $$\overline{X} > \tfrac{1}{6} - 2.05 \frac{\sqrt{5}}{6\sqrt{600}} = 0.1667 - (2.05)(0.0152)$$
   $\overline{X} > 0.1355$ (Compare with Problem 8, Exercises 11.2.)
   b. $\overline{X} > 0.1667 - (2.33)(0.0152)$; $\overline{X} > 0.1313$

3. Note that for the case of 72 out of 600 throws, we reject the hypothesis for the two-sided test but would accept the hypothesis for the one-sided test given the same significance level. The two-tailed test is more conservative in that it is not as likely to accept the hypothesis.

4. For the one-tailed test, since $\sigma = \sqrt{\frac{35}{6}} = 2.42$,

$$\text{at the 5\% level, } |\bar{X} - 7| > \frac{1.64(2.42)}{8} \doteq 0.496$$

$$\text{at the 2\% level, } |\bar{X} - 7| > \frac{2.05(2.42)}{8} \doteq 0.620$$

$$\text{at the 1\% level, } |\bar{X} - 7| > \frac{2.33(2.42)}{8} \doteq 0.705$$

Since $|7.65 - 7| = 0.65$, the hypothesis is rejected between the 2% and the 1% levels. (We do not reject at the 1% level.)

Exercises 11.4 (Page 275)

1. $\alpha = 1 - P_{0.5}(X = 4, 5, 6) = 1 - (0.2051 + 0.2461 + 0.2051)$
$= 0.3437$ (Table B)

3. $\alpha = 1 - P_{0.5}(39.5 \leq X \leq 60.5)$; $\mu = p = 0.5$, $\sigma = 0.5$, $\frac{\sigma}{\sqrt{n}} = 0.05$,
$\frac{1}{2}(60.5 - 39.5) = 10.5$; $\frac{10.5}{100} = 0.105$. Using normal approximation,

$$P\left(\left|\frac{\bar{X} - 0.5}{0.05}\right| < \frac{0.105}{0.05}\right) = P(|Z| \leq 2.1) = 0.9821 \quad \text{(Table N)}$$

So $\alpha \doteq 1 - 0.9821 \doteq 0.0179$.

5. $\alpha = P_{0.1}(X = 3) = (0.1)^3 = 0.001$

6. $\alpha = P_{0.1}(X = 2, 3) = \binom{3}{2}(0.1)^2(0.9) + (0.1)^3 = 0.028$

7. $\alpha = P_{0.1}(X \neq 0) = 1 - P_{0.1}(X = 0) = 1 - (0.9)^3 = 0.271$

9. $P\left(\frac{\bar{x} - 70}{0.5} > \frac{0.6}{0.5}\right) = P(Z > 1.2) = 0.1151$

Exercises 11.5 (Pages 277–278)

1. $\beta = P_{0.6}(\text{accept H}) = P_{0.6}(X < 3) = 1 - P_{0.6}(X = 3)$
$= 1 - (0.6)^3 = 1 - 0.216 = 0.784$

2. $\beta = P_{0.6}(X = 0, 1) = (0.4)^3 + \binom{3}{1}(0.6)(0.4)^2 = 0.352$

3. $\beta = P_{0.6}(X = 0) = (0.4)^3 = 0.064$

4. $\beta = P_A(\text{accept H}) = P_A(\text{reject A}) = 1$, since A is certain to be rejected, even if it is true.

5. Reject H regardless of experimental evidence. Then $\beta = P_A(\text{reject A}) = 0$, since we would never reject A. $\alpha = P_H(\text{reject H}) = 1$, since H is rejected even if it is true.

6. $\beta = P_0(\text{accept H}) = P_0(X = 1) + P_0(X = 2) = 0$

7. $\beta = P_{0.1}(X = 1) + P_{0.1}(X = 2) = \binom{3}{1}(0.1)(0.9)^2 + \binom{3}{1}(0.1)^2(0.9)$
$= 0.270$

8. $\beta = P_{0.2}(X = 1) + P_{0.2}(X = 2) = \binom{3}{1}(0.2)(0.8)^2 + \binom{3}{2}(0.2)^2(0.8)$
$= 0.480$

9. $\beta = P_{0.3}(X = 1) + P_{0.3}(X = 2)$
$= \binom{3}{1}(0.3)(0.7)^2 + \binom{3}{2}(0.3)^2(0.7) = 0.630$

10. $\beta = P_{0.4}(X = 1) + P_{0.4}(X = 2)$
$= \binom{3}{1}(0.4)(0.6)^2 + \binom{3}{2}(0.4)^2(0.6) = 0.720$

11. A: $p = 0.6$, $\beta = 0.720$ A: $p = 0.9$, $\beta = 0.270$
A: $p = 0.7$, $\beta = 0.630$ A: $p = 1$, $\beta = 0$
A: $p = 0.8$, $\beta = 0.480$

12. P_p(accept H: $p = 0.5$)

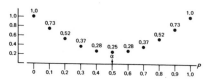

13. By making only three trials, our test allows us to decide between, say, $p = 0.5$ and $p = 0.1$ rather reliably. We would not expect three trials to enable us to distinguish $p = 0.5$ from $p = 0.4$ with much confidence.

15. The test would be more powerful (against the various values of $p \neq 0.5$).

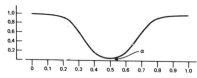

Chapter Review   (Pages 279–280)

1. $\mu_X = \mu_Y$ or $\mu_X - \mu_Y = 0$; that is, the null hypothesis is that the population means are the same.

3. $\alpha \doteq 0.438$ **5.** $\alpha \doteq 0.067$, $\beta \doteq 0.556$ **7.** $\alpha \doteq 0.002$

9. Our test is: reject the null hypothesis if the sample means differ by more than 2.21.

Table B Binomial Distributions $b(n, p; x)$

n	x	.05	.10	.20	p .25	.30	.40	.50
1	0	.9500	.9000	.8000	.7500	.7000	.6000	.5000
	1	.0500	.1000	.2000	.2500	.3000	.4000	.5000
2	0	.9025	.8100	.6400	.5625	.4900	.3600	.2500
	1	.0950	.1800	.3200	.3750	.4200	.4800	.5000
	2	.0025	.0100	.0400	.0625	.0900	.1600	.2500
3	0	.8574	.7290	.5120	.4219	.3430	.2160	.1250
	1	.1354	.2430	.3840	.4219	.4410	.4320	.3750
	2	.0071	.0270	.0960	.1406	.1890	.2880	.3750
	3	.0001	.0010	.0080	.0156	.0270	.0640	.1250
4	0	.8145	.6561	.4096	.3164	.2401	.1296	.0625
	1	.1715	.2916	.4096	.4219	.4116	.3456	.2500
	2	.0135	.0486	.1536	.2109	.2646	.3456	.3750
	3	.0005	.0036	.0256	.0469	.0756	.1536	.2500
	4	.0000	.0001	.0016	.0039	.0081	.0256	.0625
5	0	.7738	.5905	.3277	.2373	.1681	.0778	.0312
	1	.2036	.3280	.4096	.3955	.3602	.2592	.1562
	2	.0214	.0729	.2048	.2637	.3087	.3456	.3125
	3	.0011	.0081	.0512	.0879	.1323	.2304	.3125
	4	.0000	.0004	.0064	.0146	.0284	.0768	.1562
	5	.0000	.0000	.0003	.0010	.0024	.0102	.0312
6	0	.7351	.5314	.2621	.1780	.1176	.0467	.0156
	1	.2321	.3543	.3932	.3560	.3025	.1866	.0938
	2	.0305	.0984	.2458	.2966	.3241	.3110	.2344
	3	.0021	.0146	.0819	.1318	.1852	.2765	.3125
	4	.0001	.0012	.0154	.0330	.0595	.1382	.2344
	5	.0000	.0001	.0015	.0044	.0102	.0369	.0938
	6	.0000	.0000	.0001	.0002	.0007	.0041	.0156
7	0	.6983	.4783	.2097	.1335	.0824	.0280	.0078
	1	.2573	.3720	.3670	.3115	.2471	.1306	.0547
	2	.0406	.1240	.2753	.3115	.3177	.2613	.1641
	3	.0036	.0230	.1147	.1730	.2269	.2903	.2734
	4	.0002	.0026	.0287	.0577	.0972	.1935	.2734
	5	.0000	.0002	.0043	.0115	.0250	.0774	.1641
	6	.0000	.0000	.0004	.0013	.0036	.0172	.0547
	7	.0000	.0000	.0000	.0001	.0002	.0016	.0078
8	0	.6634	.4305	.1678	.1001	.0576	.0168	.0039
	1	.2793	.3826	.3355	.2670	.1977	.0896	.0312
	2	.0515	.1488	.2936	.3115	.2965	.2090	.1094
	3	.0054	.0331	.1468	.2076	.2541	.2787	.2188
	4	.0004	.0046	.0459	.0865	.1361	.2322	.2734
	5	.0000	.0004	.0092	.0231	.0467	.1239	.2188
	6	.0000	.0000	.0011	.0038	.0100	.0413	.1094
	7	.0000	.0000	.0001	.0004	.0012	.0079	.0312
	8	.0000	.0000	.0000	.0000	.0001	.0007	.0039

Table B Binomial Distributions (continued)

n	x	.05	.10	.20	p .25	.30	.40	.50
9	0	.6302	.3874	.1342	.0751	.0404	.0101	.0020
	1	.2985	.3874	.3020	.2253	.1556	.0605	.0176
	2	.0629	.1722	.3020	.3003	.2668	.1612	.0703
	3	.0077	.0446	.1762	.2336	.2668	.2508	.1641
	4	.0006	.0074	.0661	.1168	.1715	.2508	.2461
	5	.0000	.0008	.0165	.0389	.0735	.1672	.2461
	6	.0000	.0001	.0028	.0087	.0210	.0743	.1641
	7	.0000	.0000	.0003	.0012	.0039	.0212	.0703
	8	.0000	.0000	.0000	.0001	.0004	.0035	.0176
	9	.0000	.0000	.0000	.0000	.0000	.0003	.0020
10	0	.5987	.3487	.1074	.0563	.0282	.0060	.0010
	1	.3151	.3874	.2684	.1877	.1211	.0403	.0098
	2	.0746	.1937	.3020	.2816	.2335	.1209	.0439
	3	.0105	.0574	.2013	.2503	.2668	.2150	.1172
	4	.0010	.0112	.0881	.1460	.2001	.2508	.2051
	5	.0001	.0015	.0264	.0584	.1029	.2007	.2461
	6	.0000	.0001	.0055	.0162	.0368	.1115	.2051
	7	.0000	.0000	.0008	.0031	.0090	.0425	.1172
	8	.0000	.0000	.0001	.0004	.0014	.0106	.0439
	9	.0000	.0000	.0000	.0000	.0001	.0016	.0098
	10	.0000	.0000	.0000	.0000	.0000	.0001	.0010
11	0	.5688	.3138	.0859	.0422	.0198	.0036	.0004
	1	.3293	.3835	.2362	.1549	.0932	.0266	.0055
	2	.0867	.2131	.2953	.2581	.1998	.0887	.0269
	3	.0137	.0710	.2215	.2581	.2568	.1774	.0806
	4	.0014	.0158	.1107	.1721	.2201	.2365	.1611
	5	.0001	.0025	.0388	.0803	.1321	.2207	.2256
	6	.0000	.0003	.0097	.0268	.0566	.1471	.2256
	7	.0000	.0000	.0017	.0064	.0173	.0701	.1611
	8	.0000	.0000	.0002	.0011	.0037	.0234	.0806
	9	.0000	.0000	.0000	.0001	.0005	.0052	.0269
	10	.0000	.0000	.0000	.0000	.0000	.0007	.0054
	11	.0000	.0000	.0000	.0000	.0000	.0000	.0005
12	0	.5404	.2824	.0687	.0317	.0138	.0022	.0002
	1	.3413	.3766	.2062	.1267	.0712	.0174	.0029
	2	.0988	.2301	.2835	.2323	.1678	.0639	.0161
	3	.0173	.0852	.2362	.2581	.2397	.1419	.0537
	4	.0021	.0213	.1329	.1936	.2311	.2128	.1208
	5	.0002	.0038	.0532	.1032	.1585	.2270	.1934
	6	.0000	.0005	.0155	.0401	.0792	.1766	.2256
	7	.0000	.0000	.0033	.0115	.0291	.1009	.1934
	8	.0000	.0000	.0005	.0024	.0078	.0420	.1208
	9	.0000	.0000	.0001	.0004	.0015	.0125	.0537
	10	.0000	.0000	.0000	.0000	.0002	.0025	.0161
	11	.0000	.0000	.0000	.0000	.0000	.0003	.0029
	12	.0000	.0000	.0000	.0000	.0000	.0000	.0002

Table B Binomial Distributions (continued)

n	x	.05	.10	.20	p .25	.30	.40	.50
13	0	.5133	.2542	.0550	.0238	.0097	.0013	.0001
	1	.3512	.3672	.1787	.1029	.0540	.0113	.0016
	2	.1109	.2448	.2680	.2059	.1388	.0453	.0095
	3	.0214	.0997	.2457	.2517	.2181	.1107	.0349
	4	.0028	.0277	.1535	.2097	.2337	.1845	.0873
	5	.0003	.0055	.0691	.1258	.1803	.2214	.1571
	6	.0000	.0008	.0230	.0559	.1030	.1968	.2095
	7	.0000	.0001	.0058	.0186	.0442	.1312	.2095
	8	.0000	.0000	.0011	.0047	.0142	.0656	.1571
	9	.0000	.0000	.0001	.0009	.0034	.0243	.0873
	10	.0000	.0000	.0000	.0001	.0006	.0065	.0349
	11	.0000	.0000	.0000	.0000	.0001	.0012	.0095
	12	.0000	.0000	.0000	.0000	.0000	.0001	.0016
	13	.0000	.0000	.0000	.0000	.0000	.0000	.0001
14	0	.4877	.2288	.0440	.0178	.0068	.0008	.0001
	1	.3593	.3559	.1539	.0832	.0407	.0073	.0009
	2	.1229	.2570	.2501	.1802	.1134	.0317	.0056
	3	.0259	.1142	.2501	.2402	.1943	.0845	.0222
	4	.0037	.0349	.1720	.2202	.2290	.1549	.0611
	5	.0004	.0078	.0860	.1468	.1963	.2066	.1222
	6	.0000	.0013	.0322	.0734	.1262	.2066	.1833
	7	.0000	.0002	.0092	.0280	.0618	.1574	.2095
	8	.0000	.0000	.0020	.0082	.0232	.0918	.1833
	9	.0000	.0000	.0003	.0018	.0066	.0408	.1222
	10	.0000	.0000	.0000	.0003	.0014	.0136	.0611
	11	.0000	.0000	.0000	.0000	.0002	.0033	.0222
	12	.0000	.0000	.0000	.0000	.0000	.0005	.0056
	13	.0000	.0000	.0000	.0000	.0000	.0001	.0009
	14	.0000	.0000	.0000	.0000	.0000	.0000	.0001
15	0	.4633	.2059	.0352	.0134	.0047	.0005	.0000
	1	.3658	.3432	.1319	.0668	.0305	.0047	.0005
	2	.1348	.2669	.2309	.1559	.0916	.0219	.0032
	3	.0307	.1285	.2501	.2252	.1700	.0634	.0139
	4	.0049	.0428	.1876	.2252	.2186	.1268	.0417
	5	.0006	.0105	.1032	.1651	.2061	.1859	.0916
	6	.0000	.0019	.0430	.0917	.1472	.2066	.1527
	7	.0000	.0003	.0138	.0393	.0811	.1771	.1964
	8	.0000	.0000	.0035	.0131	.0348	.1181	.1964
	9	.0000	.0000	.0007	.0034	.0116	.0612	.1527
	10	.0000	.0000	.0001	.0007	.0030	.0245	.0916
	11	.0000	.0000	.0000	.0001	.0006	.0074	.0417
	12	.0000	.0000	.0000	.0000	.0001	.0016	.0139
	13	.0000	.0000	.0000	.0000	.0000	.0003	.0032
	14	.0000	.0000	.0000	.0000	.0000	.0000	.0005
	15	.0000	.0000	.0000	.0000	.0000	.0000	.0000

Table B Binomial Distributions (continued)

n	x	.05	.10	.20	p .25	.30	.40	.50
16	0	.4401	.1853	.0281	.0100	.0033	.0003	.0000
	1	.3706	.3294	.1126	.0535	.0228	.0030	.0002
	2	.1463	.2745	.2111	.1336	.0732	.0150	.0018
	3	.0359	.1423	.2463	.2079	.1465	.0468	.0085
	4	.0061	.0514	.2001	.2252	.2040	.1014	.0278
	5	.0008	.0137	.1201	.1802	.2099	.1623	.0667
	6	.0001	.0028	.0550	.1101	.1649	.1983	.1222
	7	.0000	.0004	.0197	.0524	.1010	.1889	.1746
	8	.0000	.0001	.0055	.0197	.0487	.1417	.1964
	9	.0000	.0000	.0012	.0058	.0185	.0840	.1746
	10	.0000	.0000	.0002	.0014	.0056	.0392	.1222
	11	.0000	.0000	.0000	.0002	.0013	.0142	.0667
	12	.0000	.0000	.0000	.0000	.0002	.0040	.0278
	13	.0000	.0000	.0000	.0000	.0000	.0008	.0085
	14	.0000	.0000	.0000	.0000	.0000	.0001	.0018
	15	.0000	.0000	.0000	.0000	.0000	.0000	.0002
	16	.0000	.0000	.0000	.0000	.0000	.0000	.0000
17	0	.4181	.1668	.0225	.0075	.0023	.0002	.0000
	1	.3741	.3150	.0957	.0426	.0169	.0019	.0001
	2	.1575	.2800	.1914	.1136	.0581	.0102	.0010
	3	.0415	.1556	.2393	.1893	.1245	.0341	.0052
	4	.9076	.0605	.2093	.2209	.1868	.0796	.0182
	5	.0010	.0175	.1361	.1914	.2081	.1379	.0472
	6	.0001	.0039	.0680	.1276	.1784	.1839	.0944
	7	.0000	.0007	.0267	.0668	.1201	.1927	.1484
	8	.0000	.0001	.0084	.0279	.0644	.1606	.1855
	9	.0000	.0000	.0021	.0093	.0276	.1070	.1855
	10	.0000	.0000	.0004	.0025	.0095	.0571	.1484
	11	.0000	.0000	.0001	.0005	.0026	.0242	.0944
	12	.0000	.0000	.0000	.0001	.0006	.0081	.0472
	13	.0000	.0000	.0000	.0000	.0001	.0021	.0182
	14	.0000	.0000	.0000	.0000	.0000	.0004	.0052
	15	.0000	.0000	.0000	.0000	.0000	.0001	.0010
	16	.0000	.0000	.0000	.0000	.0000	.0000	.0001
	17	.0000	.0000	.0000	.0000	.0000	.0000	.0000
18	0	.3972	.1501	.0180	.0056	.0016	.0001	.0000
	1	.3763	.3002	.0811	.0338	.0126	.0012	.0001
	2	.1683	.2835	.1723	.0958	.0458	.0069	.0006
	3	.0473	.1680	.2297	.1704	.1046	.0246	.0031
	4	.0093	.0700	.2153	.2130	.1681	.0614	.0117
	5	.0014	.0218	.1507	.1988	.2017	.1146	.0327
	6	.0002	.0052	.0816	.1436	.1873	.1655	.0708
	7	.0000	.0010	.0350	.0820	.1376	.1892	.1214
	8	.0000	.0002	.0120	.0376	.0811	.1734	.1669
	9	.0000	.0000	.0033	.0139	.0386	.1284	.1855
	10	.0000	.0000	.0008	.0042	.0149	.0771	.1669
	11	.0000	.0000	.0001	.0010	.0046	.0374	.1214

Table B Binomial Distributions (continued)

n	x	.05	.10	.20	.25	.30	.40	.50
18	12	.0000	.0000	.0000	.0002	.0012	.0145	.0708
	13	.0000	.0000	.0000	.0000	.0002	.0045	.0327
	14	.0000	.0000	.0000	.0000	.0000	.0011	.0117
	15	.0000	.0000	.0000	.0000	.0000	.0002	.0031
	16	.0000	.0000	.0000	.0000	.0000	.0000	.0006
	17	.0000	.0000	.0000	.0000	.0000	.0000	.0001
	18	.0000	.0000	.0000	.0000	.0000	.0000	.0000
19	0	.3774	.1351	.0144	.0042	.0011	.0001	.0000
	1	.3774	.2852	.0685	.0268	.0093	.0008	.0000
	2	.1787	.2852	.1540	.0803	.0358	.0046	.0003
	3	.0533	.1796	.2182	.1517	.0869	.0175	.0018
	4	.0112	.0798	.2182	.2023	.1491	.0467	.0074
	5	.0018	.0266	.1636	.2023	.1916	.0933	.0222
	6	.0002	.0069	.0955	.1574	.1916	.1451	.0518
	7	.0000	.0014	.0443	.0974	.1525	.1797	.0961
	8	.0000	.0002	.0166	.0487	.0981	.1797	.1442
	9	.0000	.0000	.0051	.0198	.0514	.1464	.1762
	10	.0000	.0000	.0013	.0066	.0220	.0976	.1762
	11	.0000	.0000	.0003	.0018	.0077	.0532	.1442
	12	.0000	.0000	.0000	.0004	.0022	.0237	.0961
	13	.0000	.0000	.0000	.0001	.0005	.0085	.0518
	14	.0000	.0000	.0000	.0000	.0001	.0024	.0222
	15	.0000	.0000	.0000	.0000	.0000	.0005	.0074
	16	.0000	.0000	.0000	.0000	.0000	.0001	.0018
	17	.0000	.0000	.0000	.0000	.0000	.0000	.0003
	18	.0000	.0000	.0000	.0000	.0000	.0000	.0000
	19	.0000	.0000	.0000	.0000	.0000	.0000	.0000
20	0	.3585	.1216	.0115	.0032	.0008	.0000	.0000
	1	.3774	.2702	.0576	.0211	.0068	.0005	.0000
	2	.1887	.2852	.1369	.0669	.0278	.0031	.0002
	3	.0596	.1901	.2054	.1339	.0716	.0123	.0011
	4	.0133	.0898	.2182	.1897	.1304	.0350	.0046
	5	.0022	.0319	.1746	.2023	.1789	.0746	.0148
	6	.0003	.0089	.1091	.1686	.1916	.1244	.0370
	7	.0000	.0020	.0545	.1124	.1643	.1659	.0739
	8	.0000	.0004	.0222	.0609	.1144	.1797	.1201
	9	.0000	.0001	.0074	.0271	.0654	.1597	.1602
	10	.0000	.0000	.0020	.0099	.0308	.1171	.1762
	11	.0000	.0000	.0005	.0030	.0120	.0710	.1602
	12	.0000	.0000	.0001	.0008	.0039	.0355	.1201
	13	.0000	.0000	.0000	.0002	.0010	.0146	.0739
	14	.0000	.0000	.0000	.0000	.0002	.0049	.0370
	15	.0000	.0000	.0000	.0000	.0000	.0013	.0148
	16	.0000	.0000	.0000	.0000	.0000	.0003	.0046
	17	.0000	.0000	.0000	.0000	.0000	.0000	.0011
	18	.0000	.0000	.0000	.0000	.0000	.0000	.0002
	19	.0000	.0000	.0000	.0000	.0000	.0000	.0000
	20	.0000	.0000	.0000	.0000	.0000	.0000	.0000

Table N The Cumulative Standard Normal Distribution

−0.09	−0.08	−0.07	−0.06	−0.05	−0.04	−0.03	−0.02	−0.01	0.00	z
									0.0013	−3.0
0.0014	0.0014	0.0015	0.0015	0.0016	0.0016	0.0017	0.0017	0.0018	0.0019	−2.9
.0019	.0020	.0021	.0021	.0022	.0023	.0023	.0024	.0025	.0026	−2.8
.0026	.0027	.0028	.0029	.0030	.0031	.0032	.0033	.0034	.0035	−2.7
.0036	.0037	.0038	.0039	.0040	.0041	.0043	.0044	.0045	.0047	−2.6
.0048	.0049	.0051	.0052	.0054	.0055	.0057	.0059	.0060	.0062	−2.5
.0064	.0066	.0068	.0069	.0071	.0073	.0075	.0078	.0080	.0082	−2.4
.0084	.0087	.0089	.0091	.0094	.0096	.0099	.0102	.0104	.0107	−2.3
.0110	.0113	.0116	.0119	.0122	.0125	.0129	.0132	.0136	.0139	−2.2
.0143	.0146	.0150	.0154	.0158	.0162	.0166	.0170	.0174	.0179	−2.1
.0183	.0188	.0192	.0197	.0202	.0207	.0212	.0217	.0222	.0227	−2.0
0.0233	0.0239	0.0244	0.0250	0.0256	0.0262	0.0268	0.0274	0.0281	0.0287	−1.9
.0294	.0301	.0307	.0314	.0322	.0329	.0336	.0344	.0351	.0359	−1.8
.0367	.0375	.0384	.0392	.0401	.0409	.0418	.0427	.0436	.0446	−1.7
.0455	.0465	.0475	.0485	.0495	.0505	.0516	.0526	.0537	.0548	−1.6
.0559	.0571	.0582	.0594	.0606	.0618	.0630	.0643	.0655	.0668	−1.5
.0681	.0694	.0708	.0721	.0735	.0749	.0764	.0778	.0793	.0808	−1.4
.0823	.0838	.0853	.0869	.0885	.0901	.0918	.0934	.0951	.0968	−1.3
.0985	.1003	.1020	.1038	.1056	.1075	.1093	.1112	.1131	.1151	−1.2
.1170	.1190	.1210	.1230	.1251	.1271	.1292	.1314	.1335	.1357	−1.1
.1379	.1401	.1423	.1446	.1469	.1492	.1515	.1539	.1562	.1587	−1.0
0.1611	0.1635	0.1660	0.1685	0.1711	0.1736	0.1762	0.1788	0.1814	0.1841	−0.9
.1867	.1894	.1921	.1949	.1977	.2005	.2033	.2061	.2090	.2119	−0.8
.2148	.2177	.2206	.2236	.2266	.2296	.2327	.2358	.2389	.2420	−0.7
.2451	.2483	.2514	.2546	.2578	.2611	.2643	.2676	.2709	.2743	−0.6
.2776	.2810	.2843	.2877	.2912	.2946	.2981	.3015	.3050	.3085	−0.5
.3121	.3156	.3192	.3228	.3264	.3300	.3336	.3372	.3409	.3446	−0.4
.3483	.3520	.3557	.3594	.3632	.3669	.3707	.3745	.3783	.3821	−0.3
.3859	.3897	.3936	.3974	.4013	.4052	.4090	.4129	.4168	.4207	−0.2
.4247	.4286	.4325	.4364	.4404	.4443	.4483	.4522	.4562	.4602	−0.1
.4641	.4681	.4721	.4761	.4801	.4840	.4880	.4920	.4960	.5000	0.0

z	0.00	0.01	0.02	0.03	0.04	0.05	0.06	0.07	0.08	0.09
0.0	0.5000	0.5040	0.5080	0.5120	0.5160	0.5199	0.5239	0.5279	0.5319	0.5359
0.1	.5398	.5438	.5478	.5517	.5557	.5596	.5636	.5675	.5714	.5753
0.2	.5793	.5832	.5871	.5910	.5948	.5987	.6026	.6064	.6103	.6141
0.3	.6179	.6217	.6255	.6293	.6331	.6368	.6406	.6443	.6480	.6517
0.4	.6554	.6591	.6628	.6664	.6700	.6736	.6772	.6808	.6844	.6879
0.5	.6915	.6950	.6985	.7019	.7054	.7088	.7123	.7157	.7190	.7224
0.6	.7257	.7291	.7324	.7357	.7389	.7422	.7454	.7486	.7517	.7549
0.7	.7580	.7611	.7642	.7673	.7704	.7734	.7764	.7794	.7823	.7852
0.8	.7881	.7910	.7939	.7967	.7995	.8023	.8051	.8079	.8106	.8133
0.9	.8159	.8186	.8212	.8238	.8264	.8289	.8315	.8340	.8365	.8389
1.0	0.8413	0.8438	0.8461	0.8485	0.8508	0.8531	0.8554	0.8577	0.8599	0.8621
1.1	.8643	.8665	.8686	.8708	.8729	.8749	.8770	.8790	.8810	.8830
1.2	.8849	.8869	.8888	.8907	.8925	.8944	.8962	.8980	.8997	.9015
1.3	.9032	.9049	.9066	.9082	.9099	.9115	.9131	.9147	.9162	.9177
1.4	.9192	.9207	.9222	.9236	.9251	.9265	.9279	.9292	.9306	.9319
1.5	.9332	.9345	.9357	.9370	.9382	.9394	.9406	.9418	.9429	.9441
1.6	.9452	.9463	.9474	.9484	.9495	.9505	.9515	.9525	.9535	.9545
1.7	.9554	.9564	.9573	.9582	.9591	.9599	.9608	.9616	.9625	.9633
1.8	.9641	.9649	.9656	.9664	.9671	.9678	.9686	.9693	.9699	.9706
1.9	.9713	.9719	.9726	.9732	.9738	.9744	.9750	.9756	.9761	.9767
2.0	0.9773	0.9778	0.9783	0.9788	0.9793	0.9798	0.9803	0.9808	0.9812	0.9817
2.1	.9821	.9826	.9830	.9834	.9838	.9842	.9846	.9850	.9854	.9857
2.2	.9861	.9864	.9868	.9871	.9875	.9878	.9881	.9884	.9887	.9890
2.3	.9893	.9896	.9898	.9901	.9904	.9906	.9909	.9911	.9913	.9916
2.4	.9918	.9920	.9922	.9925	.9927	.9929	.9931	.9932	.9934	.9936
2.5	.9938	.9940	.9941	.9943	.9945	.9946	.9948	.9949	.9951	.9952
2.6	.9953	.9955	.9956	.9957	.9959	.9960	.9961	.9962	.9963	.9964
2.7	.9965	.9966	.9967	.9968	.9969	.9970	.9971	.9972	.9973	.9974
2.8	.9974	.9975	.9976	.9977	.9977	.9978	.9979	.9979	.9980	.9981
2.9	.9981	.9982	.9983	.9983	.9984	.9984	.9985	.9985	.9986	.9986
3.0	0.9987									

Table P   Poisson Distributions

Entries in this table are the values of

$$\frac{e^{-\lambda} \lambda^x}{x!}$$

for the indicated values of x and λ.

x \ λ	0.1	0.2	0.3	0.4	0.5	0.6	0.7	0.8	0.9	1.0
0	.9048	.8187	.7408	.6703	.6065	.5488	.4966	.4493	.4066	.3679
1	.0905	.1637	.2222	.2681	.3033	.3293	.3476	.3595	.3659	.3679
2	.0045	.0164	.0333	.0536	.0758	.0988	.1217	.1438	.1647	.1839
3	.0002	.0011	.0033	.0072	.0126	.0198	.0284	.0383	.0494	.0613
4	.0000	.0001	.0002	.0007	.0016	.0030	.0050	.0077	.0111	.0153
5	.0000	.0000	.0000	.0001	.0002	.0004	.0007	.0012	.0020	.0031
6	.0000	.0000	.0000	.0000	.0000	.0000	.0001	.0002	.0003	.0005
7	.0000	.0000	.0000	.0000	.0000	.0000	.0000	.0000	.0000	.0001

x \ λ	1.1	1.2	1.3	1.4	1.5	1.6	1.7	1.8	1.9	2.0
0	.3329	.3012	.2725	.2466	.2231	.2019	.1827	.1653	.1496	.1353
1	.3662	.3614	.3543	.3452	.3347	.3230	.3106	.2975	.2842	.2707
2	.2014	.2169	.2303	.2417	.2510	.2584	.2640	.2678	.2700	.2707
3	.0738	.0867	.0998	.1128	.1255	.1378	.1496	.1607	.1710	.1804
4	.0203	.0260	.0324	.0395	.0471	.0551	.0636	.0723	.0812	.0902
5	.0045	.0062	.0084	.0111	.0141	.0176	.0216	.0260	.0309	.0361
6	.0008	.0012	.0018	.0026	.0035	.0047	.0061	.0078	.0098	.0120
7	.0001	.0002	.0003	.0005	.0008	.0011	.0015	.0020	.0027	.0034
8	.0000	.0000	.0001	.0001	.0001	.0002	.0003	.0005	.0006	.0009
9	.0000	.0000	.0000	.0000	.0000	.0000	.0001	.0001	.0001	.0002

x \ λ	2.1	2.2	2.3	2.4	2.5	2.6	2.7	2.8	2.9	3.0
0	.1225	.1108	.1003	.0907	.0821	.0743	.0672	.0608	.0550	.0498
1	.2572	.2438	.2306	.2177	.2052	.1931	.1815	.1703	.1596	.1494
2	.2700	.2681	.2652	.2613	.2565	.2510	.2450	.2384	.2314	.2240
3	.1890	.1966	.2033	.2090	.2138	.2176	.2205	.2225	.2237	.2240
4	.0992	.1082	.1169	.1254	.1336	.1414	.1488	.1557	.1622	.1680
5	.0417	.0476	.0538	.0602	.0668	.0735	.0804	.0872	.0940	.1008
6	.0146	.0174	.0206	.0241	.0278	.0319	.0362	.0407	.0455	.0504
7	.0044	.0055	.0068	.0083	.0099	.0118	.0139	.0163	.0188	.0216
8	.0011	.0015	.0019	.0025	.0031	.0038	.0047	.0057	.0068	.0081
9	.0003	.0004	.0005	.0007	.0009	.0011	.0014	.0018	.0022	.0027
10	.0001	.0001	.0001	.0002	.0002	.0003	.0004	.0005	.0006	.0008
11	.0000	.0000	.0000	.0000	.0000	.0001	.0001	.0001	.0002	.0002
12	.0000	.0000	.0000	.0000	.0000	.0000	.0000	.0000	.0000	.0001

Table P Poisson Distributions (continued)

λ

x	3.1	3.2	3.3	3.4	3.5	3.6	3.7	3.8	3.9	4.0
0	.0450	.0408	.0369	.0334	.0302	.0273	.0247	.0224	.0202	.0183
1	.1397	.1304	.1217	.1135	.1057	.0984	.0915	.0850	.0789	.0733
2	.2165	.2087	.2008	.1929	.1850	.1771	.1692	.1615	.1539	.1465
3	.2237	.2226	.2209	.2186	.2158	.2125	.2087	.2046	.2001	.1954
4	.1734	.1781	.1823	.1858	.1888	.1912	.1931	.1944	.1951	.1954
5	.1075	.1140	.1203	.1264	.1322	.1377	.1429	.1477	.1522	.1563
6	.0555	.0608	.0662	.0716	.0771	.0826	.0881	.0936	.0989	.1042
7	.0246	.0278	.0312	.0348	.0385	.0425	.0466	.0508	.0551	.0595
8	.0095	.0111	.0129	.0148	.0169	.0191	.0215	.0241	.0269	.0298
9	.0033	.0040	.0047	.0056	.0066	.0076	.0089	.0102	.0116	.0132
10	.0010	.0013	.0016	.0019	.0023	.0028	.0033	.0039	.0045	.0053
11	.0003	.0004	.0005	.0006	.0007	.0009	.0011	.0013	.0016	.0019
12	.0001	.0001	.0001	.0002	.0002	.0003	.0003	.0004	.0005	.0006
13	.0000	.0000	.0000	.0000	.0001	.0001	.0001	.0001	.0002	.0002
14	.0000	.0000	.0000	.0000	.0000	.0000	.0000	.0000	.0000	.0001

x	4.1	4.2	4.3	4.4	4.5	4.6	4.7	4.8	4.9	5.0
0	.0166	.0150	.0136	.0123	.0111	.0101	.0091	.0082	.0074	.0067
1	.0679	.0630	.0583	.0540	.0500	.0462	.0427	.0395	.0365	.0337
2	.1393	.1323	.1254	.1188	.1125	.1063	.1005	.0948	.0894	.0842
3	.1904	.1852	.1798	.1743	.1687	.1631	.1574	.1517	.1460	.1404
4	.1951	.1944	.1933	.1917	.1898	.1875	.1849	.1820	.1789	.1755
5	.1600	.1633	.1662	.1687	.1708	.1725	.1738	.1747	.1753	.1755
6	.1093	.1143	.1191	.1237	.1281	.1323	.1362	.1398	.1432	.1462
7	.0640	.0686	.0732	.0778	.0824	.0869	.0914	.0959	.1002	.1044
8	.0328	.0360	.0393	.0428	.0463	.0500	.0537	.0575	.0614	.0653
9	.0150	.0168	.0188	.0209	.0232	.0255	.0280	.0307	.0334	.0363
10	.0061	.0071	.0081	.0092	.0104	.0118	.0132	.0147	.0164	.0181
11	.0023	.0027	.0032	.0037	.0043	.0049	.0056	.0064	.0073	.0082
12	.0008	.0009	.0011	.0014	.0016	.0019	.0022	.0026	.0030	.0034
13	.0002	.0003	.0004	.0005	.0006	.0007	.0008	.0009	.0011	.0013
14	.0001	.0001	.0001	.0001	.0002	.0002	.0003	.0003	.0004	.0005
15	.0000	.0000	.0000	.0000	.0001	.0001	.0001	.0001	.0001	.0002

The preceding tables (B, N, and P) have been adapted, with permission, from corresponding tables in *CRC Handbook of Tables for Mathematics*, 4th ed. (Cleveland, Ohio: The Chemical Rubber Company, 1970).

Table R Squares and Square Roots

N	N²	√N	√10N	N	N²	√N	√10N	N	N²	√N	√10N
1.00	1.0000	1.00000	3.16228	**1.60**	2.5600	1.26491	4.00000	**2.20**	4.8400	1.48324	4.69042
1.01	1.0201	1.00499	3.17805	1.61	2.5921	1.26886	4.01248	2.21	4.8841	1.48661	4.70106
1.02	1.0404	1.00995	3.19374	1.62	2.6244	1.27279	4.02492	2.22	4.9284	1.48997	4.71169
1.03	1.0609	1.01489	3.20936	1.63	2.6569	1.27671	4.03733	2.23	4.9729	1.49332	4.72229
1.04	1.0816	1.01980	3.22490	1.64	2.6896	1.28062	4.04969	2.24	5.0176	1.49666	4.73286
1.05	1.1025	1.02470	3.24037	**1.65**	2.7225	1.28452	4.06202	**2.25**	5.0625	1.50000	4.74342
1.06	1.1236	1.02956	3.25576	1.66	2.7556	1.28841	4.07431	2.26	5.1076	1.50333	4.75395
1.07	1.1449	1.03441	3.27109	1.67	2.7889	1.29228	4.08656	2.27	5.1529	1.50665	4.76445
1.08	1.1664	1.03923	3.28634	1.68	2.8224	1.29615	4.09878	2.28	5.1984	1.50997	4.77493
1.09	1.1881	1.04403	3.30151	1.69	2.8561	1.30000	4.11096	2.29	5.2441	1.51327	4.78539
1.10	1.2100	1.04881	3.31662	**1.70**	2.8900	1.30384	4.12311	**2.30**	5.2900	1.51658	4.79583
1.11	1.2321	1.05357	3.33167	1.71	2.9241	1.30767	4.13521	2.31	5.3361	1.51987	4.80625
1.12	1.2544	1.05830	3.34664	1.72	2.9584	1.31149	4.14729	2.32	5.3824	1.52315	4.81664
1.13	1.2769	1.06301	3.36155	1.73	2.9929	1.31529	4.15933	2.33	5.4289	1.52643	4.82701
1.14	1.2996	1.06771	3.37639	1.74	3.0276	1.31909	4.17133	2.34	5.4756	1.52971	4.83735
1.15	1.3225	1.07238	3.39116	**1.75**	3.0625	1.32288	4.18330	**2.35**	5.5225	1.53297	4.84768
1.16	1.3456	1.07703	3.40588	1.76	3.0976	1.32665	4.19524	2.36	5.5696	1.53623	4.85798
1.17	1.3689	1.08167	3.42053	1.77	3.1329	1.33041	4.20714	2.37	5.6169	1.53948	4.86826
1.18	1.3924	1.08628	3.43511	1.78	3.1684	1.33417	4.21900	2.38	5.6644	1.54272	4.87852
1.19	1.4161	1.09087	3.44964	1.79	3.2041	1.33791	4.23084	2.39	5.7121	1.54596	4.88876
1.20	1.4400	1.09545	3.46410	**1.80**	3.2400	1.34164	4.24264	**2.40**	5.7600	1.54919	4.89898
1.21	1.4641	1.10000	3.47851	1.81	3.2761	1.34536	4.25441	2.41	5.8081	1.55242	4.90918
1.22	1.4884	1.10454	3.49285	1.82	3.3124	1.34907	4.26615	2.42	5.8564	1.55563	4.91935
1.23	1.5129	1.10905	3.50714	1.83	3.3489	1.35277	4.27785	2.43	5.9049	1.55885	4.92950
1.24	1.5376	1.11355	3.52136	1.84	3.3856	1.35647	4.28952	2.44	5.9536	1.56205	4.93964
1.25	1.5625	1.11803	3.53553	**1.85**	3.4225	1.36015	4.30116	**2.45**	6.0025	1.56525	4.94975
1.26	1.5876	1.12250	3.54965	1.86	3.4596	1.36382	4.31277	2.46	6.0516	1.56844	4.95984
1.27	1.6129	1.12694	3.56371	1.87	3.4969	1.36748	4.32435	2.47	6.1009	1.57162	4.96991
1.28	1.6384	1.13137	3.57771	1.88	3.5344	1.37113	4.33590	2.48	6.1504	1.57480	4.97996
1.29	1.6641	1.13578	3.59166	1.89	3.5721	1.37477	4.34741	2.49	6.2001	1.57797	4.98999
1.30	1.6900	1.14018	3.60555	**1.90**	3.6100	1.37840	4.35890	**2.50**	6.2500	1.58114	5.00000
1.31	1.7161	1.14455	3.61939	1.91	3.6481	1.38203	4.37035	2.51	6.3001	1.58430	5.00999
1.32	1.7424	1.14891	3.63318	1.92	3.6864	1.38564	4.38178	2.52	6.3504	1.58745	5.01996
1.33	1.7689	1.15326	3.64692	1.93	3.7249	1.38924	4.39318	2.53	6.4009	1.59060	5.02991
1.34	1.7956	1.15758	3.66060	1.94	3.7636	1.39284	4.40454	2.54	6.4516	1.59374	5.03984
1.35	1.8225	1.16190	3.67423	**1.95**	3.8025	1.39642	4.41588	**2.55**	6.5025	1.59687	5.04975
1.36	1.8496	1.16619	3.68782	1.96	3.8416	1.40000	4.42719	2.56	6.5536	1.60000	5.05964
1.37	1.8769	1.17047	3.70135	1.97	3.8809	1.40357	4.43847	2.57	6.6049	1.60312	5.06952
1.38	1.9044	1.17473	3.71484	1.98	3.9204	1.40712	4.44972	2.58	6.6564	1.60624	5.07937
1.39	1.9321	1.17898	3.72827	1.99	3.9601	1.41067	4.46094	2.59	6.7081	1.60935	5.08920
1.40	1.9600	1.18322	3.74166	**2.00**	4.0000	1.41421	4.47214	**2.60**	6.7600	1.61245	5.09902
1.41	1.9881	1.18743	3.75500	2.01	4.0401	1.41774	4.48330	2.61	6.8121	1.61555	5.10882
1.42	2.0164	1.19164	3.76829	2.02	4.0804	1.42127	4.49444	2.62	6.8644	1.61864	5.11859
1.43	2.0449	1.19583	3.78153	2.03	4.1209	1.42478	4.50555	2.63	6.9169	1.62173	5.12835
1.44	2.0736	1.20000	3.79473	2.04	4.1616	1.42829	4.51664	2.64	6.9696	1.62481	5.13809
1.45	2.1025	1.20416	3.80789	**2.05**	4.2025	1.43178	4.52769	**2.65**	7.0225	1.62788	5.14782
1.46	2.1316	1.20830	3.82099	2.06	4.2436	1.43527	4.53872	2.66	7.0756	1.63095	5.15752
1.47	2.1609	1.21244	3.83406	2.07	4.2849	1.43875	4.54973	2.67	7.1289	1.63401	5.16720
1.48	2.1904	1.21655	3.84708	2.08	4.3264	1.44222	4.56070	2.68	7.1824	1.63707	5.17687
1.49	2.2201	1.22066	3.86005	2.09	4.3681	1.44568	4.57165	2.69	7.2361	1.64012	5.18652
1.50	2.2500	1.22474	3.87298	**2.10**	4.4100	1.44914	4.58258	**2.70**	7.2900	1.64317	5.19615
1.51	2.2801	1.22882	3.88587	2.11	4.4521	1.45258	4.59347	2.71	7.3441	1.64621	5.20577
1.52	2.3104	1.23288	3.89872	2.12	4.4944	1.45602	4.60435	2.72	7.3984	1.64924	5.21536
1.53	2.3409	1.23693	3.91152	2.13	4.5369	1.45945	4.61519	2.73	7.4529	1.65227	5.22494
1.54	2.3716	1.24097	3.92428	2.14	4.5796	1.46287	4.62601	2.74	7.5076	1.65529	5.23450
1.55	2.4025	1.24499	3.93700	**2.15**	4.6225	1.46629	4.63681	**2.75**	7.5625	1.65831	5.24404
1.56	2.4336	1.24900	3.94968	2.16	4.6656	1.46969	4.64758	2.76	7.6176	1.66132	5.25357
1.57	2.4649	1.25300	3.96232	2.17	4.7089	1.47309	4.65833	2.77	7.6729	1.66433	5.26308
1.58	2.4964	1.25698	3.97492	2.18	4.7524	1.47648	4.66905	2.78	7.7284	1.66733	5.27257
1.59	2.5281	1.26095	3.98748	2.19	4.7961	1.47986	4.67974	2.79	7.7841	1.67033	5.28205
1.60	2.5600	1.26491	4.00000	**2.20**	4.8400	1.48324	4.69042	**2.80**	7.8400	1.67332	5.29150
N	N²	√N	√10N	N	N²	√N	√10N	N	N²	√N	√10N

Reprinted by permission of the publisher from W. L. Hart's *Modern Plane Trigonometry* (Lexington, Mass.: D. C. Heath and Company, 1961).

Table R Squares and Square Roots (continued)

N	N^2	$\sqrt{N}$	$\sqrt{10N}$	N	N^2	$\sqrt{N}$	$\sqrt{10N}$	N	N^2	$\sqrt{N}$	$\sqrt{10N}$
2.80	7.8400	1.67332	5.29150	3.40	11.5600	1.84391	5.83095	4.00	16.0000	2.00000	6.32456
2.81	7.8961	1.67631	5.30094	3.41	11.6281	1.84662	5.83952	4.01	16.0801	2.00250	6.33246
2.82	7.9524	1.67929	5.31037	3.42	11.6964	1.84932	5.84808	4.02	16.1604	2.00499	6.34035
2.83	8.0089	1.68226	5.31977	3.43	11.7649	1.85203	5.85662	4.03	16.2409	2.00749	6.34823
2.84	8.0656	1.68523	5.32917	3.44	11.8336	1.85472	5.86515	4.04	16.3216	2.00998	6.35610
2.85	8.1225	1.68819	5.33854	3.45	11.9025	1.85742	5.87367	4.05	16.4025	2.01246	6.36396
2.86	8.1796	1.69115	5.34790	3.46	11.9716	1.86011	5.88218	4.06	16.4836	2.01494	6.37181
2.87	8.2369	1.69411	5.35724	3.47	12.0409	1.86279	5.89067	4.07	16.5649	2.01742	6.37966
2.88	8.2944	1.69706	5.36656	3.48	12.1104	1.86548	5.89915	4.08	16.6464	2.01990	6.38749
2.89	8.3521	1.70000	5.37587	3.49	12.1801	1.86815	5.90762	4.09	16.7281	2.02237	6.39531
2.90	8.4100	1.70294	5.38516	3.50	12.2500	1.87083	5.91608	4.10	16.8100	2.02485	6.40312
2.91	8.4681	1.70587	5.39444	3.51	12.3201	1.87350	5.92453	4.11	16.8921	2.02731	6.41093
2.92	8.5264	1.70880	5.40370	3.52	12.3904	1.87617	5.93296	4.12	16.9744	2.02978	6.41872
2.93	8.5849	1.71172	5.41295	3.53	12.4609	1.87883	5.94138	4.13	17.0569	2.03224	6.42651
2.94	8.6436	1.71464	5.42218	3.54	12.5316	1.88149	5.94979	4.14	17.1396	2.03470	6.43428
2.95	8.7025	1.71756	5.43139	3.55	12.6025	1.88414	5.95819	4.15	17.2225	2.03715	6.44205
2.96	8.7616	1.72047	5.44059	3.56	12.6736	1.88680	5.96657	4.16	17.3056	2.03961	6.44981
2.97	8.8209	1.72337	5.44977	3.57	12.7449	1.88944	5.97495	4.17	17.3889	2.04206	6.45755
2.98	8.8804	1.72627	5.45894	3.58	12.8164	1.89209	5.98331	4.18	17.4724	2.04450	6.46529
2.99	8.9401	1.72916	5.46809	3.59	12.8881	1.89473	5.99166	4.19	17.5561	2.04695	6.47302
3.00	9.0000	1.73205	5.47723	3.60	12.9600	1.89737	6.00000	4.20	17.6400	2.04939	6.48074
3.01	9.0601	1.73494	5.48635	3.61	13.0321	1.90000	6.00833	4.21	17.7241	2.05183	6.48845
3.02	9.1204	1.73781	5.49545	3.62	13.1044	1.90263	6.01664	4.22	17.8084	2.05426	6.49615
3.03	9.1809	1.74069	5.50454	3.63	13.1769	1.90526	6.02495	4.23	17.8929	2.05670	6.50384
3.04	9.2416	1.74356	5.51362	3.64	13.2496	1.90788	6.03324	4.24	17.9776	2.05913	6.51153
3.05	9.3025	1.74642	5.52268	3.65	13.3225	1.91050	6.04152	4.25	18.0625	2.06155	6.51920
3.06	9.3636	1.74929	5.53173	3.66	13.3956	1.91311	6.04979	4.26	18.1476	2.06398	6.52687
3.07	9.4249	1.75214	5.54076	3.67	13.4689	1.91572	6.05805	4.27	18.2329	2.06640	6.53452
3.08	9.4864	1.75499	5.54977	3.68	13.5424	1.91833	6.06630	4.28	18.3184	2.06882	6.54217
3.09	9.5481	1.75784	5.55878	3.69	13.6161	1.92094	6.07454	4.29	18.4041	2.07123	6.54981
3.10	9.6100	1.76068	5.56776	3.70	13.6900	1.92354	6.08276	4.30	18.4900	2.07364	6.55744
3.11	9.6721	1.76352	5.57674	3.71	13.7641	1.92614	6.09098	4.31	18.5761	2.07605	6.56506
3.12	9.7344	1.76635	5.58570	3.72	13.8384	1.92873	6.09918	4.32	18.6624	2.07846	6.57267
3.13	9.7969	1.76918	5.59464	3.73	13.9129	1.93132	6.10737	4.33	18.7489	2.08087	6.58027
3.14	9.8596	1.77200	5.60357	3.74	13.9876	1.93391	6.11555	4.34	18.8356	2.08327	6.58787
3.15	9.9225	1.77482	5.61249	3.75	14.0625	1.93649	6.12372	4.35	18.9225	2.08567	6.59545
3.16	9.9856	1.77764	5.62139	3.76	14.1376	1.93907	6.13188	4.36	19.0096	2.08806	6.60303
3.17	10.0489	1.78045	5.63028	3.77	14.2129	1.94165	6.14003	4.37	19.0969	2.09045	6.61060
3.18	10.1124	1.78326	5.63915	3.78	14.2884	1.94422	6.14817	4.38	19.1844	2.09284	6.61816
3.19	10.1761	1.78606	5.64801	3.79	14.3641	1.94679	6.15630	4.39	19.2721	2.09523	6.62571
3.20	10.2400	1.78885	5.65685	3.80	14.4400	1.94936	6.16441	4.40	19.3600	2.09762	6.63325
3.21	10.3041	1.79165	5.66569	3.81	14.5161	1.95192	6.17252	4.41	19.4481	2.10000	6.64078
3.22	10.3684	1.79444	5.67450	3.82	14.5924	1.95448	6.18061	4.42	19.5364	2.10238	6.64831
3.23	10.4329	1.79722	5.68331	3.83	14.6689	1.95704	6.18870	4.43	19.6249	2.10476	6.65582
3.24	10.4976	1.80000	5.69210	3.84	14.7456	1.95959	6.19677	4.44	19.7136	2.10713	6.66333
3.25	10.5625	1.80278	5.70088	3.85	14.8225	1.96214	6.20484	4.45	19.8025	2.10950	6.67083
3.26	10.6276	1.80555	5.70964	3.86	14.8996	1.96469	6.21289	4.46	19.8916	2.11187	6.67832
3.27	10.6929	1.80831	5.71839	3.87	14.9769	1.96723	6.22093	4.47	19.9809	2.11424	6.68581
3.28	10.7584	1.81108	5.72713	3.88	15.0544	1.96977	6.22896	4.48	20.0704	2.11660	6.69328
3.29	10.8241	1.81384	5.73585	3.89	15.1321	1.97231	6.23699	4.49	20.1601	2.11896	6.70075
3.30	10.8900	1.81659	5.74456	3.90	15.2100	1.97484	6.24500	4.50	20.2500	2.12132	6.70820
3.31	10.9561	1.81934	5.75326	3.91	15.2881	1.97737	6.25300	4.51	20.3401	2.12368	6.71565
3.32	11.0224	1.82209	5.76194	3.92	15.3664	1.97990	6.26099	4.52	20.4304	2.12603	6.72309
3.33	11.0889	1.82483	5.77062	3.93	15.4449	1.98242	6.26897	4.53	20.5209	2.12838	6.73053
3.34	11.1556	1.82757	5.77927	3.94	15.5236	1.98494	6.27694	4.54	20.6116	2.13073	6.73795
3.35	11.2225	1.83030	5.78792	3.95	15.6025	1.98746	6.28490	4.55	20.7025	2.13307	6.74537
3.36	11.2896	1.83303	5.79655	3.96	15.6816	1.98997	6.29285	4.56	20.7936	2.13542	6.75278
3.37	11.3569	1.83576	5.80517	3.97	15.7609	1.99249	6.30079	4.57	20.8849	2.13776	6.76018
3.38	11.4244	1.83848	5.81378	3.98	15.8404	1.99499	6.30872	4.58	20.9764	2.14009	6.76757
3.39	11.4921	1.84120	5.82237	3.99	15.9201	1.99750	6.31664	4.59	21.0681	2.14243	6.77495
3.40	11.5600	1.84391	5.83095	4.00	16.0000	2.00000	6.32456	4.60	21.1600	2.14476	6.78233
N	N^2	$\sqrt{N}$	$\sqrt{10N}$	N	N^2	$\sqrt{N}$	$\sqrt{10N}$	N	N^2	$\sqrt{N}$	$\sqrt{10N}$

Table R Squares and Square Roots (continued)

N	N^2	$\sqrt{N}$	$\sqrt{10N}$	N	N^2	$\sqrt{N}$	$\sqrt{10N}$	N	N^2	$\sqrt{N}$	$\sqrt{10N}$
4.60	21.1600	2.14476	6.78233	5.20	27.0400	2.28035	7.21110	5.80	33.6400	2.40832	7.61577
4.61	21.2521	2.14709	6.78970	5.21	27.1441	2.28254	7.21803	5.81	33.7561	2.41039	7.62234
4.62	21.3444	2.14942	6.79706	5.22	27.2484	2.28473	7.22496	5.82	33.8724	2.41247	7.62889
4.63	21.4369	2.15174	6.80441	5.23	27.3529	2.28692	7.23187	5.83	33.9889	2.41454	7.63544
4.64	21.5296	2.15407	6.81175	5.24	27.4576	2.28910	7.23878	5.84	34.1056	2.41661	7.64199
4.65	21.6225	2.15639	6.81909	5.25	27.5625	2.29129	7.24569	5.85	34.2225	2.41868	7.64853
4.66	21.7156	2.15870	6.82642	5.26	27.6676	2.29347	7.25259	5.86	34.3396	2.42074	7.65506
4.67	21.8089	2.16102	6.83374	5.27	27.7729	2.29565	7.25948	5.87	34.4569	2.42281	7.66159
4.68	21.9024	2.16333	6.84105	5.28	27.8784	2.29783	7.26636	5.88	34.5744	2.42487	7.66812
4.69	21.9961	2.16564	6.84836	5.29	27.9841	2.30000	7.27324	5.89	34.6921	2.42693	7.67463
4.70	22.0900	2.16795	6.85565	5.30	28.0900	2.30217	7.28011	5.90	34.8100	2.42899	7.68115
4.71	22.1841	2.17025	6.86294	5.31	28.1961	2.30434	7.28697	5.91	34.9281	2.43105	7.68765
4.72	22.2784	2.17256	6.87023	5.32	28.3024	2.30651	7.29383	5.92	35.0464	2.43311	7.69415
4.73	22.3729	2.17486	6.87750	5.33	28.4089	2.30868	7.30068	5.93	35.1649	2.43516	7.70065
4.74	22.4676	2.17715	6.88477	5.34	28.5156	2.31084	7.30753	5.94	35.2836	2.43721	7.70714
4.75	22.5625	2.17945	6.89202	5.35	28.6225	2.31301	7.31437	5.95	35.4025	2.43926	7.71362
4.76	22.6576	2.18174	6.89928	5.36	28.7296	2.31517	7.32120	5.96	35.5216	2.44131	7.72010
4.77	22.7529	2.18403	6.90652	5.37	28.8369	2.31733	7.32803	5.97	35.6409	2.44336	7.72658
4.78	22.8484	2.18632	6.91375	5.38	28.9444	2.31948	7.33485	5.98	35.7604	2.44540	7.73305
4.79	22.9441	2.18861	6.92098	5.39	29.0521	2.32164	7.34166	5.99	35.8801	2.44745	7.73951
4.80	23.0400	2.19089	6.92820	5.40	29.1600	2.32379	7.34847	6.00	36.0000	2.44949	7.74597
4.81	23.1361	2.19317	6.93542	5.41	29.2681	2.32594	7.35527	6.01	36.1201	2.45153	7.75242
4.82	23.2324	2.19545	6.94262	5.42	29.3764	2.32809	7.36206	6.02	36.2404	2.45357	7.75887
4.83	23.3289	2.19773	6.94982	5.43	29.4849	2.33024	7.36885	6.03	36.3609	2.45561	7.76531
4.84	23.4256	2.20000	6.95701	5.44	29.5936	2.33238	7.37564	6.04	36.4816	2.45764	7.77174
4.85	23.5225	2.20227	6.96419	5.45	29.7025	2.33452	7.38241	6.05	36.6025	2.45967	7.77817
4.86	23.6196	2.20454	6.97137	5.46	29.8116	2.33666	7.38918	6.06	36.7236	2.46171	7.78460
4.87	23.7169	2.20681	6.97854	5.47	29.9209	2.33880	7.39594	6.07	36.8449	2.46374	7.79102
4.88	23.8144	2.20907	6.98570	5.48	30.0304	2.34094	7.40270	6.08	36.9664	2.46577	7.79744
4.89	23.9121	2.21133	6.99285	5.49	30.1401	2.34307	7.40945	6.09	37.0881	2.46779	7.80385
4.90	24.0100	2.21359	7.00000	5.50	30.2500	2.34521	7.41620	6.10	37.2100	2.46982	7.81025
4.91	24.1081	2.21585	7.00714	5.51	30.3601	2.34734	7.42294	6.11	37.3321	2.47184	7.81665
4.92	24.2064	2.21811	7.01427	5.52	30.4704	2.34947	7.42967	6.12	37.4544	2.47386	7.82304
4.93	24.3049	2.22036	7.02140	5.53	30.5809	2.35160	7.43640	6.13	37.5769	2.47588	7.82943
4.94	24.4036	2.22261	7.02851	5.54	30.6916	2.35372	7.44312	6.14	37.6996	2.47790	7.83582
4.95	24.5025	2.22486	7.03562	5.55	30.8025	2.35584	7.44983	6.15	37.8225	2.47992	7.84219
4.96	24.6016	2.22711	7.04273	5.56	30.9136	2.35797	7.45654	6.16	37.9456	2.48193	7.84857
4.97	24.7009	2.22935	7.04982	5.57	31.0249	2.36008	7.46323	6.17	38.0689	2.48395	7.85493
4.98	24.8004	2.23159	7.05691	5.58	31.1364	2.36220	7.46994	6.18	38.1924	2.48596	7.86130
4.99	24.9001	2.23383	7.06399	5.59	31.2481	2.36432	7.47663	6.19	38.3161	2.48797	7.86766
5.00	25.0000	2.23607	7.07107	5.60	31.3600	2.36643	7.48331	6.20	38.4400	2.48998	7.87401
5.01	25.1001	2.23830	7.07814	5.61	31.4721	2.36854	7.48999	6.21	38.5641	2.49199	7.88036
5.02	25.2004	2.24054	7.08520	5.62	31.5844	2.37065	7.49667	6.22	38.6884	2.49399	7.88670
5.03	25.3009	2.24277	7.09225	5.63	31.6969	2.37276	7.50333	6.23	38.8129	2.49600	7.89303
5.04	25.4016	2.24499	7.09930	5.64	31.8096	2.37487	7.50999	6.24	38.9376	2.49800	7.89937
5.05	25.5025	2.24722	7.10634	5.65	31.9225	2.37697	7.51665	6.25	39.0625	2.50000	7.90569
5.06	25.6036	2.24944	7.11337	5.66	32.0356	2.37908	7.52330	6.26	39.1876	2.50200	7.91202
5.07	25.7049	2.25167	7.12039	5.67	32.1489	2.38118	7.52994	6.27	39.3129	2.50400	7.91833
5.08	25.8064	2.25389	7.12741	5.68	32.2624	2.38328	7.53658	6.28	39.4384	2.50599	7.92465
5.09	25.9081	2.25610	7.13442	5.69	32.3761	2.38537	7.54321	6.29	39.5641	2.50799	7.93095
5.10	26.0100	2.25832	7.14143	5.70	32.4900	2.38747	7.54983	6.30	39.6900	2.50998	7.93725
5.11	26.1121	2.26053	7.14843	5.71	32.6041	2.38956	7.55645	6.31	39.8161	2.51197	7.94355
5.12	26.2144	2.26274	7.15542	5.72	32.7184	2.39165	7.56307	6.32	39.9424	2.51396	7.94984
5.13	26.3169	2.26495	7.16240	5.73	32.8329	2.39374	7.56968	6.33	40.0689	2.51595	7.95613
5.14	26.4196	2.26716	7.16938	5.74	32.9476	2.39583	7.57628	6.34	40.1956	2.51794	7.96241
5.15	26.5225	2.26936	7.17635	5.75	33.0625	2.39792	7.58288	6.35	40.3225	2.51992	7.96869
5.16	26.6256	2.27156	7.18331	5.76	33.1776	2.40000	7.58947	6.36	40.4496	2.52190	7.97496
5.17	26.7289	2.27376	7.19027	5.77	33.2929	2.40208	7.59605	6.37	40.5769	2.52389	7.98123
5.18	26.8324	2.27596	7.19722	5.78	33.4084	2.40416	7.60263	6.38	40.7044	2.52587	7.98749
5.19	26.9361	2.27816	7.20417	5.79	33.5241	2.40624	7.60920	6.39	40.8321	2.52784	7.99375
5.20	27.0400	2.28035	7.21110	5.80	33.6400	2.40832	7.61577	6.40	40.9600	2.52982	8.00000
N	N^2	$\sqrt{N}$	$\sqrt{10N}$	N	N^2	$\sqrt{N}$	$\sqrt{10N}$	N	N^2	$\sqrt{N}$	$\sqrt{10N}$

Table R Squares and Square Roots (continued)

N	N²	√N	√10N	N	N²	√N	√10N	N	N²	√N	√10N
6.40	40.9600	2.52982	8.00000	7.00	49.0000	2.64575	8.36660	7.60	57.7600	2.75681	8.71780
6.41	41.0881	2.53180	8.00625	7.01	49.1401	2.64764	8.37257	7.61	57.9121	2.75862	8.72353
6.42	41.2164	2.53377	8.01249	7.02	49.2804	2.64953	8.37854	7.62	58.0644	2.76043	8.72926
6.43	41.3449	2.53574	8.01873	7.03	49.4209	2.65141	8.38451	7.63	58.2169	2.76225	8.73499
6.44	41.4736	2.53772	8.02496	7.04	49.5616	2.65330	8.39047	7.64	58.3696	2.76405	8.74071
6.45	41.6025	2.53969	8.03119	7.05	49.7025	2.65518	8.39643	7.65	58.5225	2.76586	8.74643
6.46	41.7316	2.54165	8.03741	7.06	49.8436	2.65707	8.40238	7.66	58.6756	2.76767	8.75214
6.47	41.8609	2.54362	8.04363	7.07	49.9849	2.65895	8.40833	7.67	58.8289	2.76948	8.75785
6.48	41.9904	2.54558	8.04984	7.08	50.1264	2.66083	8.41427	7.68	58.9824	2.77128	8.76356
6.49	42.1201	2.54755	8.05605	7.09	50.2681	2.66271	8.42021	7.69	59.1361	2.77308	8.76926
6.50	42.2500	2.54951	8.06226	7.10	50.4100	2.66458	8.42615	7.70	59.2900	2.77489	8.77496
6.51	42.3801	2.55147	8.06846	7.11	50.5521	2.66646	8.43208	7.71	59.4441	2.77669	8.78066
6.52	42.5104	2.55343	8.07465	7.12	50.6944	2.66833	8.43801	7.72	59.5984	2.77849	8.78635
6.53	42.6409	2.55539	8.08084	7.13	50.8369	2.67021	8.44393	7.73	59.7529	2.78029	8.79204
6.54	42.7716	2.55734	8.08703	7.14	50.9796	2.67208	8.44985	7.74	59.9076	2.78209	8.79773
6.55	42.9025	2.55930	8.09321	7.15	51.1225	2.67395	8.45577	7.75	60.0625	2.78388	8.80341
6.56	43.0336	2.56125	8.09938	7.16	51.2656	2.67582	8.46168	7.76	60.2176	2.78568	8.80909
6.57	43.1649	2.56320	8.10555	7.17	51.4089	2.67769	8.46759	7.77	60.3729	2.78747	8.81476
6.58	43.2964	2.56515	8.11172	7.18	51.5524	2.67955	8.47349	7.78	60.5284	2.78927	8.82043
6.59	43.4281	2.56710	8.11788	7.19	51.6961	2.68142	8.47939	7.79	60.6841	2.79106	8.82610
6.60	43.5600	2.56905	8.12404	7.20	51.8400	2.68328	8.48528	7.80	60.8400	2.79285	8.83176
6.61	43.6921	2.57099	8.13019	7.21	51.9841	2.68514	8.49117	7.81	60.9961	2.79464	8.83742
6.62	43.8244	2.57294	8.13634	7.22	52.1284	2.68701	8.49706	7.82	61.1524	2.79643	8.84308
6.63	43.9569	2.57488	8.14248	7.23	52.2729	2.68887	8.50294	7.83	61.3089	2.79821	8.84873
6.64	44.0896	2.57682	8.14862	7.24	52.4176	2.69072	8.50882	7.84	61.4656	2.80000	8.85438
6.65	44.2225	2.57876	8.15475	7.25	52.5625	2.69258	8.51469	7.85	61.6225	2.80179	8.86002
6.66	44.3556	2.58070	8.16088	7.26	52.7076	2.69444	8.52056	7.86	61.7796	2.80357	8.86566
6.67	44.4889	2.58263	8.16701	7.27	52.8529	2.69629	8.52643	7.87	61.9369	2.80535	8.87130
6.68	44.6224	2.58457	8.17313	7.28	52.9984	2.69815	8.53229	7.88	62.0944	2.80713	8.87694
6.69	44.7561	2.58650	8.17924	7.29	53.1441	2.70000	8.53815	7.89	62.2521	2.80891	8.88257
6.70	44.8900	2.58844	8.18535	7.30	53.2900	2.70185	8.54400	7.90	62.4100	2.81069	8.88819
6.71	45.0241	2.59037	8.19146	7.31	53.4361	2.70370	8.54985	7.91	62.5681	2.81247	8.89382
6.72	45.1584	2.59230	8.19756	7.32	53.5824	2.70555	8.55570	7.92	62.7264	2.81425	8.89944
6.73	45.2929	2.59422	8.20366	7.33	53.7289	2.70740	8.56154	7.93	62.8849	2.81603	8.90505
6.74	45.4276	2.59615	8.20975	7.34	53.8756	2.70924	8.56738	7.94	63.0436	2.81780	8.91067
6.75	45.5625	2.59808	8.21584	7.35	54.0225	2.71109	8.57321	7.95	63.2025	2.81957	8.91628
6.76	45.6976	2.60000	8.22192	7.36	54.1696	2.71293	8.57904	7.96	63.3616	2.82135	8.92188
6.77	45.8329	2.60192	8.22800	7.37	54.3169	2.71477	8.58487	7.97	63.5209	2.82312	8.92749
6.78	45.9684	2.60384	8.23408	7.38	54.4644	2.71662	8.59069	7.98	63.6804	2.82489	8.93308
6.79	46.1041	2.60576	8.24015	7.39	54.6121	2.71846	8.59651	7.99	63.8401	2.82666	8.93868
6.80	46.2400	2.60768	8.24621	7.40	54.7600	2.72029	8.60233	8.00	64.0000	2.82843	8.94427
6.81	46.3761	2.60960	8.25227	7.41	54.9081	2.72213	8.60814	8.01	64.1601	2.83019	8.94986
6.82	46.5124	2.61151	8.25833	7.42	55.0564	2.72397	8.61394	8.02	64.3204	2.83196	8.95545
6.83	46.6489	2.61343	8.26438	7.43	55.2049	2.72580	8.61974	8.03	64.4809	2.83373	8.96103
6.84	46.7856	2.61534	8.27043	7.44	55.3536	2.72764	8.62554	8.04	64.6416	2.83549	8.96660
6.85	46.9225	2.61725	8.27647	7.45	55.5025	2.72947	8.63134	8.05	64.8025	2.83725	8.97218
6.86	47.0596	2.61916	8.28251	7.46	55.6516	2.73130	8.63713	8.06	64.9636	2.83901	8.97775
6.87	47.1969	2.62107	8.28855	7.47	55.8009	2.73313	8.64292	8.07	65.1249	2.84077	8.98332
6.88	47.3344	2.62298	8.29458	7.48	55.9504	2.73496	8.64870	8.08	65.2864	2.84253	8.98888
6.89	47.4721	2.62488	8.30060	7.49	56.1001	2.73679	8.65448	8.09	65.4481	2.84429	8.99444
6.90	47.6100	2.62679	8.30662	7.50	56.2500	2.73861	8.66025	8.10	65.6100	2.84605	9.00000
6.91	47.7481	2.62869	8.31264	7.51	56.4001	2.74044	8.66603	8.11	65.7721	2.84781	9.00555
6.92	47.8864	2.63059	8.31865	7.52	56.5504	2.74226	8.67179	8.12	65.9344	2.84956	9.01110
6.93	48.0249	2.63249	8.32466	7.53	56.7009	2.74408	8.67756	8.13	66.0969	2.85132	9.01665
6.94	48.1636	2.63439	8.33067	7.54	56.8516	2.74591	8.68332	8.14	66.2596	2.85307	9.02219
6.95	48.3025	2.63629	8.33667	7.55	57.0025	2.74773	8.68907	8.15	66.4225	2.85482	9.02774
6.96	48.4416	2.63818	8.34266	7.56	57.1536	2.74955	8.69483	8.16	66.5856	2.85657	9.03327
6.97	48.5809	2.64008	8.34865	7.57	57.3049	2.75136	8.70057	8.17	66.7489	2.85832	9.03881
6.98	48.7204	2.64197	8.35464	7.58	57.4564	2.75318	8.70632	8.18	66.9124	2.86007	9.04434
6.99	48.8601	2.64386	8.36062	7.59	57.6081	2.75500	8.71206	8.19	67.0761	2.86182	9.04986
7.00	49.0000	2.64575	8.36660	7.60	57.7600	2.75681	8.71780	8.20	67.2400	2.86356	9.05539
N	N²	√N	√10N	N	N²	√N	√10N	N	N²	√N	√10N

Table R Squares and Square Roots (continued)

N	N²	√N	√10N	N	N²	√N	√10N	N	N²	√N	√10N
8.20	67.2400	2.86356	9.05539	**8.80**	77.4400	2.96648	9.38083	**9.40**	88.3600	3.06594	9.69536
8.21	67.4041	2.86531	9.06091	8.81	77.6161	2.96816	9.38616	9.41	88.5481	3.06757	9.70052
8.22	67.5684	2.86705	9.06642	8.82	77.7924	2.96985	9.39149	9.42	88.7364	3.06920	9.70567
8.23	67.7329	2.86880	9.07193	8.83	77.9689	2.97153	9.39681	9.43	88.9249	3.07083	9.71082
8.24	67.8976	2.87054	9.07744	8.84	78.1456	2.97321	9.40213	9.44	89.1136	3.07246	9.71597
8.25	68.0625	2.87228	9.08295	**8.85**	78.3225	2.97489	9.40744	**9.45**	89.3025	3.07409	9.72111
8.26	68.2276	2.87402	9.08845	8.86	78.4996	2.97658	9.41276	9.46	89.4916	3.07571	9.72625
8.27	68.3929	2.87576	9.09395	8.87	78.6769	2.97825	9.41807	9.47	89.6809	3.07734	9.73139
8.28	68.5584	2.87750	9.09945	8.88	78.8544	2.97993	9.42338	9.48	89.8704	3.07896	9.73653
8.29	68.7241	2.87924	9.10494	8.89	79.0321	2.98161	9.42868	9.49	90.0601	3.08058	9.74166
8.30	68.8900	2.88097	9.11043	**8.90**	79.2100	2.98329	9.43398	**9.50**	90.2500	3.08221	9.74679
8.31	69.0561	2.88271	9.11592	8.91	79.3881	2.98496	9.43928	9.51	90.4401	3.08383	9.75192
8.32	69.2224	2.88444	9.12140	8.92	79.5664	2.98664	9.44458	9.52	90.6304	3.08545	9.75705
8.33	69.3889	2.88617	9.12688	8.93	79.7449	2.98831	9.44987	9.53	90.8209	3.08707	9.76217
8.34	69.5556	2.88791	9.13236	8.94	79.9236	2.98998	9.45516	9.54	91.0116	3.08869	9.76729
8.35	69.7225	2.88964	9.13783	**8.95**	80.1025	2.99166	9.46044	**9.55**	91.2025	3.09031	9.77241
8.36	69.8896	2.89137	9.14330	8.96	80.2816	2.99333	9.46573	9.56	91.3936	3.09192	9.77753
8.37	70.0569	2.89310	9.14877	8.97	80.4609	2.99500	9.47101	9.57	91.5849	3.09354	9.78264
8.38	70.2244	2.89482	9.15423	8.98	80.6404	2.99666	9.47629	9.58	91.7764	3.09516	9.78775
8.39	70.3921	2.89655	9.15969	8.99	80.8201	2.99833	9.48156	9.59	91.9681	3.09677	9.79285
8.40	70.5600	2.89828	9.16515	**9.00**	81.0000	3.00000	9.48683	**9.60**	92.1600	3.09839	9.79796
8.41	70.7281	2.90000	9.17061	9.01	81.1801	3.00167	9.49210	9.61	92.3521	3.10000	9.80306
8.42	70.8964	2.90172	9.17606	9.02	81.3604	3.00333	9.49737	9.62	92.5444	3.10161	9.80816
8.43	71.0649	2.90345	9.18150	9.03	81.5409	3.00500	9.50263	9.63	92.7369	3.10322	9.81326
8.44	71.2336	2.90517	9.18695	9.04	81.7216	3.00666	9.50789	9.64	92.9296	3.10483	9.81835
8.45	71.4025	2.90689	9.19239	**9.05**	81.9025	3.00832	9.51315	**9.65**	93.1225	3.10644	9.82344
8.46	71.5716	2.90861	9.19783	9.06	82.0836	3.00998	9.51840	9.66	93.3156	3.10805	9.82853
8.47	71.7409	2.91033	9.20326	9.07	82.2649	3.01164	9.52365	9.67	93.5089	3.10966	9.83362
8.48	71.9104	2.91204	9.20869	9.08	82.4464	3.01330	9.52890	9.68	93.7024	3.11127	9.83870
8.49	72.0801	2.91376	9.21412	9.09	82.6281	3.01496	9.53415	9.69	93.8961	3.11288	9.84378
8.50	72.2500	2.91548	9.21954	**9.10**	82.8100	3.01662	9.53939	**9.70**	94.0900	3.11448	9.84886
8.51	72.4201	2.91719	9.22497	9.11	82.9921	3.01828	9.54463	9.71	94.2841	3.11609	9.85393
8.52	72.5904	2.91890	9.23038	9.12	83.1744	3.01993	9.54987	9.72	94.4784	3.11769	9.85901
8.53	72.7609	2.92062	9.23580	9.13	83.3569	3.02159	9.55510	9.73	94.6729	3.11929	9.86408
8.54	72.9316	2.92233	9.24121	9.14	83.5396	3.02324	9.56033	9.74	94.8676	3.12090	9.86914
8.55	73.1025	2.92404	9.24662	**9.15**	83.7225	3.02490	9.56556	**9.75**	95.0625	3.12250	9.87421
8.56	73.2736	2.92575	9.25203	9.16	83.9056	3.02655	9.57079	9.76	95.2576	3.12410	9.87927
8.57	73.4449	2.92746	9.25743	9.17	84.0889	3.02820	9.57601	9.77	95.4529	3.12570	9.88433
8.58	73.6164	2.92916	9.26283	9.18	84.2724	3.02985	9.58123	9.78	95.6484	3.12730	9.88939
8.59	73.7881	2.93087	9.26823	9.19	84.4561	3.03150	9.58645	9.79	95.8441	3.12890	9.89444
8.60	73.9600	2.93258	9.27362	**9.20**	84.6400	3.03315	9.59166	**9.80**	96.0400	3.13050	9.89949
8.61	74.1321	2.93428	9.27901	9.21	84.8241	3.03480	9.59687	9.81	96.2361	3.13209	9.90454
8.62	74.3044	2.93598	9.28440	9.22	85.0084	3.03645	9.60208	9.82	96.4324	3.13369	9.90959
8.63	74.4769	2.93769	9.28978	9.23	85.1929	3.03809	9.60729	9.83	96.6289	3.13528	9.91464
8.64	74.6496	2.93939	9.29516	9.24	85.3776	3.03974	9.61249	9.84	96.8256	3.13688	9.91968
8.65	74.8225	2.94109	9.30054	**9.25**	85.5625	3.04138	9.61769	**9.85**	97.0225	3.13847	9.92472
8.66	74.9956	2.94279	9.30591	9.26	85.7476	3.04302	9.62289	9.86	97.2196	3.14006	9.92975
8.67	75.1689	2.94449	9.31128	9.27	85.9329	3.04467	9.62808	9.87	97.4169	3.14166	9.93479
8.68	75.3424	2.94618	9.31665	9.28	86.1184	3.04631	9.63328	9.88	97.6144	3.14325	9.93982
8.69	75.5161	2.94788	9.32202	9.29	86.3041	3.04795	9.63846	9.89	97.8121	3.14484	9.94485
8.70	75.6900	2.94958	9.32738	**9.30**	86.4900	3.04959	9.64365	**9.90**	98.0100	3.14643	9.94987
8.71	75.8641	2.95127	9.33274	9.31	86.6761	3.05123	9.64883	9.91	98.2081	3.14802	9.95490
8.72	76.0384	2.95296	9.33809	9.32	86.8624	3.05287	9.65401	9.92	98.4064	3.14960	9.95992
8.73	76.2129	2.95466	9.34345	9.33	87.0489	3.05450	9.65919	9.93	98.6049	3.15119	9.96494
8.74	76.3876	2.95635	9.34880	9.34	87.2356	3.05614	9.66437	9.94	98.8036	3.15278	9.96995
8.75	76.5625	2.95804	9.35414	**9.35**	87.4225	3.05778	9.66954	**9.95**	99.0025	3.15436	9.97497
8.76	76.7376	2.95973	9.35949	9.36	87.6096	3.05941	9.67471	9.96	99.2016	3.15595	9.97998
8.77	76.9129	2.96142	9.36483	9.37	87.7969	3.06105	9.67988	9.97	99.4009	3.15753	9.98499
8.78	77.0884	2.96311	9.37017	9.38	87.9844	3.06268	9.68504	9.98	99.6004	3.15911	9.98999
8.79	77.2641	2.96479	9.37550	9.39	88.1721	3.06431	9.69020	9.99	99.8001	3.16070	9.99500
8.80	77.4400	2.96648	9.38083	**9.40**	88.3600	3.06594	9.69536	**10.00**	100.000	3.16228	10.0000
N	N²	√N	√10N	N	N²	√N	√10N	N	N²	√N	√10N

Index

Approximating a binomial distribution, 206–211
A priori probability, 15, 22
A priori reasoning, 15
Area over a given interval, 121
Average(s), 92, 105
  law of, 42, 230
  weighted, 129

Bar graph, 84
Bayes, Thomas, 74
Bayes' Formula, 73–77
  statement of, 74
Bell-shaped curve, 125
Bernoulli distribution, 116
  standardization of, 161
Bernoulli, Jacques, 3
Bernoulli random variable, 114, 153
  expected value of, 130
Bernoulli trials, 71, 177
Biased sample, 224
Binomial coefficients, 184, 190–192
  definition of, 192
  related tree diagram, 192
Binomial distribution(s), 177–184, 193–197
  approximation of, 206–211
  definition of, 183
  parameters of, 183
  related graphs, 196, 197

Cardano, Girolamo, 3
Cards, playing, 3
Cartesian product, 70
Cell of a histogram, 90
Central Limit Theorem, 233–234
  statement of, 233
Change of variable, 159
Chebyshev, P. L., 148
Chebyshev's Inequality, 148, 149, 230–231
  problems involving, 232
Coefficient(s)
  binomial, 184, 190–192
  correlation, 165

Combination(s), 186–188
Complement of a set, 31
Complementary events, 31
Computer investigations, 103–104, 173, 189–190, 212, 253
Conditional probability, 47–50, 56
  definition of, 50
Confidence interval(s), 244–247
  use of, 248–251
Confidence limits, 246
Continuous data, 11, 89, 120
Correlated random variables, 163
Correlation coefficient, 165
Covariance of random variables, 163
Critical ratio, 255

Data
  continuous, 11
  discrete, 11, 89
  experimental, 8
  grouped, 88–89
  historical, 9
  numerical, 10, 11
Decision making, 2, 278–279
De Moivre, Abraham, 211
De Moivre's Theorem, 211, 214
  restatement of, 233
Die (dice), 3
  fair, 3
Discontinuities of a probability function, 117
Discrete data, 11, 89, 117
Discrete probability distribution, 117
Discrete probability function, 117
Discrete random variable, 117
Domain of a probability function, 32, 109
Dot-frequency graph, 84
Drawing
  at random, 12
  with replacement, 12
  without replacement, 12

Elementary event, 28
Empirical probability, 16

Empty set, 31
Error
   Type I, 273
   Type II, 273, 275
Event(s), 26–28
   complementary, 31
   definition of, 27
   elementary, 28
   for a continuous random variable, 121–122
   independent, 64–68, 155
   special, 31
Expected value
   of a Bernoulli random variable, 130
   of a random variable, 128–131
      definition, 129
Experimental data, 8

Fair coin, 3
Fair die, 3
Fair game, 3, 130
Fermat, Pierre de, 3
Finite population, 221
Frequency, 82
   relative, 82–83, 93
Frequency curve, 90
Frequency distribution, 83
   grouped, 89
Frequency polygon, 90
Frequency table, 82
Function(s), 24
   linear, 143
   power, 278
   probability, 32
   quadratic, 143

Game, 3
   fair, 3, 130
Games of chance, 2, 3
Graph(s)
   bar, 84
   line-segment, 84
   of probability functions, 114–118
   relating to binomial distribution, 196, 197
Grouped data, 88–89
Grouped frequency distribution, 89

Histogram, 85, 89–90
   cell of, 90
Historical data, 9
Huygens, Christian, 3
Hypothesis testing, 16

Independent Bernoulli trial(s), special case of, 154

Independent events, 64–68, 155
   definition of, 67
Infinite population, 221
Inputs, See *Domain of probability function*
Interval(s), 88
   confidence, 244–247
   modal, 105
   one-sided, 256

Laplace, Pierre S., 3
Law of averages, 42
Law of Large Numbers, 230–232
   statement of, 231
Limits, confidence, 246
Linear function, 143
Line-segment graph, 84

Mean, 92–93
   formulas for, 94–95
   of a particular sample, 226
   of a sample, 226
Median, 105
Modal interval, 105
Mode, 105
Moseley, Henry G., 86
Multiplication principle, 20
Mutually exclusive sets, 35

Normal curve, 203
Normal distribution, 204
   standard, 125–127, 150–151, 161
   standardization of, 203–205
Null hypothesis, 265
Numerical data, 10, 11
   organization of, 10

Odds, 34
One-sided interval, 256
One-sided statistical tests, 270–272
Outputs, see *Range of probability function*

Parameter(s), 243, 244
   of a binomial distribution, 183
Particular sample, 222
   mean of, 226
Pascal, Blaise, 3, 190
Pascal's triangle, 190
   example of, 182
Percentiles, 106
Permutation(s), 185–186
Point estimate, 243
Poisson distributions, 213–214